THE ECCENTRIC UNIVERSE

ROB LOWE

Bright Pen

Visit us online at www.authorsonline.co.uk

A Bright Pen Book

Cover design by Tim Medway-Smith ©
Cover photography by Lily Coxon-Lewis ©

British Library Cataloguing in Publication Data.
A catalogue record for this book is available from the British Library.

ISBN 978-0-7552-1402-0

Authors OnLine Ltd
19 The Cinques
Gamlingay, Sandy
Bedfordshire SG19 3NU
England

This book is also available in e-book format, details of which are available at www.authorsonline.co.uk

To my long-suffering parents,
and my three very special children,
Alice, Mary and Oliver,
who would like to remain anonymous!

If readers would like to contact the author directly they may do so by emailing:

rob.lowe@theeccentricuniverse.com

To my great friends with whom I have had such fantastic times:

Felicity, Tom, Lily, Coco, Bretia, Paul, Lucinda, Tipsy, Sheila, Edward Rupert Boisier Pernicolas Du Bois Tomlinson, Joy, Millie, Madeline, Molly, Stan, Dick, Neil, Will, Rosie, Stephanie, Peter, Jon, Mick, Allan, Andy, Gary, Nancy, Pete, Roger, Jacob, Tim, Dara, Toby, Jack, Liz, Phil, Charlie, Jason, Paul, Boycie, Ross, Simon, Chris, Raggy, Gadget, Helen, Archie, Phil, Steve, Gary, John, Jonathan, Julie, Ann, Anna, Monty, Gilbert, Mark, Clair, Kate, Jo, Fishy, Matt, Carl, Luke, Nigel, Mike, Chris, David, Paul, John, Rachel, Jez, Leah, Dale, Nicky, Gordon, Jeremy, Richard, Anita, Miq, Charles, Phil, Andrew, Ian, Charlie, Patrick, Warwick, Imants, Emmanuel, Kirsty, Ray, Lord Montagu, HRH The Prince Philip, Sir Patrick Moore, Jonathan Davis, Steve Hardwick, Lewis, Colin, Oliver, Gerald, Stuart, Scott, James, Elaine, Kevin, Scott, Nita, Sharron, Bill, Nigel, Wendy, Harold, Kay, Su, Carl, Tom, Lorraine, Jacqui, Katherine, Sarah, Di, James, Nick, Fiona, Simon, Sarah, Alison, Anton, Ben, Guy, Clint, Pip, Penny, Jan, Dick, Liz, Steve, Charlotte, Becky, Glen, Chris, Sal, Luke, Emma, Margaret, Bill, Felix, Fiona, Freddi, Gareth, Giles, Robert, Gordon, Graham, Pete, Alan, Karen, Mick, Claire, Ted, Patricia, Denise, Paul, Martin, Nicky, Alison, Pat, Harry, Chris, John, Maria, Mark, Hal, Karron, Dave, Mel, Steve, Adam, Jason, Kelly, Andrew, Rupert, Phil, Janik, Gary, Tim, Bill, Kelly, Tim, Daragh, Craig, Aileen, Basher, Jason, Tim, Archie, The Pint Pot and all the fantastic people of Great Britain who I met during my bicycle ride around the coastline of the British Isles.

For further information about this book and an interactive forum for readers' thoughts, views, comments, ideas and questions please visit:

Website: **www.theeccentricuniverse.com**
Social debate: **www.facebook.com/theeccentricuniverse**

Alternatively, email the author directly at:
rob.lowe@theeccentricuniverse.com

To Mollie and Adrian,

I do so hope that you enjoy this read.

Onwards and Upwards

Best wishes

Rob Lowe

The Eccentric Universe

17/02/2012

THE ECCENTRIC UNIVERSE

The Universe is extreme – its breathtaking size and structure has fascinated Mankind throughout the ages. The more discoveries that are made, the more people are confronted with an ever-increasing assortment of inexplicable marvels. This book explores these wonders, mysteries and secrets, and the desperate struggle to explain them.

A number of extraordinary individuals have been able to show-the-way by deciphering a handful of the Universe's secrets. These geniuses made breakthroughs in knowledge for a multitude of reasons. This book explores these along with many other successful and sometimes less successful innovations of nature and the Universe. These innovations could easily at one time have been construed as slightly freakish, weird, bizarre or altogether rather eccentric.

Making headway within science and technology has never been easy. Vast swathes of the Universe remain a complete mystery which highlights just how much more there is to discover and learn. The mysteries of the Universe are comparable with an enormously complex jigsaw puzzle that also succeeds in hiding its pieces to closely guard its secrets. After making mistakes and returning to the drawing board, occasionally a stroke of genius provides Mankind with some progress.

When looking at the World and the Universe around us it becomes clear that the success of one freak invention, discovery or natural wonder ultimately develops into normality. We celebrate the eccentric nature of these successes and analyse the questions and answers to: what, why, how, and by whom?

We take a look at the somewhat baffling aspects of our world and how humanity is just beginning to get a hold on precisely how weird it is.

This book explores the way humans try to form an understanding of the Universe and how it is becoming increasingly important for eccentric views to be tolerated and taken seriously if progress is to be made: The Eccentric Universe.

CONTENTS

A Note from the Author

This aim of this book is to encourage readers to open their minds to the wider events and outcomes that happen in our world and throughout life. A number of profound topics are covered which people often think deeply about, but very rarely resolve.

I have researched hard to bring to you in plain language some of the most peculiar, astonishing and eccentric features of the Universe which our habitual, isolated and relatively stagnant lives do not normally permit us to consider. Where possible, I have studied long and hard to bring you the most plausible answers.

There are numerous references attributed to eminent philosophers or scientists, where such references are not made then the views are mine. Quotations from historical figures are included as reference within the text. There is also an index, a summary of quotations and a list of questions arising at the end of the book. All my views stem from true facts and are designed to provoke thought and encourage discussion.

I truly welcome feedback relating to the topics covered. Within various sections I ask the reader searching questions regarding a particular topic – at these points, please feel free to email me your thoughts.

Please email any thoughts, views, comments, ideas or questions you may have to **rob.lowe@theeccentricuniverse.com** and I will do my utmost to reply with a suitably thought-provoking response.

If you are inquisitive by nature, have a sense of humour, and are seeking answers to life's mysteries – then this book is for you.

Please also visit:
Website: **www.theeccentricuniverse.com**
Social debate: **www.facebook.com/theeccentricuniverse**

Rob Lowe

A NEW EPOCH

Where we learn how the Universe's mysterious and astonishing marvels can be perceived and interpreted in so many different ways. How we must think differently to make progress to understand the myriad of mind-boggling events. We take a look at some of the mistakes Mankind has made and how mistakes seem part of natural progression.

The Universe's genius, creativity and intelligence

- *"Either nothing is a miracle or everything is a miracle" – Albert Einstein.*
- *Many phenomena are unexplainable, unfathomable and exceed the limits of human understanding.*
- *The characteristics of the Universe are just as mystifying as its creation.*

The more we learn about our Universe, the more fascinating we find its constituent parts. Of the countless discoveries Mankind has made, many have no obvious explanation.

The Universe's creative genius is extraordinarily versatile, from nature's oddly guarded secrets of time, space and matter, through to its mysterious forces. With such a vast expanse of space and time which appears almost infinite, our minds are overwhelmed with the incalculable possibilities of what could exist within it.

We can only begin to appreciate the Universe's genius by observing phenomena closer to home. Some of them are truly fascinating, so let us take a swift look at a few of the Earth's remarkable offerings.

Magicians are good at bamboozling us, but did you know that matter has a natural ability to exist in two places at once?

Astonishingly, observations have also been made that prove matter can interact with itself when in two places at the same time!

Many of us dream of an eternal life, but you may not be aware there is a mysterious jellyfish within the oceans of our planet which achieves this by perpetually bypassing death!

Humans have developed and evolved language, but we fail to communicate coherently with the other ten million different species on the planet. Some can sing in frequencies we cannot hear, detect colours that our eyes cannot see and send messages in worlds of scent, electricity and polarised light that we cannot begin to interpret!

Experiences over time help creatures learn, but there are birds such as the Manx shearwater which leave their chicks behind in burrowed nests on British islands whilst they fly off to South America. Miraculously, sometime later the young birds follow, and, unaided, meet up with their parents six thousand miles away!

All current forms of communication are limited by the speed of light making them exceedingly slow when compared to the size of the Universe. However, nature has remarkably exhibited a mechanism for instantaneous communication over limitless distances!

Well there you are!

You are inextricably linked to the eccentricities of the Universe which are compiled here for your appreciation and enjoyment.

The genius of the Universe has produced billions of diverse inexplicable objects and breathtaking cosmic events. But for ourselves its most amazing achievement has been providing us with life.

Is the provision of life an ultimate objective of the Universe?

Is life a natural phenomenon?

Is life just a by-product of the existence of the Universe?

Total turmoil, ultimate tranquillity and complete balance are all experienced within the Universe thanks to its mastermind of creativity – but no one knows what underlying secrets the Universe holds!

Why is the Universe here?

How did the Universe come about?

What fuels the Universe?

How big is the Universe?

Will the Universe ever end?

We can only speculate the answers to these and other similar questions. Exactly how the Universe came about is a true mystery, leaving scientists to endlessly theorise the way in which it materialised.

The creation of life should be perceived as the Universe's outward expression of its own intelligence and its creative impulse. Its behaviour and practices are observable but often incomprehensible as it stems from an entity so original that it cannot be compared alongside anything else.

It is perhaps with this in mind that prompted Albert Einstein to say, "There are two ways to live: you can live as if nothing is a miracle, or you can live as if everything is a miracle."

The Universe is artistic, talented, practical and original. It is not bound by limitations, and presents to us a seemingly endless void of space. Its enormous appeal invites us to explore its countless galaxies, but at the same time inhibits us from doing so by its sheer size. It only allows us to contemplate potential theories from our limited viewpoint here on planet Earth. We attempt to decipher its purpose through personal experience and study its wonders via exploration, art, science and philosophy.

Those who seek the Universe's knowledge and experience are rewarded with a satisfying mental picture of a truly remarkable place. Whilst here, we should all try our best to perceive its wonder. We may not get a second chance!

The Universe has many unusual and mystifying traits to its existence. It appears to have an unfathomable behaviour, but upon closer inspection, individual characteristics become understandable with everything blending together.

Could it be that everything in the Universe is intimately connected and has a purpose?

With all this in mind, the Universe resembles a profound eccentric with genius, creativity, expression and intelligence. It is for this reason I have entitled this book "The Eccentric Universe".

Thankfully we have a planet upon which we may live and thrive, but our comprehension of what we observe is in its infancy. We can relate to what is food, we can make out shapes and how certain things function, but as to where all these things came from originally and how everything initially constructed itself – we have a great deal to learn.

We know so little about where we live.

There is so much more to discover

- *The more we learn about our Universe, the more we realise there is to discover.*
- *Occasionally we get an innovative person who makes great headway with a truly groundbreaking discovery.*
- *It has dawned upon humanity just how truly weird our Universe is.*

The fact that the Universe exists means one of two things. Either, at one point in the distant past, there was an incredible sequence of events that forced the Universe to materialise, or it has always been in existence. If the Universe has always been here then the creation of complex life would be slightly easier to comprehend. However, if the Universe is just a few billion years old then the creation of complex life may have required some type of fast-track into existence.

Scientists believe from the red-shift observations made within light from distant stars and the discovery of cosmic microwave background radiation left over from the Big Bang that the Universe came into existence roughly thirteen point seven billion years ago. What existed before this time is a total mystery, and where the energy came from to create it is completely baffling.

The human race's understanding of the Universe has progressed considerably over the centuries, but there is still so much more to discover. It was only in the early 1990's that other planets orbiting other stars were detected. Prior to this, scientists were still debating whether the Sun possessed the only planetary solar system throughout the whole Universe. Now that astronomers have observed an abundance of planets orbiting other stars, it seems daft that the topic was debated at all. It makes great sense that there are other planets. It also makes great sense that some of these are habitable.

With the discovery of other planets orbiting other stars, in an identical way it feels only natural to consider life as a Universe-wide wonder. Although extraterrestrial life has not been detected yet, many people have a gut feeling that it is out there. Just like those other planets proved to be, this too seems to make great sense.

The panspermia hypothesis was proposed as far back as the fifth century BC by the Greek Philosopher Anaxagoras, and has been upheld by some of the most revered scientists including Sir Fred Hoyle 1915 – 2001.

When considering how life could propagate Universe-wide, panspermia comes to the rescue. The panspermia hypothesis suggests that life-bearing spores are the carriers of life from one star system to another. It is not beyond the realm of biological possibility for suitably evolved spores to travel dormant for billions of years through deep space. Ultimately the spores would populate throughout the whole Universe.

Asteroids that collide with spore infested planets will send spores spiralling into outer space, just like the wind carries seeds in the air. The spores that land successfully on suitable planets would then relinquish their dormant state to evolve and diversify. Over time they may effortlessly evolve into a variety of different plants and creatures.

The more that life-forms are studied, the more it is understood that there are complex relationships with others. Most interestingly, many have finely tuned adaptations to the Universe's forces at large – aerodynamics, gravity, electricity and magnetism are all productively utilised by many forms of life. There is absolutely no reason why certain species of spore could not develop similar finely tuned adaptations regarding the fundamental workings of the whole Universe, thus evolving to tackle the complexities of intergalactic travel.

Could the inherent panspermia-style knowledge within a spore be the mysterious way that the Universe spreads life across the cosmos?

Perhaps the debate will continue for a little while yet.

It would make sense to think that when life evolves, it evolves according to the physical rules of the Universe. However, it may be a little more advanced than this. It may also evolve according to the knowledge of the expanse of space and an awareness of the distances between star systems.

We are all unexpectedly born into this miraculous Universe then rapidly confronted with its mysteries. As soon as we arrive here, we cannot walk, we cannot talk, we cannot look after ourselves and the majority of people we see make funny noises and pull funny faces at us. From this time we must make sense of the immediate world around us, determine who our mother is and cry at the appropriate moments to get what we want!

The 'way of the world' does not really change much from this point!

As we get older we come to realise we live upon a planet that has been moulded and conjured from historical cosmic events and the wishes and desires of our ancestors; all we can do is seamlessly blend into it. Frankly, life really deserves an accompanying specialist survival handbook of sorts, but alas our ancestors failed to provide us with one! Rather, we have been inadvertently set on the road to planetary destruction which the majority of us seem to be content to follow – even to this day!

Unfortunately many of us still appear quite happy to sign up to work for logging companies, fuel-burning power stations and tuna-netting enterprises or other equally destructive organisations. Furthermore, people cannot easily visualise what the planet was like prior to man's influence upon it, so it is very difficult to comprehend how devastating our presence has been.

Some extremely large areas of land have long since been subject to deforestation and handed over to complex agricultural practices which are now well established. Huge towns and cities have been built where thriving forests once stood. From these already deforested areas where massive populations now live, people learn of modern day deforestation activities around the globe and consider it atrocious!

In 1820, half the United States of America was covered in forest; now there is virtually none left. Developing countries see that the United States benefited greatly from this deforestation and view any attempt to deny developing countries the same opportunities as hypocritical. They argue that the poor should not have to bear the cost of preservation when the rich created the problem in the first instance.

For those currently cutting the trees down, they only see it as a means of economic survival – continuing what human beings have been doing for years. However, there are planetary limits to this activity – especially when we have technological improvements making it easier to cut the trees down.

It is thought that if deforestation continues – which by all accounts it will – there will be very little forest left in just two or three decades. To put all this very sad news into perspective, it has been estimated that a fifth of the entire world's rainforest was destroyed between 1960 and 1990. Rainforest once covered fourteen per cent of the world's land surface, but within a fifty-year period from 1960 this was reduced to just six per cent – and so this trend still continues.

Unfortunately, human beings have an insatiable appetite for decimating forests – the trees are cut down for several reasons. According to the United Nations Framework Convention on Climate Change, UNFCCC, subsistence farming is responsible for forty-eight per cent of deforestation, commercial agriculture is responsible for thirty-two per cent of deforestation, logging is responsible for fourteen per cent of deforestation and the need for fuel woods is responsible for five per cent.

The general consensus of opinion is that it is too late to reverse the devastation inflicted upon our planet and as time goes by the situation seems to just get progressively worse.

Governments have left addressing the difficulties of the planet for far longer than is sensible. Their compromise to keep the age old status quo in tandem with introducing environmental policies has in the main left these policies suffering greatly. Within just two hundred years, the Earth's human population has increased dramatically and we have reduced our wonderful planet, teeming with diverse life, into one which is struggling. Jeopardised species, unsustainable food chains, mass pollution, our garbage culture, the plundering of the planet's resources and dreadful wars are but a few of our difficulties.

But there is hope!

In any discipline or walk of life there are always one or two people who lead the way. This may be because they are the most innovative, the first to discover something, the founder of an idea, the fastest, or the person with the most knowledge. Within this book we will identify some of the more successful and sometimes the less successful innovations of people, species, nature and the

Universe. All these at one time could have been construed as slightly freakish, weird, bizarre or altogether rather eccentric spectacles.

When looking at the World and the Universe around us it becomes clear that the success of one freak invention, discovery or natural wonder ultimately develops into normality. This book celebrates the eccentric nature of these successes and analyses the questions and answers to, "What, why, how, and by whom?"

We will take a general look at the somewhat baffling aspects of the Universe and how humanity is just beginning to get a hold on precisely how weird it is.

It would appear that the more we learn about our Universe, the more we find there is to discover. This book reveals the way in which humans try to form an understanding of the Universe and how it is becoming increasingly important for eccentric views to be tolerated and taken seriously if progress is to be made.

For the survival of the planet and to continue the advancement of human knowledge, we must all adopt a healthier and more compassionate approach towards the Earth's resources. Some people and organisations like Greenpeace have begun to get environmental issues onto world government agendas. For those that think this is a step in the right direction, unfortunately there may not be enough people determined to make the appropriate impact – but we shall soon see.

Alternatively, perhaps the world is heading in the direction that it is supposed to be at this time. The people living in their little bubbles that are controlled by governments are perhaps in for a real shock – again, we shall soon see.

Maybe the Universe has seen enough – had we treated the planet better, perhaps more of its secrets would have been revealed to us and the human race would be free to explore the Universe properly. It would make sense that intelligent races have to prove themselves before being granted the ability to go forth and multiply. If the Universe nurtures a prolific creature that is greedy and destructive then it will have failed in its mission.

Whilst most of us deep down know that the planet is being treated horrendously by humanity, we do very little about it.

Unfortunately the solution to the situation is a collective one rather than an individual one.

To prevent further destruction of the planet and to progress our understanding of the Universe, we must think in a very different way.

We must think differently

- *We all see things in a different way, one person may see true beauty in an animal, and another will see it as food.*
- *We must think differently otherwise we will never get to understand the inner secrets of the Universe.*
- *Perhaps a new approach will begin to solve the many unanswered questions that have arisen, bringing about a new epoch of thought.*

There are roughly ten million species of creature on Earth, of these the human race is the only one that can leave lasting legacies. Our ability to build bridges, houses, skyscrapers, stadiums and dams has excelled and etched lasting constructions all over the world from which future generations may benefit. This is a revolutionary ability that no other creature, not even the dinosaurs managed to achieve.

Many people wish to leave their mark on the planet, whether this is to be an architect of a great cathedral, a famous actor, an author of a book or any other noteworthy achievement. Desires, wishes, experiences, looks, confidence, character, motivation and attitude are just some of the attributes that make people who they are.

The Universe's astonishing phenomena can be interpreted by different people in countless ways. An artist may look upon a supernova as a beautiful subject for its colourful detail and intriguing shapes, whereas an astronomer may be more interested in cataloguing it, determining its age and establishing its distance from Earth. There is no right or wrong way to think, but according to Albert Einstein it would appear we need to think differently if we are to understand more and make significant progress.

Despite us all living within the same intriguing Universe, our individual cravings for knowledge differs considerably. Some people do not seem to care about the intricacies of the Universe about us, and others will dedicate their entire lives to it; from the poor orphans in the back of beyond who are too desperate to contemplate any deep questions, to Albert Einstein who worked out all the answers, then told us what the questions were!

We all have our position within this realm of potential thought and it is about time we knew a touch more. It is interesting, it can be riveting and we can all contribute.

Imagine how differently an eccentric professor may be thinking when in his dimly-lit garage producing crazy inventions that might change the world. He shakes with excitement as he switches on the machine for the first time, his hair unkempt, and sparks fly as the contraption whirs into life. We know he has been thinking differently!

Alas, 'hair unkempt and sparks flying' is not me! But I am told my original views and eccentric thoughts may well be of interest to many people, and one day may be of value to society. I have therefore documented them here for everyone's judgement.

No matter how difficult a problem is, with the appropriate amount of analysis and dedication there should surely be a way forward. But with some problems, to achieve this, we need to think differently.

Much of our understanding of the Universe has transpired as a result of addressing topics in seemingly counter-intuitive ways. Discoveries are made by looking at things in different ways, ignoring what may appear as common sense solutions and developing less plausible solutions in an almost eccentric fashion. As you will see, eccentric ideas often strike home as the ultimate way forward.

Einstein, referring to problems, famously said, "We can't solve problems by using the same kind of thinking we used when we created them."

Therefore we need a new approach to tackle the unknown. A new train of thought is required to determine answers to our unanswered questions. This will mean looking at topics from different perspectives, extrapolating knowledge in directions previously alien to us, considering numerous alternatives and not being worried about being wrong.

Einstein also said, "The thinking that we are has brought us to where we have already been. In order to go somewhere else, we must think in a different way."

He is simply saying that we have arrived at where we are in the world today by thinking in the fashion we have in the past, but in order to make future progress with knowledge, we must think differently.

There are numerous ways of looking at everything. For example, when looking closely at an insect, your mind can conjure up two extremes. The first extreme is that you see a rather strange-looking creature which you cannot wait to squash because you think it might bite you and suck your blood. Anyway, it gives you the creeps. The second extreme is that you see a beautifully-latticed creature with a myriad of mind-boggling survival tactics, an unimaginable process of reproduction and an inconceivable point of creation.

In a way, Albert Einstein was alluding towards this type of thought.

Do not let your inner inherent initial instinct become prevalent within you, allowing you to grab the leather fly swat and "Squish". You must think deeper, more logically and more constructively about the beauty within the Universe and then the answers will flow. Just like your inner instinct when looking at a creepy crawly insect, let your mind grasp the whole as best it can – rather than some derisory initial thought.

Incidentally, for those of you who reach for the leather fly swat to exterminate the insect, perhaps next time you should consider the billions of years it took for the insect to evolve before simply swiping it.

We must all adopt a fresh approach to solve the unknown.

Are we about to enter a time with a new epoch of thought?

Perhaps it will be better to contemplate new views, ideologies and concepts with a modified approach.

Maybe we should take a lesson from airline pilots – so long as the approach is correct, we will have a much better chance of success!

It only takes a special second look at something by someone who thinks differently to spot that a concept or idea is not quite right. For years people thought horse's legs moved in pairs, as a consequence artists drew hunting scenes with horses splayed out like wide-open stepladders. Some thought the world was flat. Many thought it would be impossible to build flying machines. Many thought that man stepping foot on the Moon was impossible.

Just by thinking differently people have made both small and giant leaps forward.

Revolutionary groundbreaking ideas and discoveries

- *Although our pace of change is currently fast, truly significant dynamic discoveries are relatively rare.*
- *Improved communication means that new theories are scrutinised, challenged and commented upon by a much larger audience.*
- *Could Gaia explain how nature is embedded deep within the fabric of the Universe?*

Despite there being billions of people upon planet Earth, truly revolutionary groundbreaking ideas and discoveries are relatively rare occurrences. Over the last few hundred years there have been thousands of discoveries, but only a handful of them are significant dynamic discoveries which have influenced and amazed everyone who has heard about them.

Within Astronomy we had heliocentrism which was first proposed as far back as the third century BC by Aristarchus of Samos.

Heliocentrism is the astronomical model which has the Earth and the other planets revolving around a stationary Sun at the centre of the solar system.

The heliocentric motion of the planets around the Sun was confirmed as a fully predictive mathematical model by Nicolaus Copernicus, 1473 – 1543. The announcement of this cosmic discovery brought about what came to be known as the 'Copernican Revolution' – a name which just shows how radical this discovery was in its day. No longer was the Earth considered the centre of the Universe.

Imagine the paradigm shift away from the Ptolemaic model of the cosmos which had the Earth positioned at the centre of the Universe. The Ptolemaic view with the Earth at the centre had been in vogue for well over a thousand years since the astronomer Claudius Ptolemaeus, AD 90 – 168, developed this widely accepted view of celestial motion.

For Copernicus to announce that the Sun was at the centre would have been a truly revolutionary proposal at the time. The announcement would have suddenly and dramatically shifted everyone's preconceived focus of where they thought they sat in the Universe to somewhere totally alien to them.

Other truly revolutionary groundbreaking discoveries include Charles Darwin's theory of natural selection, the discovery that the world is not flat, the discovery and understanding of DNA, the discovery of the atom, the development of medicines and the creation of vaccines. Many of these innovations took decades, if not centuries, to devise and successfully implement globally.

As soon as any budding innovator proposes a new theory it is scrutinised and challenged by a large scientific community. However, over the centuries the scientific community has proved to be less than receptive to new revolutionary groundbreaking ideas – it has often taken decades to embrace new ideologies. Classic examples have been the acceptances – that the Earth is curved; that the Earth is not at the centre of the Universe; evolution; and that climate change is a real problem.

The trends associated with the historical acknowledgement of innovation and discovery made me wonder whether there are any theories currently being proposed that are having difficulties being accepted. I took some time to research such proposals and came across the intricacies of Gaia.

The concept of Gaia is fascinating, but just like other revolutionary theories it has had its critics over the years. Gaia was first formulated in the 1960's by the independent research scientist Doctor James Lovelock. He initially published his Gaia hypothesis within journal articles in the early 1970's. This was followed by a popular 1979 book called, 'Gaia: A new look at life on Earth'.

In Doctor James Lovelock's own words he describes Gaia as being, "A complex entity involving the Earth's biosphere, atmosphere, oceans, and soil; the totality constituting a feedback or cybernetic system which seeks an optimal physical and chemical environment for life on this planet."

Gaia is becoming a relatively popular scientific belief providing people with a theoretical insight into how the Earth is capable of self-regulating in a state of homeostasis.

Human beings possess homeostatic capabilities, sweating in hot weather to keep their bodies cool and conserving energy in cold weather to raise body heat. In this way humans are able to maintain a stable, constant temperature.

Termites are perfectionists at homeostasis and are splendid examples of Gaia on a small scale. Termites have evolved to build one of the wonders of the natural world, creating enormous tower-like nests which hold up to several million individuals. These nests are a maze of tunnels and galleries which provide air conditioning as well as the control of carbon dioxide and oxygen. Within these nests the termites cultivate fungal gardens which benefit from the closely controlled climate, and although the external temperature will vary greatly the nests are capable of maintaining a constant temperature to within one degree centigrade.

Gaia has been likened to viewing the Earth as a single organism in much the same way as a human being is considered a single organism. Just as human beings are a multitude of independently coexisting cells, bacteria, viruses and organs, all the living entities upon Earth may be viewed as a whole. It appears that the proximity of all these living entities together complement and nurture the existence of other living entities in a positive and harmonious way. This harmonious relationship between living entities balances out so perfectly that the Earth maintains a constant equilibrium ideal for life.

Lovelock highlights the relationships between diverse creatures, one example being that sea creatures produce sulphur and iodine in approximately the same quantities as required by land creatures. His hypothesis claims the existence of a global control system of surface temperature, atmosphere composition and ocean salt content.

Lovelock's main arguments for Gaia were firstly that the global surface temperature of the Earth has remained constant despite an increase in the energy provided by the Sun. Secondly that the Earth's atmospheric composition remains constant, even though it should be unstable. Finally, the ocean salt content remains at a constant level, although there is no single regulating mechanism.

The Gaia hypothesis was championed by certain environmentalists and climate scientists, but was rejected by many others, both within scientific circles and outside. Gaia has however been backed by some very notable academics such as Sir Crispin Tickell and is now recognised by the Geological Society of London.

One of the major revolutions coming from the Gaia hypothesis is that Lovelock expects human civilisation to be hard-pressed to survive. Lovelock believes the challenges will be similar to the Paleocene-Eocene thermal maximum when atmospheric concentration of carbon dioxide was very high at four hundred and fifty parts per million. At this time the majority of land was thought to be mostly scrub and desert, the Arctic Ocean was estimated to be around twenty-three degrees centigrade and had crocodiles swimming in it.

Gaia has yet to be truly proven one way or the other and so is currently categorised into the 'brand new, not sure yet' box. I personally love the concept and believe it to be a logical step forward in an approach to understanding life on planet Earth in a way that has never been attempted previously.

Any new revolutionary concept is bound to come up against fierce opposition.

Why would the scientists of the world suddenly embrace a new concept that would hit their current working practice so hard that they would have to change the way they think, the way they conduct their analysis and force them to revisit all their results?

They will undoubtedly give it a very hard time first.

So we shall see how Gaia develops in the future. I personally like the thought of Intergalactic Gaia. Perhaps the heat from stars and the cold throughout the deep of space interacts with life; maybe the universal cosmic cycles conform to some type of predetermined schedule that is known to promote life. I feel that if

Gaia is a correct theory then it must spread further than the Earth, certainly within our solar system.

All of this implies an inherent knowledge embedded somewhere deep within the fabric of the Universe. Something which, ostensibly, should be no more astounding than our own existence!

Our ancestors have done a tremendous job to pave the way forward, defining our diction and documenting knowledge. However, mistakes are often made! Tackling the unknown is made that much trickier, as when you get older, you realise that we who have been fortunate enough to be born with good brains and provided with a decent education, know there may not be such a thing as 'absolute truth'.

There is no reason why we should accept explanations devised years ago when our knowledge and understanding of the world was nothing like it is today. We are in jeopardy of believing that the zebra got its stripes due to some tale about a bet with a lion, when really it evolved this way to confuse the tsetse fly.

So let us now take a look at some of the mistakes and ambiguities we have found at the forefront of our knowledge, giving rise to some most interesting and amusing eccentricities.

It is beginning to dawn on us that much of our knowledge within this world could be based upon ancient eccentric views.

Oh no!

Inaccuracies and ambiguities at the forefront of knowledge

- *Just like seismic fault lines and cracks on the Earth's surface, historical human errors often make their presence known.*
- *It is sometimes difficult to define and make sense of all that we discover. Only time resolves the inaccuracies that are occasionally made.*
- *There have been blunders of enormous magnitude that have given the human race environmental challenges of epic proportion.*

At the forefront of human knowledge, especially science and technology, we often discover amusing and entertaining inaccuracies and ambiguities. Some of them are outright absurd and others are weird or mystifying.

Inaccuracies and ambiguities appear like cracks at the surface of our knowledge. We only need to look at the number of rival theories, factual inaccuracies, genuine mistakes, financial blunders and differences of opinion to see that these somewhat unavoidable eccentricities of life persist throughout. Let me share some fabulous instances of these.

One splendid illustration is how the English language with its many dialects, has evolved. There are many differences of opinion and the language has yet to settle on the definitive spellings of many words.

Here are a small number of them which are worth focussing or even focusing upon! It is fascinating that they have never settled down to a single spelling in such a pedantic world – and incidentally these are not alternative American spellings:

'adrenalin' or 'adrenaline'; 'caster sugar' or 'castor sugar'; 'connection' or 'connexion'; 'connecter' or 'connector'; 'dietician' or 'dietitian'; 'disassociate' or 'dissociate'; 'encyclopaedia' or 'encyclopedia'; 'gipsy' or gypsy'; 'granddad' or 'grandad'; 'jewelry' or 'jewellery'; 'judgement' or 'judgment'; 'knowledgable' or 'knowledgeable'; 'likable' or 'likeable'; 'lovable' or 'loveable'; 'mediaeval' or 'medieval'; 'millepede' or 'millipede'; "nosy' or 'nosey'; 'ratable' or 'rateable'; 'unmistakable' or 'unmistakeable'; 'veranda' or 'verandah'.

To name but a few!

In a similar way that certain words are unresolved, there are some fundamental features of our lives which have had equally tricky times establishing themselves. A tremendous example is how Mankind has attempted to establish the precise dates within our yearly calendar.

One major inaccuracy with far-reaching implications was a very noticeable misalignment with the phases of the Moon, the equinoxes and how Easter was calculated. It became clear that a remedy to the calendar was required to get everything back into alignment. Doctor Aloysius Lilius, 1510 – 1576, proposed a solution to this significant problem which had seen an extra day being built up every three hundred and ten years. When Doctor Aloysius Lilius's solution was brought to the attention of Pope

Gregory XIII, the Pope was so concerned about the problem that he decreed the immediate introduction of the Gregorian calendar.

The proposed solution recommended that the new calendar was phased in over a period of time, therefore keeping disruption to a minimum. As ten days needed to be clawed back, it was pointed out that these ten days could be removed by omitting the leap day, 29th February, on each of its occurrences over a forty-year period. Although this seemed a sensible approach and a relatively straightforward solution to return the heavens back into alignment, Pope Gregory XIII decided upon the new calendar's immediate introduction. This meant the sudden loss of ten days, so the day after Thursday 4th October 1582 became Friday 15th October 1582.

The introduction of the Gregorian calendar, which also brought with it three fewer leap years every four hundred years, performed the necessary alteration to bring March 21st back in line with the equinox. Pope Gregory XIII was instantly satisfied as this was precisely how, in year 325, the council of Christian bishops, called the First Council of Nicaea, had intended the calendar to be. The deletion of these ten days from the calendar immediately accomplished the required alignment and the yearly schedule of ceremonies returned to being as it had been originally envisaged.

Mankind has produced a constant trickle of mistakes, both large and small. Assumptions about the Titanic being unsinkable, building the tower of Pisa on soft boggy ground, believing our solar system was the only one in the Universe and Coca-Cola once turning down the purchase of Pepsi-Cola for a mere one thousand dollars. There have also been endless ecological, sociological, biological, environmental, political and financial blunders that have left people reeling in disbelief.

One of the most amusing debates to follow is how scientists change their recommendation of what constitutes a healthy diet on almost a daily basis!

Having taken millions of years to evolve to where we are today, anyone would have thought that we should know what our diet consists of by now!

High-profile mistakes of today include expensive project failures; poor government financial management; security breaches; horrific transport crashes; and environmental disasters.

People with a genuine desire to make major scientific breakthroughs, in an attempt to develop the leading edge of our understanding, often conclude with incorrect information, anomalies and wide-ranging inaccuracies.

One widely accepted incorrect conclusion in the middle-ages was that everyone believed the Earth was at the centre of the solar system – and of course many actually thought the world was flat. In the last year of his life Nicolaus Copernicus published his epochal book "De Revolutionibus Orbium Coelestium", or in English, "On the Revolutions of the Celestial Spheres". His death was timely as he would have been arrested within this era for his heretical views.

In 1633, most bizarrely, Galileo Galilei, 1564 – 1642, one of the most revered and respected scientists, was convicted of heresy simply because he became a supporter of the views of Copernicus. To follow the theories of Copernicus was seen as contrary to the true sense and authority of the Holy Scripture – Galileo was subsequently placed under house arrest for the remainder of his life.

How foolishly narrow-minded of the authorities of that era!

If only the authorities had known of Arthur Schopenhauer's famous quotation, "All truth passes through three stages. First, it is ridiculed. Second, it is violently opposed. Third, it is accepted as being self-evident."

It is one thing for someone to challenge knowledge, but to get that challenge correct and be locked away for life as a consequence is astonishing. The attitude of the authorities at this time inflicted hardship and grief upon Galileo, which over the passage of time, not only proved to be totally unreasonable but was without doubt counter-productive in terms of astronomical progress.

We cannot be expected to get everything right first time, but undoubtedly some decisions and ideas in the past have defied all logic. We have to admire some of our ancestors who tried their utmost to be unique in their thought, but just got it wrong!

At least they were having a go!

Limited worldly knowledge will result in limited worldly concepts and ideas. The past was subsequently littered with inaccurate assumptions, opinions and beliefs because our ancestors were restricted in their ability to reason, deduce, extrapolate and surmise due to their limited knowledge.

"Oh, that looks a little like a swarm of insects on the Moon – it therefore cannot be anything else other than a swarm of insects!"

Life would be unexciting without inaccuracies and ambiguities!

Swarms of insects and postage stamps on the Moon

- *Not all theories survive the test of time.*
- *The commander of Apollo 15 had a few surprise deliveries for the inhabitants of the Moon, much to the disbelief of NASA.*
- *Scientific predictions about the future have proved very difficult to establish and it is a good job they are sometimes wrong.*

Without the appropriate level of scientific knowledge, backup and support, it is quite easy to develop incorrect theories. But in some respects it is more important that an incorrect theory exists, rather than there being no theory at all. At least we can improve an incorrect theory, whilst in the absence of any theory whatsoever we may lose sight of the topic completely. For example, if Edwin Hubble, 1889 – 1953, had never proposed the theory that the Universe was expanding, perhaps we would be oblivious to another key theory that stems directly from this discovery.

The theory that the Universe was expanding came as a major revolution to the world and was eventually confirmed with observational experiments. However, in 1998 it was realised that Hubble's revelation was only partially correct – it was revealed that not only was the Universe expanding, but the rate of expansion was increasing. This meant that for the first time we could visualise the fate of the Universe – by all accounts it looks like it will expand forever!

Edwin Hubble came up with his original theory based upon the study of the apparent red-shift of light observed within the glow emitted from distant stars.

Many theories have survived the test of time; but others have not.

Aristotle, 384 BC – 322 BC, erroneously taught that the human heart was responsible for the sensation of feeling and was the centre of personal intelligence. He also taught that living creatures could be spontaneously created without parents or ancestors. I am not sure what could have brought him to that weird conclusion!

One of my favourite erroneous beliefs has to be telegony associated with pregnancy. This was the amazing theory that animal and human offspring could inherit the characteristics of a previous mate of the female parent. This meant that the child of a widowed or remarried woman might have traits that resembled those of a previous husband. What is equally amazing about this belief was that it was widely understood as 'the way of the world' until the late nineteenth century!

This ridiculous belief would have had an enormous impact regarding the way people, especially women, conducted their lives. Imagine the long term guilt and despair that would be felt by any person having a secret affair with someone. For the remainder of her childbearing days, any woman would live in fear of giving birth to a child resembling their secret lover, or even the postman!

This theory prevented women living promiscuous lives – the worry of having an early affair, then later in life giving birth to a child that resembled the earlier lover would be too much to endure in society. Alternatively, this belief could have been brought about by men denying affairs – to such an extent that it became a mainstream belief.

Aristotle incorrectly deduced that heavier objects fall faster than lighter objects; we needed Sir Isaac Newton, 1643 – 1727, to correct this view.

However, gravity was also misconstrued by Newton at the start of his research. Like Aristotle, Newton originally thought that heavier objects fell faster than lighter ones – until he eventually revised his thoughts. He corrected his views by stating that everything fell at the same speed. To state that everything fell at the same speed was revolutionary, especially when here on Earth a feather and a hammer are seen to fall at totally different speeds. Newton was right – the only reason why objects fall at different speeds on Earth is due to their air resistance.

Leonardo da Vinci also mistook how a falling object behaved; he stated that a falling object's speed increased the further it fell, when in actual fact it increases the longer it falls. He had confused time with distance.

Newton's revolutionary predictions were officially and dramatically proved in the absence of an atmosphere on the Moon in 1971. Commander David Scott famously dropped a hammer and a feather at the same time for them both to hit the Moon's surface simultaneously during a moonwalk on the Apollo 15 mission.

Interestingly, on the same mission, Commander David Scott and the other Apollo astronauts had very strangely decided to take three hundred and ninety-eight commemorative postage stamp covers to the Moon without the permission of NASA!

Upon the safe return of the envelopes, one hundred of them were sold to a German stamp dealer before NASA realised what the astronauts had done. The astronauts were disciplined and were never to fly again.

Extremely strange but true!

The value of these commemorative postage stamp covers have increased over the years, with one selling in January 2008 for eighteen thousand dollars.

Nobody at mission control knew the three hundred and ninety-eight commemorative envelopes had been taken to the Moon, so the extra weight within the lunar module could have easily caused problems. The secretly stashed payload could have had catastrophic consequences, becoming an extremely high price to pay for the comparative profits that would be obtained by the astronauts upon their safe return. If the weight of these envelopes had prevented the successful take-off of the lunar module from the Moon's surface, or its successful docking with the mother ship, there would have been plenty of questions raised.

It was an extremely strange and penny-pinching action to attempt when participating in such a costly and truly momentous occasion in history.

Incidentally, the astronauts had secretly decided to take four hundred envelopes to the Moon, but two were damaged before take-off – hence the figure of three hundred and ninety-eight.

The forefront of technology has always been susceptible to mistakes made by revered scientists. Doctor Dionysius Lardner, 1793 – 1859, was an Irish scientific writer who popularised science and technology. He authored numerous mathematical and physical reference books on such subjects as algebraic geometry, differential and integral calculus, the steam engine and natural philosophy, but is most famous as the editor of Lardner's Cabinet Cyclopaedia.

You would have thought Lardner would have been forward-thinking and adventurous in his outlook. However, he very inaccurately announced to an extremely influential audience that a steamship would never be able to sail across the Atlantic because it would not be able to carry enough coal. It was rather awkward for Doctor Lardner when in 1838 his credibility was in question after Isambard Brunel's SS Great Western successfully docked in New York, completing the first steam powered Atlantic crossing. Considering that steam engines were one of his areas of specialisation – it made people of the time wonder how much else of his influential publication was inaccurate!

Doctor Dionysius Lardner was also responsible for stating that if railway trains went too fast then the passengers would all die of asphyxiation.

It is perhaps a good thing that sometimes the best scientific minds are revealed to be wrong!

The famous Austrian physicist and philosopher Ernst Mach, 1842 – 1907, believed both the concepts of matter being made up of atomic structures and the theory of relativity to be completely false and dogmatic.

Percival Lowell, 1855 – 1916, mapped and announced the existence of over five hundred Martian canals which later turned out to be just an optical illusion.

The reputable astronomer William Pickering, 1858 – 1938, glibly announced that a number of dark spots within the Eratosthenes crater on the Moon were swarms of insects or herds of small animals. "Oh, that looks a little like a swarm of insects on the Moon – it therefore cannot be anything else other than a swarm of insects!"

Young children and babies learn by mistakes, scientists seem no different!

We can all remember the accidents we had in our childhood that gave us aches and pains from banged heads, burnt fingers and grazed knees. In a similar way we see scientists get red faces and bruised egos as they witness failed projects, incorrectly proposed theories, poorly thought-out predictions and scientifically initiated environmental disasters.

Within our highly automated and consumer oriented world, Mankind's future mistakes may well come at a much higher financial and environmental cost.

All of this does not conjure a particularly pretty picture!

A lesson in mistakes

- *People and animals are not infallible – all manner of errors and repeated mistakes are made.*
- *Time can be spent worrying about an important occasion – when the day arrives it is totally unlike you expected and much more pleasant.*
- *We seem unable to cope with the vast amount of refuse produced – we create the need for power, but have difficulties sustaining the level required.*

We all make mistakes. People have design faults that inevitably lead to slip-ups. It is very true, and being on the receiving end can be extremely frustrating. Try as we might, we continue to repeat the same simple, preventable errors every day.

Humans seem pre-programmed to make mistakes. We look without seeing, forget things in seconds, are notorious for falling off ladders and think we are way above average in most things. Human error is consistent, so there must be reasons behind all these mistakes.

We slap our foreheads as we forget names and facts, say the wrong things, forget computer passwords, fall for optical illusions and forget where we stashed our jewellery. It is definitely down to human design rather than personality or intelligence. The way we see, think and remember sets us up for mistakes.

We are subconsciously biased, quick to judge by appearances, prone to sticking with old strategies that work poorly in new situations and overconfident of our own abilities. We seem to think we can get tasks done in a shorter time than is actually possible, know we are right when we are wrong and often misjudge our physical limitations.

In a world where a large majority of people have a preference for the number seven, favourite colour as blue and when asked to name a vegetable, say, "Carrot," no wonder we make mistakes!

Very often we make the very same silly mistakes!

So, why do we make mistakes?

The Universe mysteriously presents matter in two very different ways. There is the world of the large where we observe objects made up of trillions of tiny particles organised into the everyday shapes such as pendants, pegs, planets, peas and people. Then there is the world of the very small where we observe these same particles individually, where they act very differently.

Within the familiar world of the large, we see cats jump on hot tin roofs, sheep accidentally fall off cliffs and humans fall off ladders. However, at the subatomic level mistakes do not seem to occur, there appears to be complete consistency.

At the subatomic level, chemicals will always react in exactly the same way, gravity will always persist and light will always travel. We are not familiar with there being any mistakes at the level of the small.

We do not get up one day to find that the light from the Sun will no longer travel to us, gravity has become stronger or magnets are suddenly twice as powerful. If we introduce two chemicals that normally react, and then one day they fail to react – this just would not make sense. This never happens.

So at which point, from tiny quarks through to human beings and blue whales, do mistakes start to be introduced? At precisely what scale of magnitude do errors get introduced? For instance, I once mistook a closed glass patio door for being open – I'm large; there is an awful lot going on within me – I made a mistake and I hurt myself. Somehow, I very much doubt that chemical elements will make mistakes by behaving differently under absolutely identical environmental conditions, even if the experiment were to be repeated trillions of times – or is this not the case?

As we begin to observe larger and larger items that depend upon trillions of particles of matter all working together, there comes a point where mistakes become possible.

The science of chemistry and physics appear totally consistent, this is different to the nature it supports. Atoms bond to become molecules and they bond to become large clumps of matter, this eventually materialises to become a living organism. There must be a level at which mistakes become a possibility.

It is like an archer taking aim. Over extremely short distances the accuracy is amazing, but over longer distances the margin of error becomes apparent.

Having thought about this; I do not think atoms make mistakes; I do not think molecules make mistakes; I do not think lumps of rock can make mistakes; but, clearly living organisms can make mistakes and it would seem the more intelligent they are, the bigger the mistakes they can make. It would seem that the only things that can make mistakes are creatures. Because creatures are a subset of the Universe then I am pretty sure that the Universe is capable of making mistakes – so does this mean that there is some kind of life inherent within the Universe?

I remember being in a foreign country with friends. Having managed to become lost from the crowd with no means of communication, I tried to find them again. When I could not find them, my mind was awash with thoughts such as, "How am I going to get back to work from here for Monday morning?"

Looking around I felt panicky but suddenly realised that no one knew I was lost.

"Why does no one know I am lost?" I thought to myself.

It must be because I look normal and I am not wearing a hat on my head saying, "I am lost".

What does a lost person look like to a person who is not lost?

Did I look lost?

I knew I was lost, but no one in the area knew this – quite simply because I had not told them. There is no way that anyone looking at me could ever know I was lost and it would have never dawned on them that I was lost. Although this is just a small thing, when many of these small things mount up, assumptions are made, and then preconceived ideas form and mistakes are made. This is the nature of a mistake.

Amazingly, I once heard that approximately one car in a hundred travelling along a motorway is going in the wrong direction having missed an earlier turn. I personally have been caught out a few times in my life and been one of these motorists travelling in the wrong direction!

Unless you knew this fact, you would automatically assume that everyone was travelling in the right direction along a motorway.

Funny old world!

Our mistakes have unfortunately found themselves manifested into issues upon our planet. Large companies, governments and greedy individuals have overseen terrible worldly atrocities.

It has been calculated that roughly nine out of ten things we worry about transpire to have been not worth worrying about in the first instance. This is a fairly useful philosophy to adopt as it truly simplifies our lives should you choose to adopt it. The philosophy implies that it is a mistake to worry about anything. So having adopted this philosophy, what happens to the occasional thing that you forget to worry about that slips through this philosophical net?

Well you have all the energy in the world to concentrate on it as you have not been concerned about anything for a while!

However, is it possible to apply this philosophy to our planet also?

Are we worrying about our planet unnecessarily?

The philosophy may not work for our precious planet Earth, it has no voice to express its views – however, it has a number of issues and it certainly has extremely unpleasant resident humans. Only those things that present an imminent threat are discussed on the global news stations, but there are quite possibly many more planetary issues we should be worried about!

The fact that the world is broken up into so many different countries creates wildly different social, political and environmental agendas – each with their own mistakes being made. This is very worrying. How can anything settle within a world that does not harmonise globally – it would appear that mistakes on a grand scale are being exacerbated.

We live within an 'invent once, everyone has' world. We need only one intelligent person to invent the television, a mobile phone or a computer, and miraculously – we can all have one. As individuals we are no more intelligent than our ancestors thousands of years ago. However, collectively our innovation seems to have improved dramatically since we have increased in number and improved global communication. By pooling differing elements of innovation and capabilities upon a global basis we find the design of new technologies turning around in weeks rather than decades.

Despite our intellect, versatility and technological abilities, we are collectively making one almighty mistake – this is our inability to plan for the future. Mountains of landfill waste, oil slicks, global warming and an uneven distribution of wealth, all mistakes that have occurred due to our poor vision – and it is getting worse!

Planning is important in every aspect of life – so it is about time we accurately planned the resources of the planet.

Blunders before getting innovations right

- *The inventors of the electric motor faced bankruptcy due to the inability of everyone to see its potential.*
- *Global disasters are very often the direct result of mistakes made in the past – how many looming disasters are yet to surface?*
- *We sustain ourselves upon the planet by eating away at the very heart of what keeps us alive.*

Anyone would have thought that the first commercially manufactured electric motor would have been a roaring success. In 1837 Tomas Davenport and his wife Emily developed and patented the electric motor for commercial use. Running up to six hundred revolutions per minute, anyone would have thought that people at the time would have been queuing up to financially back their groundbreaking invention – especially as it had an accompanying patent.

The venture should have been a licence to print money!

However, people failed to see the scope and versatility of the electric motor. This, coupled with the decision to use zinc electrodes, which were extremely expensive at the time, placed the Davenport's into financial difficulty. The Davenports were forced into bankruptcy despite their electric motor being easily capable of powering printing presses and tools.

I sincerely hope that aliens were not monitoring the human race at this time – they would have been astounded!

The inability for such a groundbreaking invention to take-off, would have given any observers of Mankind very grave doubts about their ability to advance!

Quite often in the field of science, theories not only crack, they often crumble and fall into an abyss. Sometimes ideas are discarded and the theories have to be totally rediscovered at a later date. Scientific theorists' arguments ensue and camps of believers and nonbelievers are drawn to their specific corners.

Both the famous physicians Nikola Tesla, 1856 – 1943, and Ernest Rutherford, 1871 – 1937, believed that nuclear power was never going to be possible.

These are most revered and respected scientists!

It brings into question exactly how much of the information written within text books is actually true.

But how many mainstream theories and concepts amongst the depths of Mankind's knowledge are inaccurate, misleading or ill-founded?

Who knows?

Nevertheless, the mistakes that are being currently made within the environment are on a huge scale – these environmental mistakes will undoubtedly cause significant problems in the future. Many of these problems have been fuelled by the advancement of science and technology. Such problems now come in the form of nuclear waste, burning of limited fossil fuels to create energy, exponential population growth, major economic difficulties, mass extinction of species, rapid climate change, deforestation, tragic wars, oil spillages and the unequal distribution of food causing regional famine. It would appear that these mistakes created by poor planning and foresight in the past have now manifested themselves as potential global disasters.

Many are hoping that science and technology can safely dig everyone out of the hole into which it appears to have got everyone.

Another of our biggest mistakes is losing focus of what is important in life. We are also bashing things out on a computer keyboard, I am sure this was never a dream for which human beings longed.

Not when I was young anyway.

However, more and more of us are being driven to it!

When we are on our death beds will we think, "Oh ... if I had only spent a little more time in front of my computer."

Perhaps the governments of the world ought to be looking very closely at assessing the long term implications of threats to our planet. The governments perhaps ought to outline an effective recovery plan which will encourage the invention of other new greener technologies. However, we all seem to be pretty much in the dark regarding the planet's ability to survive into the future.

There are many conflicting views, but it is impossible to think that humans can continue to sustain themselves on the planet whilst simultaneously contributing towards its demise. It is like biting the hand that feeds you or sawing away at a branch on which you are standing!

It would appear that our natural scientific evolutionary path often witnesses mistakes before we get things right; this has been and always will be a hallmark trait at the forefront of technology.

Where pride is at stake, people can get seriously carried away. Just like gamblers who persistently gamble more and more trying to recoup their losses, scientists can prefabricate their research findings to create a sensational story – digging a deep hole in the process. There are many examples of failures and hoaxes in the past that have deliberately or accidentally fooled everyone, placing people in compromised positions as far as the truth goes.

If we developed the elixir of life, we would probably make a mess of using it!

The truth about the peppered moth

- *It is sometimes difficult to differentiate between old wives' tales and the truth.*
- *Bogus information has been taught to tens of millions of students throughout the world.*
- *A bizarrely inaccurate nature documentary was produced, revealing to the unsuspecting public what one individual thought happened in the wild.*

In a similar way that old wives' tales contain untruths and have been handed down over the years, our present-day knowledge can easily become interspersed with flawed theories.

Taking centre stage, a truly historic blunder that taught a major mistaken theory was the documentary that reported upon the evolution of the peppered moth.

The difficulties all began when a famous entomologist called James William Tutt, 1858 – 1911, suggested that birds could influence the rise and fall of species by natural selection.

The peppered moth was then earmarked as an example species to explain to the world how its natural selection had been dramatically and rapidly influenced by birds. The peppered moth's normal habitat and camouflage had been identified as the light-coloured lichen growing on the bark of trees.

During the industrial revolution the tree bark had reportedly become black with soot. This in turn had the effect of making the light-coloured peppered moth accelerate evolution to become dark in colour. In a very short number of successive generations its camouflage appeared to miraculously change to accommodate the rapid environmental change – only the darker moths were reported to survive and breed successfully.

This fact was documented within text books and taught to tens of millions of students throughout the world as an example of the pace at which evolution could operate.

The truth was that the peppered moth was already split into two sub-species prior to the industrial revolution; one was light-coloured and one was dark-coloured. The change brought about by the industrial revolution never had the ability to influence the moth so rapidly and so dramatically. It transpired that the light-coloured peppered moth that mimicked the colour and texture of light-coloured lichen, very rarely landed on the lichen anyway – choosing to spend the majority of its time in the canopy of the trees.

The whole concept behind the documentary seemed flawed to the film crew – despite this the producers insisted on getting the shots they wanted to show the public.

As the photographers were unable to get the desired pictures of the moths on the trees, the producers went to great lengths to get what they had been told happened in the wild.

Because the documentary to report on the demise of the lighter coloured moth had been commissioned, the producers felt compelled to deliver the end result – after all, they would not get paid otherwise!

The production crew consistently hit major problems, but rather than be deterred they soldiered on – they decided to get the footage whatever the cost.

The documentary was a success – it clearly shows a flock of birds hungrily eating the light-coloured peppered moths as they rested without camouflage on the soot covered tree bark. Darker moths were filmed in the same situation, but the birds seemed to ignore them – how amazing!

This beautifully-shot footage more than convinced viewers about the plight of this remarkable moth and its remarkably rapid evolutionary path to become 'dark' so quickly.

But then, some years later, the truth leaked out!

It transpired that the peppered moths used in the documentary had been reared in a laboratory. To get the footage required the film crew had chosen to film the feeding of these moths to captured birds inside an aviary!

The film showed the birds having a wonderful time eating these lighter coloured moths as they became visible when they landed on the dark soot-covered tree bark. However, during filming they discovered that the moths never landed on the tree bark, so to combat this, the moths were glued to the trees before filming!

Still photographs of these pictures found their way into text books too!

The whole story had been bizarrely prefabricated based upon one man's incorrect beliefs – it makes me chortle with laughter!

What is even worse, once this type of information is in the public domain, it takes years for such twaddle to vanish. Amazingly another documentary highlighting this very same erroneous story about the peppered moth was aired on British television as recently as March 2010. It would appear that the nature programme research team had come across this documented information, and chose to include these erroneous facts in their commentary – truly breathtaking.

It can take decades to eradicate inaccuracies from public knowledge – after all, we all hear people mentioning old wives' tales.

We will continue to witness erroneous teaching practices – but they are now relayed to the masses by the powerful medium of television, text books and the Internet. The lesson we need to take away from this is that we may well exist with erroneous or missing knowledge from the past and we must do all we can to identify these things, challenge them, and remedy them.

Although some things are clearly hoaxes or a prefabrication of the truth, there are other things which appear as illusions and have no explanation. As a young child might ask their parents, "Where did I come from?"

Some parents may make up a cockeyed story about a stork whilst others may duck the question. However, some parents may be inclined to blurt out the inconceivable truth to the poor unsuspecting child.

This world is full of objects and events we witness but have no logical explanation for. Therefore, in a similar way to people having been misinformed by the authoritative report about the peppered moth, the young child may well believe the cockeyed story about the stork.

We all have to be careful about what we casually believe!

Black and white mysteriously making colour

- *When scientists cannot provide any satisfactory explanation to the outcome of an experiment, they may choose to ignore it.*
- *Incredibly amazing and truly baffling results are produced when two black and white slides are overlapped.*
- *Our brains seem capable of adjusting and manipulating images remarkably well.*

Whereas the peppered moth story portrays a saga of wide-ranging inaccuracy and misrepresentation, scientists may choose to totally ignore an experimental outcome, purely because there is no logical explanation. Scientists may provide no explanation and no reference to their experiment simply to 'save face'.

A fabulous illustration of an experiment with no logical explanation regarding its outcome is one which involves mysterious phenomenon associated with light. It is a great example of how our understanding of light may be incorrect – or at least the experiment reveals our understanding of light to appear incomplete.

Edwin Land, 1909 – 1991, scientist and inventor, best known as the co-founder of the Polaroid Corporation, conducted a considerable amount of experiments with filtered light and image projection. One such image projection experiment produced baffling results using two black and white slides. One of the slides in the experiment was a simple black and white image of a colourful scene, and the other slide was a black and white image of the same colourful scene but taken through a red filter. Both black and white slides could be projected individually through a projector to produce a black and white image on a screen. The results on the screen were clearly seen as pure black and white images.

Now here comes the weird bit.

Edwin Land then proceeded to project both images together from separate projectors. He made the images overlap each other on the screen in precisely the same place. Unsurprisingly, he saw a combined black and white image produced from the two images, just as you would expect. But something very strange happened when he inserted a red filter in front of the projector that was producing the black and white image taken through the red filter. Miraculously and inexplicably, the picture became colour!

He saw exactly the same colours that were present in the original scene; colours became visible upon the display screen where he should be seeing a slightly red tinge of black and white!

He saw green, blue, yellow, brown, orange, purple, all shades of white and shades of black as well as reds!

How could this be possible?

How could it be that two black and white slides can create colour in this way?

Amazingly – Edwin Land never worked it out, and to this day no one really knows!

Upon very close analysis of the actual image projected on the screen, it has been found that it is made up solely of shades of black, white and red. After all this is all it can be as no colour is coming from the projector other than shades of red.

The only explanation is that the human brain is imagining and witnessing the colour due to an optical illusion – the colours that are seen do not exist in reality!

This experiment astounds everyone and it boils down to there being a lack of interpretation regarding scientists' understanding of the human being's concept of light. Although the setup is fairly simple to explain, the reason for it happening has totally eluded and baffled scientists.

The closest anyone has got to a plausible explanation of this secret of human vision is to imagine what a green apple looks like at different times of the day. When you look at the same green apple at dusk, midday and sunrise, the brain seems to know it is the same shade of green being viewed – even though the ambient light is dramatically different. There is a complex colour adjusting process that takes place in our brains that ensures the continuity of colour through different strengths of daylight.

This effect is believed to be a feature of human colour perception called 'colour constancy'. In 1971, Edwin Land developed his own 'retinex theory' to explain the observations he made. He formed the word 'retinex' from 'retina' and 'cortex' as he believed both the eye and the brain are involved in the process.

This effect has never been explained by scientists. There is no reason why this happens, or where on earth the illusion of colour comes from. Light's behaviour under these circumstances is most mysterious and because there is no explanation whatsoever it is strangely omitted from text books!

The question remains: How do our minds manage to interpret the original colours that were in the original scene – especially when we know that these colours have not been captured within the black and white slides?

The illusion cannot be caused by association – for example thinking that lettuce can only be green, as the illusion happens just the same with randomly coloured book ends or a display of colours on swatches.

The truth about this phenomenon is that it appears to be just an illusion. The difficulty is that even when you know it is an illusion, and you have zoomed very closely into the image and seen there are only shades of red, black and white, this still does not explain why your brain actually thinks it sees the colours.

The frustrating part is that you actually see the colour which you now know is not really there.

It would appear the illusion is in your brain, not in the image itself.

There is no current explanation for this phenomenon known as 'colour constancy' other than the earlier explanation of comparing a green apple in natural light at different times of the day!

A multitude of other illusions could be clouding our view of the world, only time will tell. If our minds are capable of distorting the images around us, then perhaps we need to re-examine everything. Our understanding of existence, consciousness, our senses and what is 'normal', could all be mistaken. Therefore we need to embark upon a new voyage of discovery to determine whether our approach is correct and get a handle on the unknown.

Goodness only knows where such an eccentric voyage of discovery may lead!

Within this chapter we have learnt a little about where we live. We know that to prevent further devastation to the planet and to progress our understanding of the Universe, we must think in a very different way. We have learnt that by thinking differently we can make significant headway. We know that much of the knowledge we possess could be based upon ancient eccentric views. We know that life would be unexciting without its idiosyncrasies – but this does not necessarily conjure a particularly pretty picture. We know that planning is an essential part of life, especially where resources are concerned. We realise that the human race often does not recognise a good thing when it is staring it in the face. We have to be careful about what we believe as we enter an era that encompasses a new voyage of discovery. The mysteries of the Universe should be beginning to unfold.

You may begin to get excited!

A NEW VOYAGE OF DISCOVERY

Where we learn how guns have provided humans the power to make their presence known upon Earth. We see how Mankind's voyage of discovery has resulted in our presence being detected in the atmosphere. We look at how mimicking nature has made progress in a world that is witnessing an enormous population boom. We also see how preconceived ideas can hamper our ability to be innovative and make progress.

One hundred million bison skins for a few dollars each

- *Mankind has clearly affected the wellbeing of plants and animals – a combination of greed, extortion and fundamental needs are to blame.*
- *Mankind has hunted many animals to extinction – the scale of which has sometimes been huge.*
- *Sanctuaries are a last-ditch attempt to prevent the extinction of many animals – it is a shame we have to resort to this measure at all.*

A significant number of animal species have had their numbers detrimentally affected by Mankind.

In the middle of the nineteenth century the bison population of America was estimated to be somewhere in the region of one hundred million – however in the 1870's this number plummeted dramatically due to unadulterated greed and corruption. The bison were hunted for their skins which were worth just a few dollars each; their carcasses were left to rot on the plains.

In the height of the hunting it was thought as many as one hundred thousand bison could have been killed each day. The hunters learnt to shoot at their lungs because their heads were so thick that the bullets often bounced off. To make matters worse for the bison, many hunters would have several guns – allowing the barrels to cool down whilst firing rapidly.

Train loads of bison skins were transported to the cities until it dawned upon the hundreds of commercial hunting teams that there were not many left.

What a surprise this must have been to them!

If this mass slaughter was not tragic enough for the bison, brace yourself for this. In 1874 a federal bill to protect the few remaining bison was vetoed by President Ulysses Grant. His reason for rejecting the protection of the small number that remained was simply because the Indigenous American Indians relied upon bison for their survival.

In an attempt to occupy the land, the Indians had been displaced and killed by the tens of thousands – but still they caused the government problems by staging battles in an effort to defend their territory.

It seemed a better idea at the time to completely eradicate all the bison and therefore totally deprive the American Indians of their source of food and clothing.

Knowing that bison were crucial to their way of life, in 1875 General Philip Sheridan, who was a close friend of President Ulysses Grant, successfully pleaded to a joint sitting of congress that the remaining herds should be slaughtered to totally deprive the native Indians. The planned extinction of bison was passed by government as the preferred option, rather than protecting the remaining few. By 1884 the government had succeeded and bison were very close to total extinction.

The remaining bison that exist today are much indebted to the efforts of a Scotsman called James Philip, 1858 – 1911, who managed to rear a small herd of five up to a reasonable number. It was only thanks to dedicated individuals such as James Philip, and a few isolated small herds that were discovered and amalgamated to improve the genetic diversity that bison actually survived as a species.

As for the American Indians, they were treated badly right the way through to 9th January 1918 at the Battle of Bear Valley which is recognised as the last of the American Indian Wars, in which the Yaquis chief died from a wound to his chest. As a race we cannot be proud of any of this whatsoever!

Elephants, lions, tigers, pandas, polar bears, rhinoceroses and hippopotami have all been treated extremely badly by the human race, and now we find it almost too late. This has meant establishing sanctuaries to attempt to prevent their total demise – a very sad state of affairs.

Although human beings are a product of nature, with all their wisdom and ability, it is difficult to see how they have created anything other than destruction and devastation. Humans picked up sticks to use as arrows to survive, but once they had invented the gun, events became somewhat different and spiralled out of hand. Guns, poaching and human greed have led to the demise of species on a cataclysmic scale.

The indigenous American Indians were very thoughtful people and believed very strongly in looking after planet Earth which they called their Mother. Whenever they took anything from the planet, they replaced it. For every tree they chopped down they planted another one, for every bison they killed, they bred another one. This way of life was decimated by the early European settlers with their insatiable appetite, their get rich quick culture, their inconsiderate nature and greed. Alcohol and guns were introduced and this perfect way of life slid away to be replaced with impiety, immorality and shamelessness.

Governments of the world still show no compassion!

Discovery of the passenger pigeon

- *Humans evolved by means of hunting and clearing trees for dwellings – the overpopulation of the plant means this has become problematic.*
- *The Earth's ecosystem was initially untouched by early man, natural vegetative recovery was possible because there were so few people.*
- *Species are becoming extinct at an alarming rate – Mankind's influence on the planet is clearly to blame.*

Our ancestors worked hard and meant well, we can be sure of this. However, they started a trend that is proving difficult to bring to a halt. Our ancestors innocently collected the Earth's resources and cleared areas for farming on a small scale. These early days proved fine, but popular trends are inclined to build up momentum.

If the trend in question is the automated production of essential food, this will no doubt develop into a major dependency – especially as the very item being produced will sustain an increasing population. The production and the increased dependency of this production will soon spiral exponentially out of control. There are clear signs of this happening now – the automated methods of farming are constantly being improved to support an ever increasing population.

Like any attack upon a finite resource, a crisis point will always be reached if it goes unmanaged. So what started off as innocent farming methods and a simple means of benefiting from resources on a small scale has escalated out of hand. With the arrival of machinery, the simple methods of the early days very quickly became much more productive with automation.

Unfortunately no one spotted that the transition to automated methods would bring a whole set of difficulties – until it was almost too late.

Having been born into a world that had already been plundered and decimated prior to our arrival – many see this mistreatment as our ancestors' fault.

But is this true?

Our ancestors did not see the situation getting out of hand decades before us. However, many now believe it is about time we all managed the situation better before the situation gets any worse.

We can only imagine the trees which used to inhabit the vast and empty plains. The problem may sadly be more severe than anyone realised – unfortunately it appears that many environmental difficulties are impossible to do much about. For example, we cannot bring back extinct creatures and we cannot replace the fossil fuels we have burnt that took the Earth hundreds of millions of years to create.

Humans have lived for millions of years by hunting – they cleared a small patch of ground for habitation which quickly grew back once they had moved on. As a consequence, early humans did not leave much of a trace; in fact they left so little we have missing links within our evolutionary path.

Unlike today!

People in the future will be able to map every minute of our current lives. They could simply dig up the millions and millions of tons of waste we have produced, and by looking at the 'sell by' dates on everything I am sure they could piece together our every moment!

Early man made no significant impact on the Earth's ecosystem; they did not seem to cause any damage quite simply because there were not that many of them.

A small group of termites can move small mounds of earth, but by adding vast numbers of termites, greater feats become possible. Therefore with enormous numbers of termites, massive mounds of earth can be moved and reshaped. Similarly, with large numbers of human beings, we are capable of chopping down all the Earth's trees!

Humans are upsetting the balance of nature extremely rapidly, which may be reflected upon in two ways. Many believe that humans are nasty, devastating creatures that must be carefully managed to prevent any further damage. Alternatively, many believe that humans are only part of nature itself and the direction the world is heading is how everything is supposed to be.

Incidentally, as far as nature is concerned, humans are the only creatures that leave behind any lasting legacy other than nests, burrows, shed skin, eggshell, bones, fossils or fossilised footprints. The billions of items of rubbish humans produce prompt me to think that we should introduce comprehensive control measures, even if this is just to err on the side of caution. After all no one knows what the future truly holds.

Respecting our planet's resources and treating the whole world like a national park may help – this should go part way to protecting the future of the planet and addressing some of the pressing environmental issues. Rather than considering the implications of our actions, we have indiscriminately disrupted the planet in so many different ways it is difficult to know whether some aspects of damage are recoverable.

When settlers in America found the passenger pigeon they were in flocks of up to two billion at a time and it was considered to be one of the most numerous of all birds throughout the world. A single flock could be a mile wide and three hundred miles long and take hours to pass overhead. These wonderfully social birds had as many as one hundred nests to a tree and migrated from the forest habitats throughout North America and Canada to Southern United States, Mexico and Cuba.

They were reported to be in infinite numbers by early settlers. They were hunted inexhaustibly by commercial hunters to be used as food for slaves, agricultural fertilizer and live targets for sport trap shooting. Eventually after this seemingly endless shooting and netting there was only one left called Martha that was kept in Cincinnati Zoo. Martha died on 1st September 1914. There are two memorials to Martha the passenger pigeon, one on the banks of the Mississippi river in Wyalusing State Park and one at Cincinnati Zoo.

Well done everyone!

Of the approximate ten thousand different species of bird on Earth, two per cent of species have become extinct within the last five hundred years, with another twelve per cent considered to be under threat of extinction. The rate at which species are entering the bracket of pending extinction is increasing at an alarming rate. Strangely, we seem to have a fraternity of human beings who are infatuated by endangered species to such an extent that they will pay good money to have them poached as pets or trophies. With more people frequenting the planet, the more widespread this will unfortunately become.

A very sorry state of affairs.

What horrible creatures we are!

The ozone layer, toxic waste and Thomas Midgley

- *Many people are not very proactive regarding environmental issues – however, resources are limited and time is ticking.*
- *The Earth is a finely-tuned planet with many critical environmental components; a change in any one of them could be devastating.*
- *Mankind's activities can now be detected in the atmosphere – the consequences are unknown.*

The energy we use, the deforestation of the planet, damage to the ozone layer, the pumping of tens of billions of tons of carbon dioxide into our precious atmosphere every year, toxic waste and the general pollution of land and sea have all got their associated tragic stories. The whole ecosystem is just not sustainable at this rate.

One awful tragedy appears to be that many human beings have no hesitation in putting themselves first before the environment. Very few people in comparison are actively doing anything to reverse this awkward situation. The people who are actively doing something about it are just a tiny proportion compared with the numbers of people responsible for its decimation.

Despite the most amazing two hundred years of technological innovation we cannot be proud of what we have achieved upon the planet. We have discovered the use of oil and now pump one hundred million cubic metres of oil from the ground every single day. If that is not bad enough, we then convert all this oil into greenhouse gases.

We have massively damaged the ozone layer. We pump thirty billion tons of carbon dioxide into the atmosphere every year which has the effect of warming the planet's surface; this is now equivalent to having increased the natural amount of carbon dioxide in the Earth's atmosphere by nearly forty per cent.

Toxic waste causes death to creatures when it finds its way into streams, rivers, lakes and the oceans. Toxic waste is often transported around the world onboard ships for dumping – quite often to less advanced countries who have little experience of handling it.

The filling of disused mines, landfill sites and deserts with billions of tons of rubbish and dumping waste into the sea are common practices.

We have done well!

We have multiplied our population by eight fold in just two hundred years and this trend looks set to continue. We have chopped down the Earth's original six million square miles of mature tropical forests at a rate of roughly an acre per second to a mere one and a half million square miles. Along with this deforestation we have lost tens of thousands of species in the process.

We occasionally have proud industry spokesmen with scientific-looking people milling around in the background, pulling the wool over our eyes by announcing such things as, "There is a decline in the rate of growth of deforestation."

This still means it will all be gone by the year 2090, except for a few national park reserves. Many are predicting an extremely sad and desperate situation very soon.

As for the depletion of our ozone layer, we have Thomas Midgley to thank for this. He was once described by a historian as being the person who had more impact on the atmosphere than any other single organism.

Thomas Midgley accrued over one hundred chemical patents including chlorofluorocarbons used in air-conditioning, aerosol spray propellants, and cleaning processes achieved by exploiting by-products of chemical processes.

Since the take-up of these patents in 1930 there has been a drastic year-upon-year depletion of the ozone layer that equates to a decade-upon-decade depletion of approximately four per cent. This is a flabbergasting depletion of the total volume of ozone within the Earth's stratosphere in such a short period of time.

The absence of an ozone layer would permit biologically harmful ultraviolet rays from the Sun to reach the Earth's surface, and is therefore critical to the wellbeing of life. Even though the problem with chlorofluorocarbons was known as early as 1978, it took the signing of an international treaty called the 'Montreal Protocol', to finally ban authorised production in 1996. However, not all nations of the world conform to the 'Montreal Protocol' and production still continues to this day. The problem is also made worse because the chlorofluorocarbons that are already in the atmosphere can remain there for many years – continuing to wreak havoc on our precious ozone layer.

We shall just have to wait and see whether this one issue alone has condemned the creatures of planet Earth!

Incidentally, chlorofluorocarbons do not get produced within the natural world which means their abundance in the atmosphere can only be attributed to the activity of human beings.

Following Thomas Midgley's invention of chlorofluorocarbons there followed an extremely ill-fated twist, in more than one respect!

At the age of fifty-one he contracted polio which left him severely disabled. Being the innovative person he was, he invented a series of ropes and pulleys to make it easier for him to get in and out of bed. On 2nd November 1944 he unfortunately became tangled up in his system of ropes and died of strangulation.

An unfortunate way to go, but at least he avoided hearing how devastating his original invention had become. The devastation his inventions had inflicted upon the planet was discovered just a few years after his death.

Thomas Midgley – what a horrifying life-form!

The population boom

- *The population of the world has undergone an enormous increase in recent times – there are no signs of the increase abating.*
- *Longer-term strategies are required if governments are to safeguard the survival of the human race for hundreds of millions of years.*
- *The increased population is nibbling away at the Earth's resources, which can only sustain so much.*

It has been estimated that the population of the world reached one billion people shortly after 1800. Quite incredibly, just over two hundred years later, there is now a population of more than seven billion people. This is a frighteningly significant increase. Advances in medicine, improved agricultural production and cleaner living have contributed towards the significant increase in our species.

Not too many people seem to be overly concerned about this enormous population increase and it is seldom a major topic in the world news. However, when we look closely at the longer-term implications regarding the planet's resources, we find there is no way the increase can be sustained.

The population boom is similar to the overpricing of houses during a housing boom – everyone knows the bubble will have to burst at some stage. However, in human terms this would equate to starvation.

There are already signs that the planet is struggling to feed everyone. The signs of this struggle have been clear to see in the poorer parts of the world. There will always be people that will suffer on the planet, even during the good times, just like there are still people that suffer during a housing boom.

It is the population bubble bursting that we have to be wary of. Although the population bubble has burst in regions such as Ethiopia in recent times, the bubble has yet to burst on a global scale.

Having researched a few relatively reputable sources regarding human population growth in the near future, I found the figures vary somewhat. However, if we apply some simple mathematics, assuming the current growth continues at the same rate – then we will have a population approaching fifty billion by the year 2200.

This is clearly unsustainable and frightening!

A large population such as this could not be supported indefinitely, especially when they have a propensity to plunder the planet's resources. We wait to see what this population boom shall bring, whether this be a shortage of food, sociological breakdown, ecological breakdown or a catastrophic war.

Let us hope it will be none of these!

It needs just one aspect of our fragile network of technological and sociological dependencies to fail and we will have a breakdown within society which becomes extremely tricky to remedy.

The recent population boom can be put into perspective by considering the lifetime of Henry Allingham, the oldest surviving World War I veteran, born 1896, died 2009. The difference in population during Henry's lifetime went from roughly one point six billion to six point seven billion. An enormous population boom during his lifetime.

What is truly incredible is that it took four billion years for the population of the planet to reach one point six billion at the time Henry was born.

Amazingly, in the last ten years of Henry's life the total population of the planet increased by an extra one point seven billion, which in itself is more than the whole population of the planet at his birth.

Let us ponder on this incredible fact for a moment.

The staggering truth is that the population of the world increased in ten years by more than the total population of the planet one hundred years earlier. With this exponential growth it is difficult to see how the over exploitation of the planet's resources can be avoided without significant intervention and control. Perhaps the governments of the world should unite and have joint common sense agendas. However, in most countries of the world, every few years the governments change.

Invariably, a new government's approach starts by complaining about the mess they have inherited from the previous government. As a result there is a terrible lack of continuity on important matters regarding the future resources of the planet and financial control.

If nations insisted that their successive governments were jointly responsible for implementing radical improvements to the way resources, waste and population are handled – perhaps the planet would have a better chance of survival.

Perhaps we should request that governments implement multimillion-year strategies, not four-year strategies!

For example, if governments truly considered what the world may look like if they continue to build roads at the pace they are today, they will realise that the whole planet will be covered in tarmac roads by 836,420 AD. As for one hundred billion years' time, we would still be here to witness what happens if we get our strategy correct today.

With vast numbers of humans on the planet there seems no end to the devastation we are capable of inflicting.

In every field of human interest people have exploited the planet to levels that are extraordinary. One such example is our need for iron which is in huge demand for a myriad of products globally. To satisfy this demand, a massive hole has been excavated called the 'Hull Rust Mahoning Open Pit Mine', at Hibbing, Minnesota. The open hole measures over three miles long, two miles wide and is over five hundred feet deep!

The total surface area of the land upon the whole of Earth is just fifty-seven million square miles; so if humans dug up the equivalent of the 'Hull rust Mahoning Open Pit Mine' every year, then it would take less than ten million years, at this pace, for human beings to dig up the whole surface of the planet!

They are progressing very well digging this hole in Hibbing with a population of just twenty thousand people. Just imagine what could be achieved if the whole population of the planet started digging!

The mine has now produced nearly one billion tons of iron ore in total, and currently produces around eight million tons of ore a year. On top of this, roughly the same amount of waste rock is produced.

If this hole is not a large enough scar on the surface of the Earth, then just take a look at the system of roads which humans have created. There are now enough roads to stretch approximately fifty million miles, which is equivalent to roughly two thousand times around the world; or one hundred times to the Moon and back!

If you think that is staggering, then consider how dreadful it is that we have enough dedicated tree-logging roads within our valuable forests to stretch four times around the world!

If our booming population has still not surprised you enough, then try reflecting on these facts and figures.

For our growing population there are approaching one billion motor cars, sixty thousand merchant ships, one and a half billion television sets, and over four billion mobile phone subscriptions. Each year we consume over twenty billion bushels of wheat, we drink half a trillion glasses of Coca-Cola, and make over a trillion plastic bags!

People will continue to affect the planet – so long as there are no incentives to stop. The insatiable appetite for human possessions will inevitably bring about disastrous times. It may unfortunately be just a matter of when.

Perhaps improved global awareness is the answer. A better method of educating people may make them more mindful of the precious resources the Earth has to offer.

Proactive environmentalists appear unable to fully project their ideas and feelings through to all those that matter in the world – namely the world's leaders.

We unfortunately still live in an era when large corporations feel they are being 'unfairly targeted', when approached by environmentalists. At present large corporations clearly have little consideration for future resources!

Perhaps condoms, allotments and a return to moral values, is the answer to the World's problems.

Biomimicry providing innovation

- *Biomimicry – the simulating of nature to develop new innovation.*
- *We look, hear, touch, taste and smell – building up our interpretation of the world about us.*
- *Discovering the unknown may result in a whole host of new biomimicry opportunities.*

Biomimicry has brought about a whole host of valuable innovations through the observation and mimicry of nature. Velcro was invented in this way by Swiss engineer George de Mestral in 1941. Whilst returning home after a hunting trip with his dog in the Alps, he took a close look under a microscope at the burrs of burdock that stuck to his clothes and his dog's fur. He noticed the hundreds of hooks that caught on anything with a loop.

We now use Velcro as fasteners on clothes, shoes, spacesuits, wall hangings and as zipper replacements on almost everything. Other biomimicry innovations include echolocation sticks for the visually impaired, inspired by bats; water purification mechanisms, inspired by marsh ecosystems; and air-conditioning systems, inspired by termites.

We all started life as babies and only saw what our eyes glanced at; heard what our ears listened to; touched what our skin felt; tasted what our tongues came into contact with; and smelt what our noses sniffed. Throughout our lives, using these senses, we have built up our interpretation of our own known universe; which is constantly forming. It enables us to familiarise ourselves with normal everyday life – comparing it with a catalogue of previous experiences.

These life experiences act as our brain's input and constitute everything we ever get to know.

We are gifted with the power to deduce; this enables us to compare experiences and objects with others. This makes it possible for us to deduce biomimicry-oriented solutions.

The power of deduction allows us to develop conclusions which may in turn become established as knowledge in our minds. However, the knowledge we deduce may be wrong!

"Oh, yes," said Fred, "I've travelled for miles and miles, everywhere seems flat – so the world must be flat."

This conclusion was so plausible, that for years it was never put into question that the world could be spherical.

Although we rely on our deduction skills to make progress, too many deductions will provide many incorrect conclusions that can lead to difficulties. Experts concluded that the Titanic was unsinkable and so put too few lifeboats aboard. Early attempts to fly were hampered by our incorrect aerodynamic conclusions – flapping wings were not enough. The introduction of certain species into parts of the world to combat pests has very often backfired – the introduced species overruns the area becoming problematic itself. Economic experts regularly believe that company finances are fine, only to find that they have failed to facture in all the indicators and the true conclusions are grim.

How many times have you been adamant about something – only to find you were wrong?

Humans only ever get truly familiar with their environmental surroundings, their friends and relatives, their brain's input and of course their conclusions based upon deductions from this input. Unfortunately, many of the conclusions that have been deduced during your life may be wrong – this is why early attempts to develop flight using biomimicry failed. The engineer's conclusions based on their powers of deduction were incorrect.

Biomimicry may not always provide the complete answer, but it can ultimately lead us to the correct solution. The fact that a sycamore seed hovers through the air, does not mean that a helicopter should be exactly the same but on a larger scale – we have to modify nature's design. The fact that a dandelion seed parachutes gracefully in the air is interesting – however, when we come to design a parachute, we find that lengths of material are better suited.

The genius of nature is a wonder to behold!

Our current innovations based upon biomimicry only extend to the known aspects of the Universe. Just imagine how innovation and technology could be bettered if we had a full understanding of all the unknown phenomena within the Universe.

Who knows what would be possible?

Perhaps we could mimic energy transfer and switch ourselves into another form of energy, then travel through the Universe at the speed of light – converting ourselves back into humans upon our arrival elsewhere.

Let us hope we can unlock the secrets of the Universe and discover the unknown very soon!

Beam me up, Scotty!

Do not run with preconceived ideas

- *Preconceived ideas often get people into difficulties.*
- *People quite often argue about the outcome of situations based upon their own blinkered view.*
- *By knowing that one of two children is a boy makes it twice as likely for the other child to be a girl – many people cannot follow this – but it is true.*

Running with preconceived ideas and views has caused significant problems throughout history. Major bridges have failed, buildings have collapsed, wars have been lost and significant projects have been abandoned, all due to beliefs that certain preconceived ideas were correct. Oversights, incorrect assumptions and flawed crucial knowledge often cause fundamental components to be overlooked or underestimated.

It therefore makes great sense for us to keep preconceived views and ideas to an absolute minimum. Failure to do so may well hinder advances in science, technology, manufacturing and construction.

Imagine how you feel when you are informed of something that contravenes knowledge you previously believed to be correct. It may be that you find a close friend turns out to be untrustworthy when your earlier belief allowed you to trust them; or you find that an important appointment was on Thursday, not Friday – and you missed it; or you discover that you have ill health when you thought all was fine. Preconceived views of this type often cause anxiety and bring difficulties to our lives.

When these types of difficulties occur we often feel it keenly and it becomes heartfelt. Our minds experience an almighty, en masse, acquaintance-shift with significant mental affairs within an instant – often placing other factors of our lives out of kilter. Perhaps you were booked to go on holiday with this 'now' untrustworthy person; perhaps by missing your appointment your job is on the line; or perhaps your ill health has meant you miss a key family occasion to which you had been looking forward.

These types of moments bring psychological pain as our minds try to readjust our lives to accommodate the shift from our perceived normality.

Preconceived ideas are used by everyone on a daily basis – just to prove how inaccurate they can be, try the following example.

What follows may be awkward or difficult to comprehend at first – but stick with it. Supposedly intelligent friends of mine have tried to follow the logic within this example and have ended up disagreeing profusely with the actual outcome. They were wrong!

This example truly helps reveal just how strongly our preconceived ideas make us believe in what we want to believe.

If you can follow this carefully, I can guarantee you will be amazed!

There is a conundrum called the 'boy or girl paradox', which involves two carefully phrased questions that has led people over the years to argue about the outcome.

The first question is:

Mr Jones has two children. The older child is a girl. What is the probability that both children are girls?

The second question is:

Mr Smith has two children. At least one of them is a boy. What is the probability that both children are boys?

This puzzle was first published by Martin Gardner in 1959; he initially gave the answers as a half and a third, respectively.

Shortly after publishing, Gardner acknowledged that the second question was a little ambiguous and the answer could be a half rather than a third depending on how you found out that one child was a boy. For instance, if you found out that the eldest of the two children was a boy, then the chances that the other child is a boy is a half. However, if all that you knew was that at least one of them is a boy and categorically nothing else about them, then the chances that the other child is a boy is just a third.

So if top mathematicians can get, what appear to be, relatively straightforward calculations wrong – what hope is there for the rest of us?

There are variants of this question with differing degrees of ambiguity.

As a result the paradox has caused a great deal of controversy since it was first proposed. Professors of Mathematics have argued strongly for both answers with a great deal of confidence – some on occasion have even scorned those taking an opposing view.

It is the answer to the second question that is problematic.

Mr Smith has two children. At least one of them is a boy. What is the probability that both children are boys?

The intuitive answer is that the probability is a half – equal fifty-fifty. However, when you begin to dissect and analyse the actual possible outcomes it becomes clear that the probability is just a third.

The logic is as follows – you only know that at least one of the two children is a boy. Knowing that one is a boy does not tell you whether it is the first child or the second child. This gives you the outcomes of girl/boy, boy/girl and boy/boy. Notice that the outcome girl/girl is missing because obviously one of them is not a boy. Now when you look at the three possible outcomes where only one of the children is a boy, you will notice that only one of the three possibilities has an outcome where the other child is a boy.

This means that by simply knowing that one of the two children is a boy, makes the other child twice as likely to be a girl.

This is not what your intuition tells you – but it is true!

Please do not lose the will to live if you have not followed this – it is tricky!

If you still cannot believe this and need further clarification then take two coins to represent the children. Decide in your mind whether heads or tails represents a boy or a girl. Toss the two coins together repeatedly. Each time, so long as one of the two coins represents a boy, as you know that at least one of them is a boy, then make a note of the other outcome. If you do this a number of times you will begin to notice that the other coin will be twice as likely to represent a girl as a boy.

This is extremely counter-intuitive as we have been taught that a flip of a coin is fifty-fifty heads or tails, and that girls and boys are born randomly in a similar fifty-fifty manner.

If you perform one hundred tosses of the two coins you will note that so long as one of them represents a boy, the coin representing a girl will show twice as often. So after one hundred successful tosses you will note something like sixty-seven girls and thirty-three boys as the gender outcome for the other child.

This has caused great arguments, but it is true!

People who run with preconceived ideas without listening to the logic of others, or without challenging long-held beliefs, may find themselves in trouble from time to time.

The reality of happenings in relation to other happenings can often be very deceptive!

More preconceptions

- *Preconceived ideas about the sex of a puppy at the pet-shop.*
- *Preconceived ideas about opening doors.*
- *Preconceived ideas about tossing coins.*

Another means of examining closely the 'boy or girl paradox' is by looking at another similar tale, but this time based around a pet-shop.

The preconception is best portrayed by imagining that a pet-shop owner tells you that there are two new puppies for sale. However, it is not known whether they are both male, both female or one of each sex. You tell the pet-shop owner that you only want a male puppy. So the pet-shop owner telephones the person who has the two puppies and asks, "Is at least one a male?"

The answer is affirmative.

"Yes," the pet-shop owner informs you with a smile, "I have a male puppy for you."

At this point, you know that one of the two puppies is a male – and this is categorically all that you know about them. The question now is, "With the probability of a puppy at birth being fifty-fifty, male or female, what is the likelihood that the other puppy is also male?"

Amazingly, when knowing the information precisely as detailed here, the sex of the other puppy is twice as likely to be a female!

"What!" I hear you cry, "This cannot be the case!"

Again – do not lose the will to live.

The explanation is a follows.

In a world where the sex of a puppy is fifty-fifty male or female, there are four possible outcomes in respect of their births which are; two females; two males; a female and a male; or a male and a female.

As we know one puppy is a male, we can safely eliminate the option where there are two female puppies. This leaves us with three possibilities, namely; two males; a female and a male; or a male and a female. Two of these three possible remaining options have the other puppy as a female. Therefore the sex of the other puppy is twice as likely to be a female.

You may take a deep breath if you wish – you may need it!

Another difficult conundrum to absorb that highlights that we often run with preconceived ideas and views, is called the Monty Hall problem. It is a puzzle based on a television game show which involves there being three doors, two of which have a booby prize behind and one which has a star prize. When the contestant has chosen a door, the game show host then opens one of the other two remaining doors revealing one of the booby prizes.

At this point the contestant is asked whether they would like to swap doors or not. Most people in this situation choose not to swap – sticking to their original gut feel and instinct.

The contestant's reason not to swap doors is further substantiated in the mind by the thought of living with the option to swap and perhaps whilst doing so altering the choice to a booby prize – after having initially chosen correctly. This would play on the contestant's mind for a long time.

As it turns out, it is twice as likely that you will win the star prize if you swap doors at this point.

Deep breaths once again!

The best way to think about this is by considering the state of the three doors prior to choosing one. There are two booby prize doors and just one star prize door, so you are twice as likely to pick a booby prize as the star prize. Once the game show host reveals one of the two remaining doors with a booby prize behind it and offers the swap, the swap should be taken because the other door now offers a two thirds chance of being the star prize – rather than the chance of one third when you originally chose the door.

If you still do not believe this, then get a friend to assist with the setup with real doors and astound yourselves.

These types of conundrum can sometimes be challenging to follow – so do not worry if you are struggling – but there is just one more great example of how your brain can run with preconceived ideas that I would like to share with you.

It is called the Lewis Carroll pillow problem.

Inside a pillow case is one counter. This one counter has an equal probability of being a black or a white counter. Into this pillow case with the one random counter you now place a white counter and shake. You reach in and pull out a white counter. What is the probability that the other counter is also white?

"Oooooooooh," I hear you cry.

Let us recap – you placed in a white counter and pulled out a white counter, leaving the random counter which you would think was fifty-fifty.

Therefore you may think it is fifty-fifty as to whether the remaining counter is black or white.

However this is not the case!

Amazingly you are twice as likely to pull out a white counter. If you do not believe this, then get a pillow and try it!

I did promise no more – but this one is just marvellous. So from the depths of the eccentric Universe I bring you the 'heads, heads, tails or heads, tails, heads' conundrum. I promise this is the last one.

Again – an interesting observation that highlights our susceptibility to believe what we want to believe and revealing some extremely strong preconceived views that we hold.

We all know that tossing a coin has a fifty-fifty outcome; it will be either heads or tails. However, when we specifically seek certain occurrences of coin tossing sequences, the results become rather confounding.

Let us determine the average number of times a coin must be tossed before we achieve the exact sequence 'heads, heads, tails'. We will also determine the average number of times a coin must be tossed before we achieve the exact sequence 'heads, tails, heads'.

Irrespective of the precise answer to each of the two sequences, most people instantly suggest that the average number of throws required to achieve both these sequences will be identical.

This conclusion is reached quite simply because we all know the outcome of a toss of a coin is fifty-fifty.

Well, it is not as simple as this – the number of throws to achieve each of the outcomes is very different!

To understand this, concentrate carefully on the following explanation.

Whether we seek the answer by repeatedly tossing a coin or work it out mathematically, we find that the number of coin tosses required before we get the sequence 'heads, tails, heads', averages to be about ten times. Whereas, the number of coin tosses required before we get the sequence 'heads, heads, tails', averages to be about only eight times.

Does your brain ache?

It all boils down to the fact that when you are tossing a coin repeatedly to seek the precise occurrence of 'heads, tails, tails', upon getting the first two tosses correct, 'heads, tails'. When your next toss is incorrect upon throwing a 'head', then at least with a 'head' you have started the sequence you are looking for again – namely, you already have the starting 'head' for your next attempt.

Whereas, when looking for 'heads, tails, heads', and almost achieve the sequence by throwing 'heads, tails, tails', then your last throw does not start the sequence you are looking for.

Quite incredible, as most people instantly think that any combination of a coin tossing outcome is just as likely as any other.

This is not the case!

We just need someone to convert this phenomenon to a roulette table using reds and blacks to represent the coin tosses.

Let us move on before our minds go completely potty!

Accidentally classifying a lump of rock as a planet!

- *Planet Pluto was named after the Roman god of the underworld who could render himself invisible! A planet one day: gone the next!*
- *We had to rely on experiments to mentally visualise matter prior to making a decision about its structure.*
- *If you tried explaining in the 1900's how the quantum world worked as we understand it today, you would be considered a complete crackpot.*

Imagine how Clyde Tombaugh, 1906 – 1997, may have felt if he had still been alive in 2006 when the planet Pluto he discovered in 1930 was reclassified as a lump of rock. Somehow, even though we are now told by astronomers that we have only eight planets rather than the preconceived nine, we still give credence to Pluto in conversation as we all find it difficult to drop its status so abruptly.

We must also feel sorry for Venetia Burney of Oxford, England, who at the age of eleven in 1930 named the new planet Pluto. She lived until April 2009 and therefore lived to witness the reclassification of Pluto. The name was chosen from a large number of suggestions because it not only depicted the Roman god of the underworld who could render himself invisible, but coincidentally the first two letters represented the initials of Percival Lowell, who had earlier predicted the existence of a planet further out than Neptune.

Unfortunately, following the discovery of the Kuiper belt in the 1990s, Pluto was no longer considered a planet orbiting alone, but as one of a group of icy bodies in a specific region orbiting around the Sun. When it came to light that at least one of this group of bodies was larger than Pluto, on 24th August 2006 the International Astronomical Union reclassified Pluto as a dwarf planet along with two other similarly-sized bodies – so we reverted back to only having eight planets rendering millions of textbooks incorrect.

Clyde Tombaugh's widow Patricia said that although Clyde may have been disappointed with the downgrading of Pluto, he would have accepted the decision had he known the reasons.

There is very often more to consider than what we observe.

If any lesson is to be learnt from this type of experience then perhaps it should be to resist forming too many conclusions utilising other knowledge and to only believe what you yourself see, hear, touch, smell and taste – not to be influenced too much by others and certainly not run with preconceived ideas!

When people are under pressure to come up with results and further the field of science, they may become overly eager to produce their findings. What is published may often be premature and ill-founded, but in fact 'correct' based upon all known information at the time. So periodically it is essential that discoveries are revisited retrospectively for their current credibility in light of new discoveries.

It would be interesting to know how many scientists have been wrestling with their own internal knowledge conflicts and anomalies as they announce a new discovery.

As we venture further into our advanced technological future, more theories and discoveries may become fact-based upon limitations inherent within previous discoveries. These inaccuracies and limitations will be used as the foundations to derive the supporting evidence for further advancements.

It is like using broken pillars to hold up a building.

Clyde Tombaugh's telescopic observations indicated a ninth planet in our solar system. However, it transpired not to be the fault of the telescope that devised the scientific view of Pluto as a planet, but our general lack of knowledge about our solar system at the time of discovery. There were too many preconceived ideas floating around at the time.

When Clyde weighed up all the facts available to him in 1930 about this new celestial body, it brought him to the conclusion that he had discovered another major planet.

It was the limited knowledge of the solar system that caused the difficulties and preconceived views about planets.

Let us hope these types of shortfalls are better managed and understood in the future to prevent scientists developing such things as ill-founded biological advancements that ultimately backfire – potentially producing calamities of enormous proportion!

With this in mind, we have to be careful when believing some scientists theories.

There is an extremely awkward difficulty facing scientists when studying the realm of the very small. They do not have microscopes powerful enough to visualise anything at this level, so they have to rely on the results of experiments to feel and guess their way forward. The make-up of small subatomic particles must be ascertained by experimentation prior to making a final decision about the physical look of a substance. There is also the difficulty that learning subject matter within the arena of subatomic particles is made increasingly awkward because the whole subject is alien and counterintuitive to the human mind.

The subatomic world of quantum physics is very unlike the everyday macro world we are so used to living within and the rules are quite different. In many respects we are subject to having to believe the findings of earlier scientists without experiencing all the preceding deterministic experimental results.

We trust in them implicitly!

Incidentally, the term 'quantum' comes from an indivisible unit of quantity that was defined by Max Planck in 1900 – an extremely tiny unit of energy, light or matter.

The word 'quantum' comes from the Latin 'quantus', meaning 'how much'.

If anyone were to speculate how the quantum world worked, no one could have guessed unless they were seriously deranged.

Perhaps if everyone in the world were deranged it would help!

Within this chapter we learnt how the governments of the world show little compassion to the plight of the planet. We have realised what horrid little creatures we are and how Thomas Midgley could perhaps have been the most horrifying life-form in the whole Universe. We realised how birth control and a return to nature may be a smart idea along with strengthened moral values. We learnt how the future could hold untold mind-blowing innovation – to such an extent that if we play our cards right we could one day benefit greatly from tapping into this very strange quantum world. We learnt how the reality of what happens in our minds can often fool us and be very deceptive. We saw how there are many ways that our everyday lives came be tricked into thinking certain things, just because we are not being truly logical. We realise that perhaps potty people are our friends – their eccentric views may truly be of value.

It is time now to look at whether our physical bodies are being deceived in some way – could it be that some things are not as they seem in the real world too!

THE MYSTERY OF THE SENSES

Where we investigate how our senses try to understand everything that surrounds us. How sight, sound, taste, smell and touch are all we have as a means of assessing the world in which we live. We look at the difficulties that arise when we attempt to decipher various phenomena. We discover how our senses do not actually interpret everything as the world truly is.

Sight – what our eyes really see

- *Do not be drawn into thinking that your vision provides you with the exact image of absolute reality!*
- *A white surface appears white because it is everything but white – a white surface appears white as it is unable to absorb white light.*
- *A yellow surface appears yellow because it rebounds all light that constitutes yellow.*

The image that your brain creates in your mind of sight, sound, touch, smell and taste are all biologically artificial illusions of the world around us. With sight giving us the pictorial representation of the world, this is what gives us the greatest knowledge of how things appear. However, do not be fooled into thinking that this is actually what these things are truly like!

For starters, take a white table and chair. In the absence of any light source, you will see absolutely nothing, as the table and chair have not been equipped with their own means of emitting photons of light. It is only when a light source such as the Sun, a light bulb or a candle appears, emitting photons that bounce about the room, that the table and chair are able to reveal themselves to us visually by means of emitting everything that it is not.

To be a white table and chair you need to be unable to absorb white light, so this means that a white surface is actually black in reality as it reflects all light away.

So straight away we can see we are being fooled by our brains into seeing white things which are actually black and black things that are actually white. Blue surfaces appear blue as they reflect everything that makes up blue, retaining everything that is not blue.

So in reality everything is the opposite colour to what we see!

We experience sight because photons collide with the lens of the eye and get detected by photoreceptive cells on the retina. There are two kinds of cells called rods and cones. The cones detect colour and the more sensitive rods detect the image without colour. It is estimated that each human eye has sixty million rod cells and three million cone cells.

The incoming light is converted into electrical signals by the rods and cones in a process called phototransduction. The rods and cones contain photoreceptive proteins with pigment molecules that differentiate between different colours. The cell density in the centre of the eye is much higher; this is why we see something we are looking at directly much more clearly than something located in the corner of our eye.

The electrical pulses are relayed along the optic fibre to the back of the brain where the visual cortex is located. Here cells isolate useful regularities in the visual data, for example one layer of cells will detect curved features, and another layer will detect lined features.

As the eye is such a complex aspect of the human anatomy, there are many features that protect it. Eyebrows prevent sweat from dripping in, eyelashes keep small dust particles away, eyelids protect by sweeping dirt from the surface, and tears are constantly injected to bathe the front of the eye keeping it moist and clean.

The function of sight therefore provides us with an image of what we think is out there in the real world. However, it reveals exactly the reverse of what it truly is.

In reality, objects are just a dark jumbled mass of bouncing particles – our brain puts order and colour to all that we look at using reflected photons, purely to assemble some order to the object in our minds. This happens even though the photons are not actually anything to do with the objects we are looking at, unless the object is a light.

To prove this just sit in a room at night and switch the light off, all you see is pitch black – this is what everything really looks like!

We have now learnt that there are severe problems within the world we 'think' we see.

Sound – what our ears really hear

- *The sound of a passing car is heard totally differently by a pedestrian compared to how it is heard by the driver.*
- *If a tree falls in a forest and no one is there to hear it, does it make a sound?*
- *Our brain is capable of interpreting vibrations into sound – the vibrations are relayed to our brain by the mechanics of our ears.*

Sound operates in a similar way to how vision works – where we actually observe the reverse of how an object appears in reality.

Is it possible that we could be interpreting sound in a similar skewed way?

Are we listening to something that a noise is not?

When a lion roars are we actually hearing what it is not roaring?

There is no reason why there should not be some kind of illusion happening in the process of hearing somewhere. The sound, after all, is just the resonance of particles within the atmosphere. The resonance of the particles must be interpreted by our brains as sound inside our heads.

Sound must also hold some type of illusionary property as we hear sound differently depending upon whether the object making it is moving or not. The best example is the pitch of a formula one car as its passes you on the track-side going from a high pitch to a low rumble in a matter of seconds, but you know the sound is constantly the same to the driver. This is called the Doppler Effect, which is applicable for all wave types, including sound and light.

We do not notice the Doppler Effect associated with light as it travels so quickly. However, as sound only travels at seven hundred and sixty-eight miles per hour in air, then a motor car travelling at one hundred miles per hour will have the sound waves noticeably stretched or squashed – depending on whether the car is coming towards you or travelling away.

Now let us theorise on other properties of sound. Can you recall the old saying, "If a tree falls in a forest and no one is there to hear it, does it make a sound?"

Thinking about the answer to this question has always fascinated people. In fact the tree has actually made the potential for someone or something to hear it – so long as a sound recording device or a creature with ears is there to listen. This is exactly the same as the fact that a radio station is available for listening to anywhere within the vicinity of the mast emitting the signal, provided you have the equipment to receive it.

Ears can be viewed as delicate pieces of equipment that start the process of converting vibrating air particles into sound within our brains. Sound will be heard when in close proximity to any event that smashes particles together. The smashing of the particles causes vibrations that get converted into distinct sounds in our brains.

Incidentally, there is another off-the-wall analogy with our ears and a radio. If a radio is off then you will not hear the radio station, just as when you are asleep you will not hear the rain on the window, people talking nearby or even someone shouting at you to wake up – as your ears are 'switched off'.

It is as if we go deaf at night within deep sleep as our brain switches off – when we wake up in the morning it is like a miracle has happened and we can hear again.

With our ears resembling delicate pieces of equipment that turn vibrating airwaves into sound within our brain – this implies that we are hearing something that is not real – only imaginary, or conjured up in our minds.

We can use an example of the initial cracking of a tree trunk in a forest as the tree falls. As the truck snaps, the actual snapping of the trunk does not create sound; the sound is formed in our brain by converting the vibrating air waves caused as a result of the particles of matter and air bashing into each other as the trunk snaps.

As the trunk snaps, this immediately affects the air and other matter around it. Having caused a momentary vacuum, there is a sudden change in pressure and the particles around rush to fill the vacuum – this causes a knock-on effect of vibration through the air and all substance within the immediate vicinity of the tree.

It is just a vibration of particles that our ears detect – there is no such thing as sound I am afraid. Sound only materialises when vibrations or pressure waves, have been converted into sound by our brains.

You may wonder how a radio is heard by us in the clear manner that it is – well this is quite simply because radios replicate the exact vibrations created by the original resonance of particles within the atmosphere.

Simple!

As you get further away from the tree other sounds interfere with the faint noise of the trunk cracking, just like a large number of ripples on a pond, there is just an indistinguishable rumble of a background noise.

What we can deduce from this is that a combination of our ears and the processing part of our brain is interpreting pressure waves in the air into sound in our brains. This implies that vibrating air molecules are the reality in the world and our sense of hearing works by converting this into something totally unrelated – sound within our brain.

So the bash of a drum in reality just causes air pressure changes that in turn get interpreted by our brain as the sound of a drum. What we hear as sound is nothing like the original vibration caused.

Because sound travels via vibrating particles within air, sound does not exist within the vacuum of space!

Again, we are confounded – sound is just a biologically artificial illusion within our minds.

It is not looking very hopeful that we live in a world that is anything like we thought!

Touch – what our fingers really feel

- *Touch appears to be rather less significant than our other senses during normal daily activities.*
- *Our minds portray sensations as we touch different objects – very hot surfaces create a painful sensation in our brain.*
- *Pleasure and discomfort can be felt in our minds as our fingers touch various objects.*

Let us now investigate the sense of touch. This has to be one of our rather weaker senses; of all the senses we talk about, the sense of touch is perhaps the one we least discuss with each other.

Touch is best assimilated to us within our brains by the tips of our fingers. We seem to be able to run them over surfaces and obtain various sensations. Sensations are normally registered as only mild during normal daily activity, through to absolute searing pain. These sensations range from the lovely sensation we get as we feel silk to the painful experience we get as we clamber over extremely jagged rocks at the seaside. What we experience in our brain is a sensation of pleasure or discomfort from whatever we are in contact with.

We know not to touch sharp objects from a very young age, to keep away from fire, and that rubbing your fingertips over a carpet rapidly is so uncomfortable it makes your hair stand on end. But we know that lying on a soft rug in front of a fire in cold weather is a good sensation. We refer to touchy-feely things, with the invention of stress-balls providing the sensation of touch with such a pleasurable sensation produced in your brain that it relieves anxiety.

But what are we really monitoring when we touch?

When fingertip atoms are in contact with the atoms of a piece of silk – our brain does not build a complete picture anything like our vision does; it does not hear anything like sound, there appears only to be an inner sensation that provides a value in our brains of extremely pleasant to extremely unpleasant.

Although, if you sit in a chair and shut your eyes, then feel the chair that you are sitting in – strangely an image of the chair you are in is available to you in your mind. It is as if you are able to feel a view of the chair and create an image in your mind, even though this image you have created is pitch black in nature. The shape is imagined.

Our brains will interpret the contact between silk atoms and our fingertip atoms as being extremely pleasant.

With touch we cannot determine its colour, what it sounds like, but what we are interpreting is its composition. The composition of the jagged rocks that you are clambering over at the seaside conjure up a sensation of extreme unpleasantness, the composition of the feel of the rocks rings warning bells of danger and alarm. Your brain then seems to be able to put your whole body on alert in case you slip and the potential outcome is damage to your whole body.

Although no actual visual image is formed, the brain seems to be able to create a type of out-of-body experience; perhaps touch is the most peripheral of all our senses. It certainly seems to occur at the actual point of contact with our body and is certainly capable of detecting heat.

Touch is therefore a degree of pleasure we get from the electrical signals from our skin being translated within our brain. It is as if we have an imaginary structure of ourselves, like a doll, in our minds. This doll monitors the touching sensations we experience and associates them to parts of our body within our minds. As it is not visual or audio in nature, we are at a bit of a loss as to how to describe touch.

There is a condition where people are born without the sense of touch called CIPA, Congenital Insensitivity to Pain with Anhidrosis. People with this condition strangely cannot sweat. This seems rather an odd link to the sense of touch, however, this is how our body operates. The condition is thankfully extremely rare, one in one hundred and twenty-five million people are born with it. The disorder affects the nervous system and prevents the sensation of pain, heat, cold and all other nerve centric sensations. This also includes the urge to urinate.

All manner of injuries are inadvertently inflicted upon people with this condition as you may imagine – without the sense of touch and pain. Mental illness sets in due to the onset of hyperthermia caused by the inability to sweat; infection and scarring of the tongue, lips and gums; infections of bones;

fractures; bodily scars; joint deformities; and all manner of other difficulties such as rubbing their eyes too hard.

Not nice!

It looks like touch evolved as an essential survival tactic. It appears that it makes sure that we do not injure ourselves and keeps us fully aware of our immediate environment.

It is interesting to see how our minds are yet again being artificially alerted to sensations that are caused by a translation between our nerves in contact with heat and surfaces, through to our brains.

A better picture of how the body works through its sense is now formulating.

Taste – what our tongues really contact

- *Taste has no shape and no colour, it appears as purely a mental measure of pleasure or discomfort.*
- *People get used to certain dishes of food at certain times of the day – a curry for breakfast would seem rather strange to most people.*
- *Beautifully cooked food brings your mouth to salivate, whereas just the sight of mouldy, maggot-infested food makes you feel sick.*

When investigating the sense of taste, we can look upon it as similar to touch in the way that the brain processes it. Simply – no visual image, no sound, just a mental composition of the pleasure obtained. We all know the sensation of tasting our least favourite food, and the feeling we experience from our favourite food. This sensation again is conjured somewhere within our brains, again no image or sound, unless you are eating something crunchy.

How is our brain analysing the vibrations of food molecules against our tongue into what we perceive as taste?

Perhaps the purpose of taste is a type of survival tactic, similar to touch, to prevent us from eating the wrong types of food. Eating a roast joint for breakfast seems a peculiar way of starting the day, but there does not appear any logical reason why a cup of coffee and a croissant is preferred other than 'this is how it is!'.

However, the mental image created when eating a roast at breakfast does not appeal in the brain. Maybe this is just habit.

But what seems to be conjured in the brain is a type of flavour palette that is working alongside sight. Looking at a steak, beautifully cooked to your liking, brings your mouth to salivate. However, looking at a mouldy maggot-infested raw steak makes you to turn away and wriggle in disgust; even though you have not tasted it. The mouldy maggot-infested raw steak has already triggered a very bad sensation within your brain's taste palette.

Studies have shown that our senses are extremely interconnected – more so than ever thought before. The vision of disgusting food making us feel sick is a good example of how our senses closely interact and influence one another.

Take an image of disgusting food we have not even tasted yet – it could just be a picture of maggot-infested food. Our vision seems to get our taste buds to imagine the sensation of eating the food just from the image. Although our sight is not taste, we are able to assimilate a taste from the image. We learn later on in life that this assimilated taste in our minds is sometimes not correct – and take to enjoying foods we thought we did not like when we were younger.

I remember not liking prunes as a young boy – the image of their wrinkly skins made me want to curl up in a ball and be sick. Now I look forward to a bowl of prunes and custard in the morning having now learnt that prunes are actually rather tasty.

The physical world we live in gets even 'less real' as we realise that taste is just another biologically artificial illusion created within our minds.

It is all getting rather intriguing and in a way somewhat scary!

Smell – what our nose really sniffs

- *With an empty stomach the smell of your favourite food will send flashes of delight sparking through your brain.*
- *It is our fond memory of the smell of our favourite food that has an amazing knack of reminding us of those fantastic vibrating molecules of food.*
- *If a luscious food molecule contacts one of our nasal sensors we instantly recognise it – lovely sensations created in our mind.*

Now we come to the sense of smell, which in a way complements taste. The smell of good food tends to get our brain invigorated in real-time depending upon your state of hunger. Nevertheless, when you have just filled your stomach with your favourite dish, you are somewhat less impressed with its smell once satisfied. Again, some kind of palette is being generated associated with smell in your brain, converting the aroma of vibrating food atoms into an interpretation of desire.

When sitting in a restaurant choosing your preferred dish from a menu, you are using your previous experiences as a guide to your brain's talent for remembering the sensation as you bite in, chew and swallow. Incredibly your memory is used to remember precisely what those vibrating molecules of food conjure in your mind as the aroma wafts up your nose.

We can safely say that taste and smell are a means of directly interacting with molecules at the most minute level. When a favourite food molecule comes into contact with a nasal sensor, 'bingo', we have a lovely result in our brain's smell palette. This is definitely the closest a human being can get to detect a molecule at an atomic level.

How exactly is this possible?

It is thought that molecules can fit into the odour receptor nerve cells, thus triggering the sensations we get.

The human tongue can distinguish among only five distinct qualities of taste, while the nose can distinguish among hundreds of substances, even in minute quantities. It is thought that these two combine to create flavour in our minds – another classic case of senses working very closely together.

More research is required as there are, at present, a number of competing theories regarding the mechanism of odour perception.

Whatever the outcome, we have an extremely special, highly-tuned body that we should be very proud of owning.

Having covered the five senses, we now have a complete picture of how we truly interpret the world. It is not what we think it is at all – what we picture in the mind is just an illusion created in our minds by the five senses we possess.

All extremely peculiar ... but this now gives us the ability to explore the world in different ways ... in ways that we have never known possible before!

Other bodily senses

- *What other types of senses could a human being possess?*
- *Is there such a thing as a sense of movement? You can get this sensation when you are in a lift.*
- *A mantis shrimp has the most advanced eyes – it sees more of the spectrum and in graphic detail.*

Most people who have studied homing pigeons believe that their homing ability is based on a 'map and compass' model. Their compass ability enables the birds to orient themselves allowing them to determine their location relative to their desired destination. The compass mechanism appears to rely on the Sun and the map mechanism is thought to rely on the bird's ability to detect the Earth's magnetic field.

The natural ability to detect a magnetic field to determine direction, altitude or location is called magnetoception. Amazingly, this ability has also been discovered in fruit flies, honeybees, turtles, bacteria, fungi, lobsters, sharks and stingrays.

Imagine if humans evolved into a race of airline pilots or a race of people who perpetually spend time flying around in aeroplanes. We may well eventually develop a sense similar to that of a homing pigeon – recognising fluctuations in the magnetic field of the Earth.

Just as we witness a decent smell, an incredible sight or a pleasing sound, it would be possible for us to witness the splendour of a 'home stretch' magnetic-field experience.

Could this experience truly rank alongside the smell of a rose or the taste of our favourite food?

Possibly!

What other senses could we possibly develop?

Nature seems to have done a particularly good job of developing some pretty interesting and useful senses. Some think there is a sixth sense that enables people to deduce otherwise unavailable information, but there are no organs within the human body to study this phenomenon. We will cover this fascinating subject in detail within the next section.

The sense of movement is interesting. When getting into a lift and then pressing the button to go up, but the lift goes down; then an almighty sense of movement is felt. Balance is affected and an unexpected giddiness can be felt. This sense is called equilibrioception and is one of the physiological senses. It prevents humans and animals from falling over when walking or standing still.

Proprioception, or the kinaesthetic sense, sends the parietal cortex of the brain information about the relative positions of parts of the body. This sense can be tested by closing your eyes and then trying to touch the tip of your nose with your index finger. If all is functioning well, then the position of the hand is fully known to the brain at all times, even though it is not being monitored by any other sense.

Nociception is the sense of pain. There are three types of pain receptors which represent the skin, bones and body organs. Pain was at one time thought to be a subjective experience, but recent studies have shown that it is registered in the anterior cingulate of the brain.

There are a number of inner senses within the human body which involve sensory receptors connected to internal organs. There are stretch receptors which are neurologically linked to the brain; these are found in such places as the bowel walls and muscles. Stretch receptors trigger various bodily functions via the brain.

There is also the chemoreceptor trigger zone, which receives input from hormones and communicates with the postrema, or vomiting centre.

There are other senses to which we refer, but are not truly senses. The sense of fear; the sense of excitement; the sense of awareness; the sense of loneliness; the sense of enjoyment; the sense of surprise; the sense of regret; the sense of delight; the sense of anger; the sense of wellbeing. These are no doubt brought about as a consequence of the other senses and an intelligent mind.

A mantis shrimp has an extremely enhanced sense of sight; it has the most sophisticated eyesight known within the animal kingdom. Its sense of sight is called hyperspectral colour vision, which means it can detect light from the ultraviolet, through visible light and into the infrared.

The mantis shrimp's eyes are mounted on stalks and move around independently of each other. Each eye is divided into three regions for precise vision and depth perception, and they have the ability to detect and analyse polarised light. Goodness only knows why it has developed such perfect eyesight, but some believe it helps them recognise dangers such as transparent predators and barracuda with their complex shimmering scales. Other suggestions are that their hunting style requires very accurate ranging information and accurate depth perception.

These mantis shrimps also have elaborate mating rituals during which they fluoresce, so colour must be an extremely important aspect of their lives. They grow to be about a foot long and are known to be highly intelligent, they perform ceremonial fighting, produce complex patterns on their bodies to provide behavioural communication and can learn and remember well to a point where it has been known for a captive mantis shrimp to be able to visually recognise their human keepers.

Irrespective of why the mantis shrimp possess such miraculous vision, we can only marvel at what it must be like to see the world through such perfect eyes – the clarity; the depth; the contrasting colours; the all-round vision; the infrared spectrum; the vision of heat; and the ability to see transparent objects. To imagine what it may be like to see the world through the eyes of a mantis shrimp – well, it could be as dramatic a change as a blind person trying to comprehend what it must be like to have normal vision.

As far as senses that nature could have developed are concerned, perhaps a sense of object density would be useful; being able to determine whether an object is light or heavy just by looking at it.

A sense that could detect the truth would be very useful, as would a sense that tells you that you have had enough to drink!

Perhaps there are other senses that we are just not aware of in other animals – but we do not do too badly with the five senses that we have.

We have to marvel at how nature has chosen to develop senses around what is detectable. Given a free reign to develop any senses whatsoever, we would be hard pressed to better those of a human.

Instinct – one of nature's best kept secrets

- *A wasp has a brain the size of a pinhead and yet is able to perform amazing feats without being taught.*
- *Viruses enter a cell and take over its genetic machinery, there is no obvious force or attraction – it is completely baffling as to why this happens.*
- *There is a great price to pay in nature for the mothering of a child, instinct can help resolve this.*

Instinct never ceases to amaze all who study it; some of the most amazing behavioural aspects of nature are handed from generation to generation in this way. One notable, instinct-related, behavioural mystery emanates from the eumenes wasp. After mating, the female wasp builds a number of hollow igloos made from mud – each igloo is designed to accommodate one of her offspring. The igloos are normally built upon a wall, or the underside of a roof. The female wasp makes the mud from her own spittle, dust and very small stones.

Having made an elaborate opening in her igloo, similar to one which resembles an ornamental vase, the wasp places a number of grubs in each igloo corresponding to the sex of the egg to be laid. The wasp stings the bugs sufficiently to stun them, but not to kill them, thus keeping the larder fresh for the grub. The egg is only laid in the igloo once the grubs have been caught, paralysed and positioned in the igloo.

There are no detectable signs regarding the sex of the wasp's egg. However, instinctively the wasp knows the sex of each egg and correctly provides the female igloos with significantly more grubs than male igloos.

The wasp displays a hidden instinct, almost like predicting the future, by providing the 'yet to be laid' eggs, with the appropriate amount of food. When an eumenes wasp lays an egg, she suspends it from the ceiling of the igloo by a fine silk thread. This ensures that the young wasps are not disturbed by the semi-paralysed bugs below. When the egg hatches, the young wasp is still suspended by its rear-end so that it can raise itself out of danger if a grub becomes too lively.

A wasp has a brain the size of a pinhead, yet the mother is able to do all of this without being taught!

She has never even seen her parents who die immediately after their eggs have been laid. All this prompts a number of intriguing questions regarding instinct.

How does the female wasp know how to make mud?

How does she know how to make an igloo?

How does she know the sex of the egg she has not yet laid?

How does she know how many grubs to catch and position for each of her eggs?

How does she know how much sting venom to inject into the grubs so they are paralysed rather than killed?

How does she know to hang her egg from the ceiling?

How does she know how to manufacture a silk thread for her egg to hang?

Most intriguingly, how did all this knowledge manage to get into the mind of the wasp?

Unlike humans, who have to learn everything, wasps and many other creatures have evolved a means of passing on their lifestyle and life characteristics. Just in the same way that a creature's bodily DNA is passed on from one generation to the next, aspects of the wasp's memory and lifestyle must be passed on in a similar fashion too.

Instinct is one of the most incredible features of the natural world. It is an invisible wonder that can power the behaviour in animals. We can debate and argue about how eumenes wasps evolved, but we cannot dispute that instinct exists as an inherent part of their lives.

Even if we look at extremely small and lowly forms of life such as viruses, we see them exhibiting incredible instinct. The viruses enter cells and take them over, turning them into virus factories. We know the stages which the viruses undergo, shedding their protein coats, invading the cell to take over the genetic machinery and then totally hijacking it to produce more viruses. However, knowing what viruses do, does not give us any indication as to the "Why?" of their action.

It is a complete mystery why these viruses enter the cell and take over the genetic machinery. There is no obvious force or apparent attraction encouraging viruses to do this – however they are somehow compelled to do it by something we have yet to fully understand. Therefore, this can only be instinct revealing itself at the tiniest level of life-form.

In some remarkable relationships within nature, instinct works between two extremely diverse life-forms. The adult female yucca plant moth emerges from the ground in June or July at the same time as the yucca plant is in flower. It mates shortly after appearing from the ground and then instantly begins another truly amazing relationship.

The female yucca moth collects the pollen from a nearby flowering yucca plant using her uniquely shaped mouthparts that could only have evolved for precisely this task. She then flies with her harvested pollen to another nearby yucca plant. Once she has selected one of the flowers, she inserts her ovipositor through the wall of the carpel and lays her eggs next to the developing ovules. She then climbs to the top of the plant's style which supports the stigma and uses her specially adapted mouthparts again to transfer the pollen to the top of the stylar canal. She repeats this a few times to ensure that the plant is well-pollinated, then drops off and dies.

The female yucca moth must make sure that the plant is properly pollinated because her young are to feed upon the seed. The seed is to be their single source of food.

But how can this benefit the yucca plant?

Surely if the young moths eat the seeds, this is totally unsupportive for the yucca plant – so why does a plant help a wasp that has the potential to destroy it?

Just think for a moment how a combination of nature and instinct may overcome this problem.

We have a moth that is specifically designed to pollinate a particular plant inside which it then lays its eggs; once the eggs hatch they feed on its developing seeds. Surely this equates to doom for both life-forms!

The young moth larvae hatch roughly eight days later and begin to feed on the developing seeds. But rather than indiscriminately eating all of the seeds, all of the moth larvae miraculously leave one uneaten!

The exact same happens with every brood of moth larvae, therefore it cannot be a coincidence – the behaviour can only be attributed to collective instinct by all the larvae.

After forty days of eating all but one seed, the larvae eat their way out of the developing fruit. They then drop to the ground using a silk thread. Having burrowed into the ground to pupate, they then repeat the same remarkable sequence of events the following year.

Considering that the young moth larvae have never seen their mother or father and are therefore unable to copy what they saw their parents do, their actions can only be explained by instinct. Their lifestyle and behaviour must somehow be programmed into their genetic make-up.

It would appear that nature is not only able to pass on a blueprint for the development of a body and organs within the genes, but also complex, adult-behavioural knowledge. It is impossible to explain some aspects of the yucca moth's life in any other way.

Instinct holds many mysteries.

How does the female yucca moth know to lay her eggs next to the yucca plant's developing ovules?

How does she know that the plant must be extremely well-pollinated in order for her offspring to survive on the seeds?

One of the most incredible yucca moth mysteries is – how do all of the newly hatched larvae know not to eat just one of the seeds to allow the plant to survive?

No one knows!

However, one thing is certain, without the yucca moth the yucca plant would become extinct, and without the yucca plant the yucca moth would become extinct. Each is totally reliant upon the other for their survival. The yucca moth can only reproduce using the yucca plant and the yucca plant cannot be pollinated by any other insect.

This is an impressive example of co-evolution coupled with extraordinary instinct. However, the instinctive mechanism by which the two diverse species behave will remain a mystery for some time yet.

So what instinctive behaviours do human beings possess?

It appears there are very few indeed. The one that seems to be most strongly argued for is the maternal instinct which is inherent within females. There is also a baby's strong handgrip at birth. After years of research these two appear be the only instincts that we can truly claim.

The gripping baby's hand possibly stems back to when gripping to the mother at birth was essential for survival. Whether a baby still has a strong enough grip to hold on to their mother's hairy back as she jumps from tree to tree like a monkey, is debatable!

Having thought long and hard about our human instinct, I wondered whether kissing could be one. How wrong could I be!

I discovered that kissing is not practised by roughly ten per cent of the global population, and was unheard of in Japan until it was introduced by Westerners just a few hundred years ago. So we cannot count this as human instinct.

Instinct occurs within species within a multitude of different ways. Just take the bird called the Manx shearwater for example, what an incredible life it leads. The adults return to their established breeding sites on a few select islands off the UK. Here the females lay their single white egg in burrows. As well as having the most elaborate greeting ceremonies, they partner for life. This is wonderful as they have been recognised as one of the longest living birds in the world, living well into their fifties.

Manx shearwaters are nocturnal creatures, choosing to feed their babies at night, quite possibly for safety reasons so nasty predators cannot see where their nesting burrows are. Amazingly, their voyages to find their favourite fish can take them up to six hundred miles in one round trip.

Their level of intelligence as a youngster is one of nature's most amazing secrets. Having been incubated for fifty two days they put on so much weight that they become larger than their parents and cannot fit through the hole of the burrow. The chicks are left behind by their parents who fly off on their migration route to South America without them. Amazingly the chicks slim down until they can scramble out of the burrow and then learn to fly on their own. Once they have accomplished this they fly six thousand miles to join their parents. No one knows how they know where to go, alone, unaided and unprompted.

Imagine a human baby at a few days old making a march a few thousand miles to somewhere it had never been before to meet up with its parents.

It is totally amazing that the Manx shearwater is born with such incredible inherent knowledge. Maybe this is something from which human beings could benefit; being born knowing how to count, add, subtract, multiply and divide. This would save the taxpayers quite a sum of money for educational needs!

If only we could tap into this mechanism it would have enormous benefits. Some people believe that previous lives and experiences are available to us within our brains – possibly placed there in the same way as instinct works. Past life regression is a technique that uses hypnosis to access what are thought to be past lives or incarnations. Having understood the incredible lives of the yucca moth, the eumenes wasp and the Manx shearwater, then the storage of our past lives does not seem that bizarre a concept.

Compare any one-year-old mammal with a one-year-old human and you will notice an enormous capability difference. Young tigers and lions will be mature and hunting proficiently without question after just one year. They will be using survival, instinct and cunning to catch their prey in a fairly organised and sophisticated manner. A one-year-old human baby cannot walk, cannot understand too much language, cannot prepare its food or recognise itself in a mirror. Interestingly, a baby cannot recognise itself in a mirror until about fifteen months old.

Sixteen years of childhood is a very long time compared with all other mammals. It is a significant proportion of a human's life.

Could this mean that taking time to learn behaviour results in better long-term intelligence?

Is instinct a requirement for animals born into a dangerous world – giving them a head start in life?

Is instinct a substitute for free-thinking intelligence in some fashion?

Who knows what the answers are – but there is plenty of research remaining.

Perhaps in the future we could harness nature's ability to pass information directly on to our children. If we were ever able to do this, what type of information would we choose to pass on?

I was once asked when talking about instinct whether I thought a bee knew that it would die when it stings. I strongly believe bees do know they will die when they sting. It is as if the whole colony within the hive is one animal; losing one bee may be as insignificant as a human losing a piece of skin. There is no reason why an animal cannot be in bits; there are stranger things out there in the Universe.

Instinct is truly a wonder to behold.

ESP and the sixth sense

- *Extrasensory perception, ESP, is about obtaining thoughts and views from the past, present and future.*
- *ESP may appear strange, but weirder happenings have been found to occur in this Universe.*
- *Just as a wine taster develops their taste buds, a psychic develops their sixth sense.*

People who seem to have a knack of predicting events or reading people's minds are said to have the sixth sense. This is the same as extrasensory perception, ESP, coined by Joseph Rhine in 1927. ESP has been described as the ability to interrelate with the Universe to acquire information which is unobtainable utilising our five senses; the information is supposedly sensed with the mind.

If this mechanism is truly effective, then we have something extremely powerful available to us.

If we consider utilising ESP on a grand scale, perhaps it is conceivable that we may be able to interact with the Universe's forces to discover all sorts of usually hidden knowledge.

An exciting prospect!

Perhaps this is why people say things like, "It came to me in a flash."

It sounds like Einstein certainly had a few of those moments – maybe he was very well connected to the fabric of the Universe!

The sixth sense is most often called upon to allow us to obtain thoughts and views from people regarding their past, present and future without the use of anything other than the mind. Some view this phenomenon occurring successfully due to an individual's paranormal abilities, pure intuition, coupled with a relaxed mind.

Some people are very sceptical of the sixth sense, but as we all know there are some who swear by it and make a living from it, with many being extremely successful.

Remote viewing, mediumship, trance work, spiritual healing, psychic readings, telekinesis and out-of-body experiences are all associated with the sixth sense.

Remote viewing is the ability to gather information from a distant source. Mediumship is a form of communicating with spirits. Trance work gets people to enter an altered state of consciousness. Spiritual healing involves treating illness with contact healing, distance healing and therapeutic touch. Psychic readings involve giving someone advice about their life through such mechanisms as tarot cards, palm reading and astronomical signs. Telekinesis is the ability to affect matter with the mind.

A friend of mine called Kevin and I once witnessed telekinesis upon a pint glass in a pub. Kevin suggested that he and I concentrated all our efforts on smashing his pint glass just using the power of our minds. At the very instant we were trying our hardest, a lady sitting beside us who was totally unaware of what we were attempting, saw her empty pint glass smash in front of her – I believe she may have just wobbled it slightly. It did not drop any distance at all – it just wobbled and smashed on the table.

The shock and surprise that Kevin and I experienced was enormous. We have not stopped talking about it since – Kevin claims to have accomplished it three times before. As far as it being a coincidence, I certainly had never tried telekinesis before and the landlord said that only roughly one pint pot per month gets broken in his pub.

Out-of-body experiences involve the sensation of floating outside your body with the ability to observe.

All these sixth sense phenomena possess a special type of mystical feeling or magic to them. Many people are fond of being associated with the mystery and the intrigue of the sixth sense; it encourages open-mindedness, involves the thrill of a mystic, introduces intriguing paranormal activities and takes the mind into a world of the unknown.

One scientific explanation I have researched that could go part way to explain how this phenomenon operates is associated with a very well-known scientific fact, but is very seldom discussed by many. It is the scientific understanding that everything came from the Universe's initial singularity. The Universe's singularity was the point in time thirteen point seven billion years ago when the Universe was born out of the Big Bang.

At this moment of the Universe's birth, all matter was in contact with all other matter within an area smaller than the size of a pinhead. From this point all particles evolved and formed the Universe we know today.

The fact that all atoms have been in contact with all other atoms in the past, may well mean that all particles in the Universe are 'entangled'. Particles which are in a state of entanglement must act in cahoots with each other and are knowledgeable about one another, irrespective of the distance they are apart.

Physically-entangled particles are known to be able to detect each other's presence, despite the distances travelled. This effectively provides the ability for every particle in the Universe to have knowledge of all other particles. The conclusion being that the knowledge of all the Universe is in contact with us at all times in the form of these entangled particles. We are also physically made up of these entangled particle-pairs which effectively makes us base-stations amongst this Universe-wide, particle-based, communications web.

This peculiar state of entanglement and other strange quantum effects will be covered in detail later in the book.

The ability to tap into all these touching entangled particles may well have the ability to give us knowledge of all other happenings within the Universe.

Quite simple really!

Experiments here on Earth have proved that pairs of entangled particles possess weird properties, including the ability to react together instantaneously at a distance. This marvel is called quantum teleportation and was achieved for the first time by scientists in 1997 – the quantum status of a particle, such as its spin direction, can be relayed instantaneously from one place to another without anything having to physically travel the intervening gap.

As you might imagine, this discovery is now being rapidly investigated by scientists to determine whether it will be possible to utilise for communications purposes. It is not fully understood how this type of instantaneous connection operates, but the fact that it does makes it exciting times for communications experts – and physicists alike!

In theory, a particle will be able to detect the state of its entangled partner even if it were billions and billions of miles away. It is extremely plausible that future communication will be based around this concept.

Scientists have reported the ability to perform quantum teleportation using entangled particles over greater and greater distances; this distance has now reached many kilometres – sufficient now to communicate between a ground station and an orbiting satellite.

Because particles are actually waves of potential awaiting detection, they only understand their true state within a Newtonian world. For clarification, the Newtonian world is the one we are familiar with, where apples fall from trees – not the quantum world where electrons dash and spin about, ignoring gravity.

As a particle emerges from its quantum system state into the Newtonian world, the wave function state that the particle was in then collapses, and it becomes what is termed a point particle. It is like a ripple on a pond suddenly turning into a pebble. It is at this point that the true identity of particles is determined.

Because of entangled particles, it is not beyond the bounds of possibility for psychic interaction to stand up to the rigours of science. One day in the future we may well see a light source, a clear mind and a willing participant, yield the most fascinating results as they inadvertently receive vibes from the billions of entangled particle-pairs that are scattered throughout the Universe.

Weirder things have been found to happen in reality than the phenomenon just described.

The Big Bang itself being one!

The Big Bang is truly weird!

Take the example of someone walking up to you at a bar and telling you, "I can read your mind," you might be a bit sceptical. However, you may be prepared to give them the benefit of the doubt and let them have a go.

If someone walked up to you at a bar and told you, "There was a massive Big Bang from nothing and the Universe and you magically appeared – isn't that wonderful."

Seriously ... this is a big one to swallow ... it may seem easier to believe a person that professes to be able to read your mind.

However, scientists are adamant about this Big Bang ... so believing that it is possible for someone to read your mind is wildly less bizarre in the whole scheme of things.

The potential that every single atom is inextricably linked to absolutely everything else in the Universe is a difficult a fact to disprove if you believe science's Big Bang theory. With entanglement being a proven feature of quantum physics, then if the Big Bang is to be believed then some method of tapping into the Universe to acquire information which is unobtainable utilising our five senses, is more likely than unlikely.

Let us now experiment with extrasensory perception. With all that we know now, we surely just have to relax and let the information come flooding into our minds. We need to interpret these waves from the Universe's entangled particles, just as we do with the waves and vibrations connected with all our other senses.

As we know with our other senses, constant use and practice provides better levels of accuracy. This is perhaps why practicing psychics become better over time.

If you are a wine taster – you will develop better taste buds; if you are a music conductor – you will develop better hearing; if you are a masseur – you will develop a better sense of touch; if you are an archer – you will develop better sight; if you are a traditional rose gardener – you will develop a better sense of smell.

It stands to reason therefore if you were a psychic, you will develop your sixth sense. Not being a psychic myself I am unable to confirm this. Although having said that, if you are a wine-loving musical conductor with a love of archery and massaging people, and also grow roses in your spare time – does it stand to reason that you would make a better psychic?

Maybe – maybe not. Strangely, another question for a psychic!

If the sixth sense is not related to the wellbeing of your other senses then perhaps there is a compromise somewhere. It could be good hearing, good sight, poor touch, poor taste and poor smell that combine to create a sixth sense that compensates for the failing senses. We will never know without an in-depth study!

Another interesting snippet of information regarding the ability to tap into the Universe's entangled particles as part of the sixth sense relates to our understanding of particles of light. If you consider particles of light as the transport medium, not the transport method, then you are beginning to understand the potential of this discovery. Do not look upon light working like an electric current travelling along a copper wire, light 'is' the copper wire, and the sixth sense i.e. the entangled particles travel over it instantaneously.

Quite breathtaking stuff!

Entangled particles of light will arrive at different destinations, as far as they are concerned, instantaneously, and it is their quantum state that determines their values and hence their ability to relay their status. It is not anything to do with the speed of light. Remember that the quantum system cannot communicate; it can only relay a status. A good psychic, therefore, may become adept at interpreting a status into a verbal communication based upon perceptions of their client, but must also possess the gift of the gab to relay this succinctly!

Extrasensory perception was first tested on volunteers by Joseph Rhine using what are called Zener-cards. Rhine used Zener-cards to test people's ability to determine one of five shapes utilising just their ESP. The shapes used were a circle, a cross, wavy lines, a square and a star.

Rhine occasionally came across exceptional subjects that far outperformed the score that would be obtained purely by chance. One such person was Adam Linzmayer who in 1931 twice scored one hundred per cent when tested with a nine-card series. When Linzmayer was tested with a longer three hundred card series, he scored thirty-nine point six per cent when chance would have predicted only twenty per cent. Over time Linzmayer's scores began to drop down much closer to, but still above, chance averages. Rhine put these dropping averages down to boredom, distraction and competing obligations.

Rhine tested another promising ESP gifted individual in 1932, Hubert Pearce, who managed to surpass Linzmayer's performance with Zener-card trials that averaged forty per cent correct, whereas chance would have been just twenty per cent.

Joseph Rhine published the first edition of his book in 1934 entitled, "Extra Sensory Perception", which was widely read over the next few decades and underwent a number of editions.

Rhine was not without his critics, and some scientists began to undermine his work saying that various methods could have been utilised to gain better results – namely discarding those results that were not favourable and worthy of publishing!

I leave you to make your opinion on this matter.

The sixth sense certainly poses some challenges for future researchers!

How hypnosis works

- *Charles Lafontaine was able to discover a natural psychophysiological mechanism that underpinned conditions experienced in the mind.*
- *A hypnotist must have a willing subject upon whom to practise, the person being hypnotised must also understand what you want them to do.*
- *The subconscious mind is accountable for most thought and determines a great deal of what a person does.*

Hypnotism is very often associated with the mysterious hypnotist figure popularized in movies, comic books and television. This respectably dressed and bearded man waves a pocket watch back and forth and easily transitions his subject into a zombie-like state. Once hypnotized, the subject is compelled to obey, no matter how strange or immoral the request. The subject mutters, “Yes, master,” and proceeds to conform and act out the evil hypnotist’s wishes.

This image of a hypnosis session is far from the truth and bears no resemblance to actual hypnotism. When a subject is under hypnosis they are never slaves to their masters and they always have absolute free will. They are not in a zombie-like, deep or semi-like sleep. They are actually in a state which is more akin to being hyper-attentive.

Hypnotherapy is psychological therapy and counselling that acts as a treatment of emotional and psychological disorders, unwanted habits and undesirable feelings. The aim of all such therapy is to assist people in finding meaningful alternatives to their present unsatisfactory ways of thinking, feeling or behaving. Therapy also tends to help clients become more accepting of both themselves and others and can be most useful in promoting personal development and unlocking inner potential.

Hypnotherapy attempts to address the subconscious mind. In practice, the Hypnotherapist requires the client to be in a relaxed state, frequently enlists the power of their own imagination and may utilise a wide range of techniques from storytelling, metaphor or symbolism to the use of direct suggestions for beneficial change.

The problems that can be addressed using hypnotherapy are many and varied – some of the most common are stress, anxiety, panic, phobias, unwanted habits, addictions, disrupted sleep patterns, lack of confidence, low self-esteem, fear of examinations, public speaking, allergies, skin disorders, migraine and irritable bowel syndrome. Hypnotherapy has also proved of value within surgery when normal anaesthetics have not been practical or possible. It has also been used in the areas of both sporting and artistic performance enhancement.

Using a procedure called hypnotic induction, a mental state of mind can be achieved using a preparatory set of instructions and suggestions. The hypnotic suggestions are delivered by the hypnotist to the subject, but it may also be self-administered, which is referred to as self-suggestion or autosuggestion. The use of hypnotism for therapeutic purposes is called hypnotherapy.

Many hypnotherapists are not too keen on stage hypnotherapists as they undermine the good that could so often be offered to cure people's problems. This frequently deters people from taking a beneficial course of hypnotherapy for fear of coming out the practice clucking like a chicken!

The Scottish surgeon James Braid, 1795 – 1860, was a pioneer of hypnotism and hypnotherapy. He became very interested in a demonstration of mesmerism presented by a travelling French mesmerist called Charles Lafontaine, 1803 – 1892. Braid was fascinated by the mesmerised subjects and deduced that they were truly in an alternate mental state. He was able to discover a natural psychophysiological mechanism that underpinned these conditions experienced in the mind. From November 1841, he began presenting a number of public lectures that conveyed his theories on this most revolutionary topic.

James Braid coined the terms 'hypnosis' and 'hypnotism' from the Greek word 'hypnos', meaning sleep. He based his findings upon the earlier research by Franz Mesmer, 1734 – 1815, who incidentally paved the way for the English term 'mesmerise'.

The latest thoughts regarding what causes hypnosis suggests that it is a wakeful state of focused attention and heightened suggestibility, with diminished peripheral awareness. James Braid was the first to write a book on this subject which he published in 1843, describing 'hypnotism' as a state of physical relaxation accompanied and induced by mental concentration.

As far as hypnotising people is concerned, it would appear that one of the key enablers is having a willing subject who understands what you want them to do. If the subject is extremely willing, then a very simple induction technique may be used. Hypnotising your subject should be done seated opposite them or at a slight angle to them. They may start with their eyes open if they wish. If they start with their eyes open they should be asked to develop a distant ten-mile stare. Whilst adopting the same posture as the subject, the hypnotist should try to breathe at the same rate as they do.

Once relaxed, the hypnotist can explain proceedings using terminology in a calm voice. A typical approach would be to say, "OK, here is what we are going to do. I am going to explain some things that lead towards hypnosis and I will ask you to do them and then we will come back to our usual state of mind. This will give you a taste of what hypnosis is like and if you like it, later we will go deeper." The subject should be encouraged to agree to all suggestions.

Then the hypnotist explains, "Once in a trance your breathing can slow down, your facial muscles can go loose and limp." This should then be demonstrated by the hypnotist. The hypnotist continues by saying, "In fact all your muscles can relax and you can become very still. Are you able to do all those things?" The subject responds, "Yes." The hypnotist may then say, "Relax … when you feel like you want to close your eyes, let them close. You may experience a lack of desire to move because you are so comfortable. OK … are you ready?" The subject responds, "Yes."

The hypnotist should at this point tell the subject, "Now we shall begin."

At this point the signs of trance are explained, such as the face relaxing, body relaxing and mind clearing. Next the subject is asked to produce these relaxed features and the hypnotist looks on calmly to observe the signs of the relaxation becoming apparent. Then each time the subject produces a notable sign of trance, such as a relaxed face, the hypnotist praises them, saying things like, "Good", "Well done", "Excellent", "Wonderful", or "That's just perfect."

At the same time the subject is entering the trance state, the hypnotist should produce the signs of trance which has the tendency to help them into a trance-like state more quickly. The hypnotist should encourage the trance by saying such things as, "Good … your shoulders can slump … your breathing can slow … that's right … wonderful … your face relaxes … wonderful …" This should continue for five to seven minutes after which the hypnotist should say, "OK … I am now going to count you out to your normal state … five ... four … three … two … one … and you are back again in your normal state."

The subject should now be asked about their experience. The more the subject practises entering the trance-like state, the faster and more deeply they can enter it. The hypnotist should be reminding the subject of the signs of the hypnotic state to engrain it in their minds.

After two or three repetitions of this simple technique you will have a good level of trance. Some subjects say that it takes a little effort to swallow afterwards because the throat has been so relaxed. Sometimes the subject's arms can become comfortably heavy.

People have been questioning, pondering and arguing over hypnosis for more than 200 years, and no one is any the wiser. The fact that scientists say that only a small percentage of the human brain appears to be used, some say only ten per cent, only adds to the mystery of this subject. Contrary to this fact, it was however announced in Scientific American on 7th February 2008, that human beings actually do use one hundred per cent of their brains. This stated that it was only ever a rumour started in a book called 'The Energies of Men', by William James. Einstein, it is thought, fuelled the rumour further when he used the fact to explain his cosmic, towering intellect.

At any given moment all of the brain's regions are not concurrently firing; brain researchers using imaging technology have shown that, like the body's muscles, most are continually active over a twenty-four hour period. John Henley, a neurologist at the Mayo Clinic in Rochester, Minnesota said, "Evidence would show over a day you use one hundred per cent of the brain."

Irrespective of the situation, there is an unchartered world of consciousness and hypnotism that has yet to be explored and understood fully from a scientific perspective. No one has come close to being able to explain the precise mechanism that the brain undergoes during hypnotism – perhaps we shall never get to know the inner workings of the brain whilst in this state, or any other state for that matter!

We can observe what a person does under hypnosis, but it is not clear why they behave as they do. It is a small piece of a much larger puzzle: how the human mind works. It is unlikely that scientists will arrive at a definitive explanation of how the mind works in the foreseeable future, so it is likely that hypnosis will remain something of a mystery for some time.

Psychiatrists have often compared hypnotism to daydreaming – very similar to the feeling of 'losing yourself' in a book or a movie. Subjects are fully conscious, but they tune out most of the stimuli around them. They begin to focus intently on the subject at hand, to the near exclusion of any other thought.

An excellent way to explain hypnosis is by simply relating it to the everyday phenomenon of self-hypnosis. Within a movie, video game or book, we enter a type of daydream where an imaginary world materializes which can often become extremely vivid and real. Amazingly we can become emotionally attached to this imaginary world of imaginary events where we then begin to experience real fear, sadness, happiness, and we may even find ourselves jolting in our seats.

Having said this, there is also the trance-like state brought on by intentional relaxation and focusing exercises. This deep hypnosis is often compared to the relaxed mental state between wakefulness and sleep. Just as you are entering sleep you may witness a drift into a relaxed state but will still be semiconscious to your surroundings – a slight noise may bring you back to realise your 'normal' state once again.

In essence, within the hyper-attentive or hypnotised state, if the hypnotist suggests that your tongue has swollen up to twice its size, you will feel a sensation in your mouth and you may have trouble talking. If the hypnotist suggests that you are drinking a raspberry milkshake, you will taste the milkshake and feel it in your mouth and throat. If the hypnotist suggests that you are afraid, you may feel panicky or start to sweat. However, throughout this time you are aware that it is all imaginary. In fact you are just pretending on an intense level in much the same way that children do.

Within this state the hypnotised person is much more open to suggestion than normal. So when the hypnotist tells them to do something, they will probably embrace the idea completely. This is what makes stage hypnotist shows so entertaining. Normally reserved, sensible adults are suddenly walking around the stage clucking like chickens or singing at the top of their voices. Fear of embarrassment seems to fly out the window. The subject's sense of safety and morality remain intact throughout the experience and the hypnotist will not be able to get them to carry out anything they do not want to do.

Hypnosis could be looked upon as a way to access a person's subconscious mind directly. A human being is only aware of thoughts within their conscious mind, thus consciously thinking things over that are immediately to the fore of their minds. This has the effect of enabling someone to choose words to speak, remember where your hat and gloves are, ponder over recent experiences and remember where it is that that you need to go next. But whilst doing all this, your conscious mind is communicating with your unconscious mind.

Your unconscious mind does all the behind-the-scenes thinking, it accesses the vast database of information stored within your brain – so when a thought comes out the blue, it is because you have already thought this unconsciously. Your subconscious is undoubtedly working for you all the time, taking care of all that you do automatically; it is this that enables you to breathe, react, move and visualise things without you having to focus upon doing so all of the time.

Your subconscious mind is responsible for most of your thought and decides a great deal of what you do. When awake, your conscious mind evaluates a great deal of thoughts, decisions, ideas, information and communicates this to the subconscious mind. However, when you are asleep, the conscious mind is placed on hold and your subconscious mind has a free reign to do as it pleases.

It is therefore thought that the deep relaxation focusing exercises of hypnotism work together to calm and suppress the conscious mind, so that it takes a less active role.

When in this state, the subject is still aware of what is happening around them, but their conscious mind takes much more of a backseat to their subconscious mind allowing the hypnotist to work directly with the subconscious.

In this chapter we learnt there are severe problems in the world that we 'think' we are within. We have seen how our senses are just biologically artificial illusions within our minds. We have seen how we do not live within a world anything like we thought we did. We have seen how the five senses complete a picture of how we interpret the world – in reality it is nothing like we think it is at all. We have seen how everything is extremely peculiar in reality, but knowing this gives us an ability to explore the world in different ways. We have seen how marvellous nature is to have chosen to develop senses in the way that it has. We have seen how the future could allow us to develop instinct capabilities by harnessing nature's secrets that allow us to pass information directly on to our children. We have seen how the sixth sense possesses some challenges to future researchers.

We know what our senses are providing us and the world it creates in our minds – so we may now take a look at how our minds operate. There are some pretty crazy things that go on in there!

MAKING SENSE OF OUR SENSES

Where we investigate precisely what our senses do for us, exploring what makes every person individually different. We take a look at our ability to possess free will and what may happen to us when we die. We also look at some obscure issues like why we cannot speak with other animals, and how our minds can influence us in all that we do.

Déjà vu and other assorted delusions

- *Déjà vu conjures a feeling of familiarity within a situation which is known to be a new experience. This is accompanied with a sense of eeriness.*
- *The human mind misconstrues a new experience by channelling forward through to our conscious minds as if it is recalling something.*
- *There are some intriguing degrees of déjà vu that begin to become rather problematic at the extreme end of the scale.*

It is difficult to find anyone who has never experienced déjà vu – the English translation is 'already seen'. It occurs within adults and children alike and is an experience that lasts from a few seconds to a minute or so, giving you a sense of familiarity coupled with a sense of eeriness, strangeness or weirdness. There are many situations within which déjà vu can be witnessed.

Travelling along a road that you have never been down before, yet it feels extremely familiar. Entering a house for the first time, but it feels like you already know the place. Meeting someone for the first time, but you feel like you already know them. A random situation which for some unknown reason makes you feel like everything that is happening has occurred to you previously. Déjà vu!

With the brain being such a complex organ and the fact that it comes at birth without any type of guarantee or warranty, it is no wonder it sometimes presents us with a number of challenges!

The human brain is one of the absolute masterpieces of the entire Universe. It converts the photons emitted by the bubbling uncertainty of matter into vision within our minds. The brain is held within a skull casement, no light can enter. Yet somehow the information arrives as electrical pulses from an image presented to our eyes in the form of electrical energy. In fact, all our senses are experienced by converting electrical energy into something that we then interpret into what we believe is the real world – however, it is only imaginary.

It is this imaginary world that we conjure, coupled with our memories that then begin to play these déjà vu tricks on us. It is this real world within our minds that we believe we have witnessed before.

The most scientific explanation of déjà vu is that it is the recollection of a series of similar sensations we have experienced in the past, coupled with our minds mistaking the situation as one that we are recalling rather that experiencing.

Déjà vu is purely a mental mistake that we make, in very much the same way that our bodies make other mistakes. Physically we may fall off a log or a ladder. Verbally we may slip up and inexplicably say the wrong word. We may misinterpret the sound of a book falling off a shelf for a car backfiring. We may mistake the touch of cold water for the touch of hot water. We may mistake a dream for being reality. We may mistake smells, tastes as well as complete situations. So the human being is fraught with potential misinterpretations and anomalies. Déjà vu is just another example.

Déjà vu was first coined by French physic researcher Emile Boirac, 1851 – 1917. He made reference to it within his book "L'Avenir des Sciences Psychiques" or "The Future of Physic Sciences", published within the last year of his life in 1917.

Interestingly, déjà vu has some other closely linked phenomena which manifest themselves with slightly different symptoms to which almost all of us can relate. There is jamais vu, which means 'never seen', and there is presque vu, which means 'almost seen'.

Jamais vu is used to describe a situation that is familiar, but for some reason it no longer seems familiar. Perhaps you get home and it feels unfamiliar and totally unlike the place you are used to. You could be confronted with a person, a word or a place that you know really well, but you have a momentary lapse of any recognition before your familiarity comes flooding back.

Presque vu is used to describe a situation that is almost remembered and in English is referred to as something that is on the tip of your tongue. People suffering from this feel that the blocked word or name is permanently on the verge of being remembered. Of all the so called 'Vus', Presque vu is the easiest to conjure up in one's mind. Just think of the image or silhouette of someone you knew in the past and then try recalling their name – within just a minute or so you should be able to get that wonderful 'presque vu' feeling.

Déjà vécu, translated to 'already lived', is known to be the ultimate in déjà vu. This is when someone feels déjà vu but is unable to determine that the situation is new to them. It is almost nightmarish in that they really believe that they have experienced the situation before, when they have not. Whereas with déjà vu, there is acknowledgement at the back of your mind that the feeling of familiarity is inaccurate, with déjà vécu the feeling of familiarity is genuine. Déjà vécu can be viewed as an intense or recurring version of déjà vu. People with déjà vécu have a total inability to understand that a situation is new, people will seem familiar and the locations they are in will appear as places well-known to them.

This can clearly be problematic!

Things get progressively worse for some poor humans as these mind tricks get progressively more severe. There is semantic satiation, where a word or phrase loses all meaning and when repeated is just a meaningless sound that cannot be visualised in any way, shape or form; it is as if they are hearing these words for the first time. There is also Capgras delusion where a person is in total belief that someone they know well has been replaced by an identical imposter. This can occur within patients with schizophrenia, brain injury or dementia. There have been several cases studied where people think that their partners have been replaced with another person and refused to be with them. Quite difficult situations!

There are a number of other more bizarre delusional disorders and conditions which are caused by the brain in a similar way to déjà vu and jamais vu, such as prosopagnosia, where people lose the ability to recognise faces. Fregoli delusion is where someone thinks that everyone they meet is the same person in disguise. Intermetamorphosis is when someone believes that people keep swapping identities whilst maintaining the same appearance. They can also get a feeling that they are being mistaken for someone else.

Subjective doubles syndrome is when someone thinks that there is someone else with their appearance. Mirrored self-misidentification is a delusion where someone believes that the reflection in a mirror is of someone else. Reduplicative paramnesia is when someone believes that something has been duplicated, for example they may believe that they are within a duplicate house within a totally different area. Delusional companions syndrome is when someone believes objects such as teddy bears, vacuum cleaners and flowers are conscious beings. Clonal pluralisation of the self is when someone thinks that there are several copies of themselves in existence, both physically and psychologically.

I pity anyone suffering from any of these; personally I am happy to just experience déjà vu from time to time!

However, be warned ... it is thought that all these misidentification delusions are related, starting with déjà vu at one end, with the extreme delusional beliefs at the other.

Everyone is unique, but it is extremely rare for any person to experience feelings of which others have no previous knowledge.

What makes a person unique?

- *Distant memories float around our minds about our younger years, but our later memories are much more vivid and precise.*
- *Our memories of childhood can often be triggered from old photographs – these photographs can easily become our memories of the past.*
- *Years of our lives can be forgotten for considerable periods of time – those days were extremely real, but slip from our minds.*

All people of the world experience the same types of feelings, emotions and sensations, otherwise psychiatrists would have identified a difference by now. Everyone grieves in much the same way; becomes happy in much the same way; becomes annoyed in much the same way; and everyone thinks in much the same way.

If we are so similar, then why are we so different as individuals?

Or are we that different?

Many people might consider you fairly eccentric if you were to ask them, "Why are you always the same person?"

However, the rationale for asking this question is directed more towards determining precisely what tethers your mind to your body, rather than anything else.

What unique components make you precisely who you are?

What events and experiences have made you who you are?

Why does your mind never experience someone else's body?

When did the moulding of your unique-self begin?

Where did you, the 'you' inside your head, come from?

Has the 'you' inside your head been here on Earth before?

At what point in time did you arrive here on planet Earth?

Did you become moulded into who you are as a consequence of your surroundings?

Did your surroundings play no part at all in moulding who you are?

Did you become who you are as a result of your life's experiences?

What precisely happened to make the conscious 'you', emerge with your thought in the body you have?

Were you once just a blank shell of a human being?

Tough questions!

Irrespective of the ultimate answers to these questions, clearly we are all unique individuals and unable to directly interface with other people's minds.

It does not require a genius to notice that during a human being's lifetime their mind does not transition from being in one person's body into another. They quite unsurprisingly remain the same person!

Perhaps you are who you are as a consequence of your surroundings, your upbringing, your parents, the positioning of the stars at birth, or perhaps it was just your turn. Alternatively, perhaps at the point of your birth your 'blank shell' of a human mind quickly filled with who you are – because it just did!

It is baffling to think that there was the Universe for all those billions of years – then miraculously, one day out of the blue 'you' came along.

Why now?

Why did it take all these years?

Where were 'you' before?

What makes 'you' any different from all those that have come before you?

What makes 'you' different from all those that are yet to come?

What makes 'you' different from everyone else in the world?

Some people may argue that being who you are is as a consequence of the blue-print defined by your parents. However, this does not explain why all children born to the same parents are all separate individuals. Even identical twins experience totally different minds. This would imply there are some other factors involved too.

Perhaps it takes a while for people to become who they are. My recollection from when I was a week old is absolutely zero. Could it be that it takes time for people to develop their minds into what they can relate to as themselves?

Perhaps peoples' minds slowly materialise one piece at a time. So a one-week-old infant is only one per cent of the eighteen-year-old person they will one day become. Perhaps during pregnancy a baby's mind is undergoing the rudimentary elements of identity, by relating to things like gravity and perhaps some unknown universal natural forces. The baby's mind then develops into an individual. However, this does not explain why you are who you are – the only true explanation can be that it was your turn. A little bit like one enormous galactic board game, it was your turn. However, this begs the question, where are all the minds of people that have not yet been born?

Are the minds of unborn people sitting somewhere upon a huge intergalactic shelf?

I very much doubt it. However, it strikes me that the way the Universe organises itself with all the infinite possibilities – it could be that there is the potential for any one of an infinite number of minds to materialise inside a new born baby. It just so happens that everyone here on planet Earth was effectively plucked from this infinite resource of possible minds – after all, all these bodies had to be occupied by someone. So we should just look upon ourselves as being very fortunate.

When fortunes change – we still remain the same person. A pauper from the worst slums in the world, who gets whisked into an affluent society, still has the memories of the past and remains the very same person.

Human beings must become a unique conscious individual either before conception, at conception, during pregnancy, at birth, after birth or progressively during all of these phases. Irrespective of which of these it is, at some stage humans mysteriously and miraculously become who they are.

What is your preferred opinion?

Let us take a look at the things that make us who we are – the things that make us always feel like the same person every day. We have our memories, we have our possessions, we have our friends and family, we have our environment, we have our bodies, we have our life's challenges, we have intelligence and we have language. If all these features that make up who we are, mysteriously changed whilst we were asleep at night, upon waking we would not know who we were. So in a way we can identify the important aspects of our minds that make us think individually as we do – but this still does not really explain why we are who we are.

If you were born on Mars and had learnt Klingon as your first language, had your arms and legs blown off in an accident when you were very young – would you still be the same person as you are today?

I believe the answer would be, "Yes."

You would be exactly the same person you are today inside your mind.

Perhaps we are missing something fundamental – perhaps it is the body and brain of a person that makes up who they are. After all, a brain could not exist in isolation. So when thinking about a human's conscious mind, perhaps we should include their body in this too.

Do all human beings generally feel 'roughly the same' in their minds?

I do not think it is too daft to believe that every human being feels more or less the same. What I mean by this is that it is very doubtful that humans witness life with the same mind experience as a worm.

The only differences between individuals in the human race are looks, speech, experiences, culture, location, abilities and parents. Perhaps everyone feels exactly the same and experiences consciousness and life in an identical way, but exists in total isolation of everyone else. This would give the impression of uniqueness when we could all actually possess an identical mind.

A worrying thought that everyone could be like Rob Lowe!

Perhaps the whole human race is the complete living entity, which is made up of many isolated identical minds – a little like how we observe life within a wasps' nest. We see the wasps' nest active and flourishing – but we cannot relate directly to what is going on because we are not a wasp. If we were a wasp, things would be very different – we would look at the nest and see things we had to do. In a similar fashion, humans are unable to detach themselves from the human race as they are too embroiled within it.

So the human mind could well be an isolated part of a larger entity known as the human race – each individual mind exists in isolation giving it a feeling of uniqueness that feels distinctly disconnected from the others.

As we get older we can look back at times when we had been totally preoccupied with significant life-changing issues. Thinking back to these times it is often difficult to relate to them, time has moved on so dramatically. We just have some distant memories of times in the past and a few thoughts pop into and out of our minds when reflecting upon our much younger years.

Our memories within the present are much clearer. It is as if we live within a window of time, we are only engrossed with the events that have taken place over the last few days and planning those that are about to come. We picture ourselves within the plans we make ... and this window passes across our lives like a magnifying glass across a page of writing.

We are unable to body-swap from one person to the next – although we do age and to begin to look very different as time goes by. We would never go to sleep in our bed at night as a highly-respected district judge and wake up the next morning in prison as a hardened, convicted criminal. However, we do go to bed at night and wake up the next morning looking slightly older – which over the passage of about thirty thousand days, changes us from a young energetic person with an active mind into an older person with a less-energetic, less-active and complacent mind.

But it is still the same mind.

Quite a good job, otherwise it would be rather confusing!

Perhaps our minds are like the bubbles within a lava lamp as they break free from the main lump. They exist as an entity in their own right for a period of time before returning back to the main bubble.

This process could provide us with a small portion of the Universe's unified field, which is the holy grail of physicists. We may be provided with an isolated piece of consciousness, part of the unified field, separate from that of the Universe for the period of our lives.

Then we have to hand it back.

We are definitely the same people we were when we were younger, even though I have heard it said that all our atoms are replaced throughout our entire bodies every five years. Our memories, our consciousness, our minds and our unique feeling of individuality are obviously carried forward. However, the influential images I personally have from my childhood are the photos I have seen – these images somehow get in the way of my true memories. My childhood memories seem to be very distant.

Perhaps you are only who you are for a moment and then become someone else the next moment, but the transition is so subtle that you do not notice. Your memory is carried forward, and the people who notice it least are your friends and family as they are also busy trying to clutch onto their identity!

Perhaps, from each moment to the next you become a different person, every moment brings with it a new sensation or view that you had not experienced before; new challenges; new decisions to make. However, when you get home in the evening from your busy day, you can relax and become familiar with your old self in the comfort of recognizable surroundings and the memories of your recent past within your home are fresh in your mind once again.

For someone who has lived within two totally different surroundings, it seems interesting that this person can quite legitimately say, "Oh, those days were terrible, I have forgotten those, in fact I have not remembered those days for years."

However, those days were nevertheless lived by that person. Perhaps their comments are not genuine, but it does seem that we can live for weeks without thinking of previous years.

What we are perhaps discovering here is that your old self is dead and your new self is here now in the present. Your old self, whom you think was you, is now stuck back in the past and can never be invoked again; effectively dead!

Only 'you', now, in this split instant is alive, your future moderately planned as far as you are concerned, a new 'you' being born into the next instant of time as time progresses. That is until one day when you are not looking properly whilst crossing the road and the number thirty-two bus runs you over.

Incidentally, I have scoured the Internet and discovered that there is no record of anyone ever having been run over by a number thirty-two bus.

It is now time to take a deep breath before you pretend to be a caveman.

Imagine you are a caveman with no mirror to look at yourself in and obviously no photos of when you were younger. As you grow older, you will only have a remote inkling of what you look like until one day the wind is not blowing and whilst not in a hurry, you catch your reflection in the still water. This is all you know about your looks.

Continue imagining yourself as this prehistoric tribe member – all day and night you only ever see your fellow tribe members, your true companions that congregate around the fire in the evening.

Your image of the world becomes one with theirs – this will be because the tribe imagines itself as the living entity under these circumstances, the individual mind is just part of the whole. Each of the individuals within the tribe view the whole tribe only slightly differently from their perspective as the only alteration that one tribe member has from another is that just one face is different from all the others.

This notion of unity within a tribe is interesting. Each member of the tribe will contribute in a small way towards the larger objective of the whole tribe. Then there is the gradual replacement of tribal members over time as they are born and die. Each time this happens, it appears only to be a slight and fairly insignificant change to the tribe as a whole. Just like the atoms within a human being change over time, but we remain the same person. The changes are similarly just as subtle within a tribe, the exact same mind-set of the tribe as a whole is carried over – well beyond the boundaries of generations.

I believe this is what the function of an individual human being is.

What do you think?

It is a good job we are all unique, else the world would be a tedious place.

Remember that next time someone annoys you!

One of the key missing attributes in today's society may well be tribal unity. We occasionally get a glimpse on television of black and white film showing a long lost tribe from yesteryear. Having now been integrated into the world's rat-race – I am sure there would be many that would much prefer the tranquil life their ancestors enjoyed. In fact, having lost our tribal values, this is one reason we may suffer from so many mental problems.

It is a shame we cannot mix the two styles of life – perhaps this is where social networking fits in!

Thought – can a thought be detected outside the brain?

- *Thought is part of every human being, but what is it made of?*
- *Our brains are extremely demanding organs – they use a disproportionately large amount of the body's energy.*
- *Decisions made by different people can be very dissimilar – one person's choice of action in a crisis situation will be very different to another's.*

Thought is by far one of the most valuable assets we have, and without it we would be just as capable of achieving anything as a lump of rock. Thought is capable of making decisions in our minds and making choices about what we prefer to do.

Imagine where we would be without thought!

A major difference we all have is our unique isolated thoughts. Irrespective of how close we get to someone, it is impossible to directly witness their thoughts. Our thoughts take place in isolation and may only be relayed to others via communication.

Thought cannot be detected like a sound wave; it cannot be picked up like a radio wave; and it does not produce ripples like on a pond. Thought cannot be observed like light and it cannot be touched, tasted or smelt. In fact our thoughts do not appear to be identifiable by anyone but ourselves. Thoughts do not penetrate anywhere other than inside of our own conscious minds – or do they?

Is a thought visible to the Universe?

Can a thought be detected outside the brain?

The notion of a thought being detectable outside the brain is fascinating for a number of reasons. Firstly, if a thought is visible to the Universe then it could be influenced by other matter. This could have an impact upon the ability for us to possess free will. Secondly, if a thought is visible and constructed of physical particles, then there is a possibility that it could be detected and interpreted. Thirdly, it would definitely help the plight of psychics.

For a thought to be visible to the Universe then it must be constructed from something physical – whether this be molecules, atoms, electrons, or bosonic carrier particles like light.

As it transpires, thoughts are initiated by a burst of electromagnetism; they become part of our conscience for a fleeting moment and then fade upon recognition. Thoughts last for a very short period of time, passing through our conscious minds as a wave and collapsing in an instant.

Proof that thoughts are created by physical particles comes from research carried out in the area of lie detection. Lie detection techniques developed since the year 2000 have resulted in the perfection of 'Brain Electrical Oscillation Signature Profiling' which may now be used in criminal investigations. The technique relies upon 'response conflict' where the subject reacts physiologically as they are verbally walked through the crime scene. If the electrical responses detected from the subject's brain conform to the patterns of 'experiential knowledge', this is attributed to remembering the crime scene in the manner being described. With no questions asked and no responses expected, this technique can be used to confirm the honestly of the subject's statements during interrogation.

The whole technique is based around the neuro signature system. It has been discovered that remembering an experience triggers a neural activation of the brain, which is different from that observed when the subject has mere knowledge or familiarity of an event. Although we can look at brainwave signatures, we cannot interpret the pattern of electrons to any word, phrase or sentence.

So as we have ascertained that thoughts are constructed from electromagnetism, we can safely say that they weigh something.

This would make a great deal of sense and explain why people often say, "I have been thinking long and hard, and it has taken a great weight off my mind!"

People often notice and comment that they use up a great deal of energy when concentrating for long periods of time – to such an extent that people can feel exhausted. The human brain is just two per cent of the body weight, but receives fifteen per cent of the cardiac output, twenty per cent of the body's oxygen burn and twenty-five per cent of the body's glucose consumption.

As thoughts possess weight, they may well have a real impact in the world.

Statistics show that home teams in sport have an advantage over away teams, but what provides this advantage is not fully understood.

Is it possible for a large crowd to give the home team an advantage at sports games by producing millions of thoughts that gather momentum?

Perhaps it is the thought of being at home that makes players perform better?

Noetic science is concerned with this type of effect, which studies mind and intuition. It seeks to discover whether there is a relationship between perceptions, beliefs and intentions upon thought and consciousness. The area of study investigates whether the human mind is capable of affecting events, it recognises that thought is not imaginary, but is bose-based or photon-based. This implies that thoughts can be measured by the same rules that govern quantum phenomena such as light waves. This is far-reaching logic. Noetic theory proposes that thoughts affect matter in an identical way to gravity, only to a smaller degree. Telekinesis is related to this area of study too.

For all the research that has been completed, it is clear that thought is an area with many uncertainties.

It occurred to me that thoughts could be kept secret to an individual due to Heisenberg's Uncertainty Principle – this is where only one property of a particle may be known at the quantum level. The sensors in the brain which monitor thoughts may quickly analyse the particles, only for them to then spiral off into the Universe holding no secrets pertaining to the thought. Thought could easily employ this natural quantum effect as a means of providing security and encryption.

Even when we are close to someone and sharing activities, we think and witness events very differently. One person may think of a stage performer as a hilarious comedian, when someone else may think of them as an embarrassingly tactless entertainer. This shows we have unique identities and think differently, thus differentiating us.

The ability to think differently, and in isolation seems to be the major difference between people – it may be this that generates that little flame inside us that burns, telling us we are an individual and that we are alive – distinguishing us from others. To establish whether there are any other fundamental differences between yourself and other people, let us investigate further with a few thought experiments.

If your mother had met with a man other than your father, would you still have been born to your mother as yourself, just as you 'think' today, but just look slightly different?

This would imply that the sense of 'you' that you feel today is brought about by your mother's influence.

Alternatively, if your father had met a woman other than your mother, would you still have been born as yourself, just as you 'think' today, but born to a different mother and look slightly different?

This would imply that the sense of 'you' that you feel today is brought about by your father's influence.

It is very difficult to imagine how either of these events may have resulted in you being born as you are today. It would appear the only way you have arrived here on Earth was because both your parents decided to make you when they did. At any other time, with any other events happening, if the phone had rung, perhaps something as simple as a car backfiring outside, there would have been someone else in your place. This would be someone else that could not be recognised by yourself as the thinking 'you'.

When we look back at the evolution of the entire Universe, the formation of all that we see was inevitable from the very beginning. Once the Big Bang took place, there was nothing to prevent the cosmos forming as we see it today. All the chemical reactions, the bumping of particles, the explosions of stars and the formation of black holes, must have been just unfolding in a predictable fashion as there was nothing to make a decision to change anything ... a world without thought ... a universe without thought.

Eventually creatures evolved that were gifted with the ability to create thoughts and make decisions. This would have begun to have an effect on the future of the previously predictable world. There must have been a moment in the history of the Universe when someone or something suddenly had the capability of making a decision and made the first alteration to physical events.

This would have been an event that exerted a physical force as a consequence of pure thought – this type of event would cause something different to occur compared to what would have happened otherwise.

When could this have taken place?

Perhaps there is something out there now in the depths of the cosmos performing thought on behalf of the Universe on a scale much larger than we can comprehend. Alternatively, for billions of years the Universe was just like a huge bag of ball-bearings – bashing predictably into each other without any interruption until creatures on Earth evolved and altered physical events using thought.

It is in many ways difficult to imagine that Earthlings were the first thinking creatures to frequent the Universe. For creatures of the Earth to consider themselves as the first thinking creatures in the Universe seems almost as bizarre as believing that 'you' were the first thinking person in the history of the world.

It is the ability to make decisions within our thoughts that differentiates one person from another; one person's sequence of decisions and choices within a particular situation will be different from someone else's. This is what makes the world interesting. Certain people are more inclined to make decisions in favour of normality, shrewdness, conformity and regularity, whereas others will perhaps favour abnormality and eccentricity.

People can be very predictable and the closer we get to someone the better we can judge their thoughts. Just because many people may make similarly normal or abnormal decisions within all types of situations – does this make them predictable?

If people are predictable then there is a possibility that all our thoughts and actions were inevitable from the beginning of the Universe – and the future will be mapped out and unsurprising. The predictable thought of people is made even more obvious when you realise that thoughts are made from particles and may be influenced and detected. As we all know, magnets can direct inert iron-filings to behave in certain ways, so could this be happening to our brains?

Perhaps our thought is being channelled along a particular route as a consequence of our surroundings.

It would be sad if this were true – so you may wish to make sure this claim is incorrect by checking the thoughts of a few of your close friends.

With our intimate knowledge of a friend, this should equip us sufficiently to determine their actions within a variety of situations – especially if thoughts are visible to the Universe.

Challenge yourself to see whether you can guess how your friends will make decisions within certain situations. Each time, write your predicted answers down before asking some questions. The answers you hear from friends you thought you knew, may well surprise you.

Try these questions:

If you were on your own in a lift without security cameras, if you saw a bank note on the floor would you put it in your pocket or hand it in at reception?

If you became knowledgeable about a slanderous secret that could make money, would you ring the newspapers to prosper?

If you entered a restaurant and there was the person you dislike most sitting at one of the tables, would you always leave in this situation or would you consider sitting down to eat ignoring that fact they were there?

If you saw a spider struggling to get out of water spilt on the floor, would you help it to survive or not bother?

If someone informed you about a fortune that you were to inherit that until now had been kept secret, would you tell no one other than essential close family and continue as if nothing had happened at all?

Try to determine whether your friends have made their decisions as a consequence of the way they have been raised, or whether they made their decisions as a consequence of how they instinctively 'are'.

Get your friends to predict what you would do in these situations; again the answers may well surprise you.

It would be a very different world if our thoughts were visible to everyone!

Free will

- *Perhaps because of the film 'Jaws' we have become a little more frightened of sharks than we would have been otherwise!*
- *Perhaps human beings are pretty much the same at birth and it is mostly our experiences that mould us into who we are.*
- *As all the events in the past have occurred precisely as they have, does it then makes sense that future moments could have been planned in some way?*

First of all – a little lesson in Determinism.

Determinism is the philosophical view that every event, including human cognition, behaviour, decision, and action, is causally determined by prior events. If there is a predetermined, unbroken chain of prior occurrences back to the origin of the universe, free will is impossible.

When a meteorite is on a collision course with a planet but is yet to make impact for another ten billion years, this collision can be accurately calculated and will definitely take place. There is nothing to prevent it happening unless a creature with free will puts their mind to it and prevents the collision.

Let us consider whether it is possible to contemplate our lives in a similar fashion.

If thoughts are visible to the Universe then it is slightly more likely that thoughts are as a result of surrounding influences – effectively thoughts are being reverse-engineered by the surroundings of an 'intelligent' being. Alternatively, if thought is invisible to the Universe, then it is more likely that future of events may be changed. This is because we are able to think in total isolation of the Universe, keeping things secret from the Universe itself.

If thoughts are invisible to the Universe, then future events can be changed, otherwise we are perhaps just living out already predetermined motions, actions and thoughts – our fate-driven lives!

Put a little thought to this yourself and see what conclusions you come to – unless of course the future has already decided for you!

As an example, I might have a thought which prompts me to reach out and throw a ball for a dog. This dog then trips someone up who consequently misses their train. Not a big deal at first glance. However, this person as a consequence did not sit next to their future partner on the train. If I had not thrown the ball when I did, this person would have had a totally different experience for the rest of their life. Consequently this would affect everyone he met and eventually change the future for everyone!

Alternatively, you could argue that the only reason I threw the ball was as a direct consequence of all the matter around me influencing me to do what I did.

So the burning question is; does a human being act as they do because they are affected by their immediate surroundings?

Put another way – does a human being and all other creatures for that matter, act in a predictable fashion just as a set of dominos would fall?

When waking up in the morning, just because the way events were, a supercomputer that is monitoring all events in the Universe could calculate that I will trip over the doormat, hit my head and miss my bus.

Are we following a series of events within a life that cannot be altered?

Is it possible that everything we do is governed by billions and billions of tiny subatomic collisions – forcing us to do what we do?

Amongst this vast Universe of calculable collisions it is difficult to judge whether life-forms with free will have come about by random events and coincidence. Life-forms may have evolved purely by the random movement of atoms and molecules over billions of years. If this is the case, we would love to know at what point was it possible for a life-form to change the future by performing an unplanned action?

Arguably, it may not be possible to determine when this happened as all life-forms may have been created by chance and do not possess free will!

It is rather a sad deduction, but it may not be possible to alter the future ourselves as there are so many interconnecting historical events that moulded us to how we are today. Perhaps we are still being influenced by these at present – so everything we do could be calculated if only we knew all the properties of every atom in the Universe.

Interestingly, our present moment becomes the past as soon as it has arrived – so as soon as we witness the moment in the present it instantly becomes the past. We can envisage ourselves perpetually living in the last moment of history; thus not giving us the ability to live in the present; hence no ability to experience free will!

As you can see, it is a tricky subject.

We are all influenced by our past. Past events can never be changed – they are fixed for eternity. For example, the film 'Jaws' has been produced and no one can take it away.

Consider if the film 'Jaws' was never screened, would we all be as petrified of sharks as most of us are today?

Possibly not!

If a film called 'Let Us Vandalise the Classroom' had become a massive hit, would we all be educationally challenged and less well-tutored as a consequence?

As it transpires, perhaps because 'Jaws' was screened, we remain a little more frightened of sharks than we would have been otherwise – this is just one tiny contribution that makes us who we are and how we react within the world. This worldly reaction may be applicable down to every individual particle within our bodies.

Let us consider all the movements of all the meteorites, planets and cosmic bodies, plus all the movements of all the creatures and plants of the Earth and ask, "If I delay scratching my head, have I changed the future of the Universe for all time?"

The answer will be 'yes', so long as we have free will.

However, the answer could be 'no'. This is assuming that I was going to delay scratching my head within this Universe anyway. Following this my pseudo-conscious thoughts may improvise to subsequently give me an impression of free will.

Some people think the former, and some think the latter.

What are your thoughts?

If delaying scratching your head changes the Universe for all time, then this prevents there having to be an infinite number of universes to accommodate peoples' individual private scratching habits!

If anyone needs this explaining further please get in contact with your psychiatrist!

Perhaps every human being is actually born similar to a blank piece of paper and miraculously each one of us materialises upon it. The experiences you undergo create you and that is what you are made up of. However – what is to prevent you becoming your brother or sister with whom you are being brought up?

Simply the fact that it is not possible!

However, your outlook on life and experiences will be pretty similar. My belief is that I am 'me' and somehow I doubt I would have been 'me' if I had been a different sperm and nurtured in my mother's womb differently before I was born. Somehow, it is very difficult to imagine that the world could be here without me in it!

Where else could I be ... at this point?

Consider the amount of creatures upon this planet with the ability to utilise free will to do whatever they wish, whenever they want. Somehow, the planet seems to have its own agenda associated with tectonic plate movements, volcano eruptions, weather patterns and its associated free annual trip around the Sun. Just looking at the several trillion living creatures on Earth, each with free will to eat its next morsel, scratch its head, have a good day, have a bad day or waggle its arms about, does this mean that every new moment of time has not been planned?

For instance, my head itches, and I decide not to scratch it for a second longer than I would have done normally. If I had not been conscious of 'not scratching it', just for the purposes of this precise experiment – I have changed the future of the Universe!

I must have changed the future of the Universe as all future events are now influenced by me not having scratched my head.

The butterfly that flapped its wings in the jungle caused a hurricane across the other side of the world. The first domino that fell caused a million more to fall also. If the Universe did not change as a consequence of me not scratching my head when I should have instinctively done so, then I would be surprised!

Would the whole Universe get to know that I had scratched my head later than expected?

This thought experiment may help.

Imagine there are two rooms next to each other, totally soundproofed. In one there is yourself sitting for an hour, hungry, reading a book. After an hour you are invited out and asked if you knew what was going on in the room next to you. Of course your answer will be, "No, not a clue". If you are then told that there was a room full of people having a party, eating your favourite food and that you missed out, does this then change your history, your present and your future?

I think it alters your present, but the very act of just getting to know what was going on right next to you has an impact on your future thoughts.

Reflect upon this for a moment.

Imagine if you are now suddenly told that the room next to you was actually empty and that the previous statement given to you about the party and your favourite food was a complete lie – also you are shown timed video evidence to prove this was the case. What effect does this then have on your history, present and future?

What this is attempting to expose is that the actions within the Universe are having no true influence upon your life – irrespective of whether or not they occur and even if they pretend to have occurred. The main point being that the trillions of creatures on the planet do what they do, either when they want to do it, or when they would have done it anyway, without control. Perhaps their consciousness is in turn interpreting the situation as if they had chosen to do it themselves anyway.

So there we have it – I 'think' the Universe is entering a new era where change to the future is being brought about by intelligent human activity.

Were you inevitable?

- *Would you be different if your mother had eaten differently during her pregnancy?*
- *Could an event as small as someone randomly swinging their arms about, change the future of the Universe?*
- *Everything throughout the history of the Universe has caused your presence on Earth – was this inevitable from the outset or just pure chance later?*

If the mind of life-forms cannot influence the future of the Universe through thought, then your presence in this world was inevitable from the beginning.

If the mind of life-forms can influence the future of the Universe through thought, then your presence in this world was random and purely as a consequence of luck.

If we believe that every action has an equal and opposite reaction, then throughout the history of the Universe it would not be possible to change events, therefore your arrival in this world would have been inevitable from the beginning. Your body's formation and your self-awareness would have been a foregone conclusion at the very start of the Universe. A little like a long line of dominoes ready to topple, as soon as you have toppled the first domino you know all the rest are going to topple too.

Following this logic; out you pop into this world, just as you were supposed to.

Everything seems to have come from an organized and calculated beginning. You were inevitable from the very start of the Universe.

An alternative view to this is that anything is possible; randomness takes over and chaos prevails, which makes every outcome throughout the Universe pot luck. When someone makes a decision within a thought and a physical action is performed as a consequence, this will have an impact within the Universe that equates to random movement. This would mean that your presence here on Earth came about as a consequence of a whole host of random events that you could interpret as coincidences.

Ultimately becoming yourself as you are now was perhaps just a random marvel, one of billions and billions or even close to an infinite number of possibilities at the moment of your arrival. Alternatively, everything was planned and events could not be altered, as everything that happens is as a consequence of other previous events and activities.

Let us for a moment imagine a mother has just this moment conceived and for the next nine months has two choices available to her. There are two doors to enter, each of which contains nine months' supply of food. She chooses a door at random and eats the food in this room for nine months and then gives birth to the baby. Would this baby be the same and also think and be conscious in exactly the same way if the mother choose to enter the other door and nurture the baby on a totally different food source?

Somehow my intuition says, "Yes, it would be the same baby." For the very same reasons you are the same person you were when you were two years old. Perhaps food choices from the moment of conception are not relevant.

We could debate that it is your DNA making you who you are, and we could imagine that the atoms that make you look as you do are just to make you visible as a human.

This is an alternative view that suggests that it really does not matter which atoms of the universe make you, so long as they are the correct ones, in the right order and in the right quantity. From my experience, I have never felt any different as a person when switching my eating habits!

Do you feel that you have been exactly the same person since your moment of conception, irrespective of what you have ingested?

Do you think you would you be someone else if you had eaten differently?

Now imagine that this lady has a choice of two doors to choose a loving partner, and following this she then has a choice of two doors to choose the food for nine months. My instinct tells me that by going through random door one, a totally different conscious being will be created than if she had gone through random door two. In fact, time will play a part inside the random room also, as it is not until the random moment of conception that things have been determined as a blueprint for a new conscious being. Feed this blueprint on anything relatively nutritious and the same conscious being may well result.

However hard it is to break out the mould of doing what you instinctively should be doing, you are never sure whether you would have done those actions anyway, even if you had not been thinking in advance about it. Perhaps a crazy non-action such as 'not scratching your head', when you really felt like scratching it, was going to happen anyway. Perhaps it is just that your mind then fills the preceding moments with the silly thought of, "Let's be silly and not scratch my head for a while, even though it is itchy", giving you the sensation of free will even though you do not have it.

Is it possible for a person on the other side of the world to swing their arms around in the air for no specific reason and not alter my life and the future of the Universe?

No – this action would have a significant impact on life and the Universe as everything would be different from this point onwards. By swinging their arms in the air they may have disturbed a fly which would have mated with another fly to produce offspring, which would have been eaten by a bird. This bird no longer gets caught by a young man to feed his family – so they suffer. This suffering is reported in the newspaper and read by millions of people, causing all their breakfasts to be eaten at different rates. As a consequence, millions of people start the day later or earlier than normal which has untold implications worldwide.

Just for a moment try to imagine that you were never conceived. You were never a foetus in your mother's womb, never born and never lived as you are today. The burning question here is – would your mind be conscious that you had never been born?

I very much doubt you would be aware that you had never been born.

This would be the exactly same for the infinite number of other people that were never born too. Having an infinite number of unborn people is becoming less strange; this to me implies that we become who we are as a result of our experiences and were just an isolated shell of a human being at birth, identical in every way to all other human beings when they were born.

Perhaps a foetus in a womb has no context or framework within the Universe until it is born and able to relate to its surroundings. A foetus might be just as bereft of consciousness as a kidney, liver or heart for a time. It may just be following genetic instructions to evolve in the defined manner until one day – it pops out and becomes an individual.

Somehow I very much doubt that anyone who has not been born would be conscious that they were not born.

So now that 'you' have been born, you have experienced living within the Universe and can appreciate it, no doubt something you quite treasure.

If you had never been born, would this mean that whatever it is that makes 'you' who 'you' are, would never have known that anything ever existed?

So, either the Universe was destined to develop as it has from the first moment it was created, which included the development of you; or there was a whole plethora of random events which miraculously and randomly made you from the chaos of the Universe.

It is one or the other.

I go with the chaos.

What are your views?

What happens when you die?

- *If someone is yet to be born, is this the same as being dead?*
- *Should we remember what it was like not to be alive, if not having been born and being dead are the same?*
- *The human being's brain waves stop – the body just then appears as a shell which used to house the person you knew when they existed.*

Some scientists believe that death begins when your heart stops and goes on for a period of time. The heart stopping is just one of the symptoms of death. If there are a number of other symptoms, there is no reason why the symptoms could not just be temporary and reverse. The heart starts beating again and life resumes normally.

Roughly fifteen per cent of people who have been brought back to life will tell us that their consciousness was still present whilst dead. There are numerous stories where dead people were able to see doctors and nurses working on them whilst they seemed to float freely about the room. This contravenes the state that is thought to affect our minds when a medical flat-line state is observed.

Science has traditionally stated that when a person's pulse ceases then they have passed away. However, it is so much trickier to determine what activity is taking place in the brain at the point of death. Is there any life left in there for a while so that you may have your final thoughts?

No one knows whether the 'out-of-body experiences' witnessed by many means that we have gained a new insight into the inner workings of the brain when someone is at death's door. Scientists do not know whether out-of-body experiences are real or if they are just a trick of the mind to ward off the process of the body shutting down.

There seems no way of finding out.

Perhaps this mystery shall remain this way ad infinitum!

However, there is an experiment being conducted in a number of cardiac units around the world. A small picture shelf has been installed near the patients' beds, the idea being that anyone having an out-of-body experience prior to coming back to life will be able to recall what they saw sitting upon it.

We wait for any resuscitated patients to surprise the world!

To answer what happens when you die is rather tricky for anyone to attempt. Taking away any preconceived ideas and removing prior knowledge from your mind is easy as there is very little known on the subject by anybody. There is one very obvious fact associated with the life and death of every human being on this planet; this is that every human being who is alive on this planet 'now' has at one time been 'not alive' before. This is if you can look upon being unborn or 'not alive', as being dead.

Question yourself as to whether there is any status difference between being 'not yet born' and being dead.

Is it necessary for someone to have had a body at one time to be considered dead?

Unless I am mistaken, if you are not alive on planet Earth, or any other planet for that matter, then you must be dead. If being dead and unborn are the same thing then we should all remember what it was like when we were not alive before.

However, my recollection of this 'not being alive' experience is non-existent. This implies that time and thought are non-existent within a dead state. People have no recollection of being dead because thought and memory are not operational at these times.

Kind of makes sense!

There have been near-death experiences recorded, but to be honest, anyone who has returned from one of these is quite categorically 'not dead', and therefore unable to recall what it is like to be truly dead.

When a person dies, people who are alive are able to view them as being in a state of death. However, from the dead person's perspective this cannot be the case as they are not even in a status; they have assumed a 'not alive' state which is the same as the unborn status from which we came. The dead person is no longer here; and their current state would appear to be of no consequence to themselves.

The unborn status from which we came cannot be too tragic a state. I somehow doubt nature would permit it to be bad. There is no logical reason why the dead state of 'nothingness' should be credited with a bad experience. Visions of peoples' souls endlessly drifting in space for eternity is generally accepted as a vision confined to science fiction films.

The body the dead person leaves behind is just a shell of their former existence and their brain waves collapse and no longer function. All knowledge of existence is lost, memories within your head are lost and consciousness is no longer a reality. As the dead person's electrical brain activity no longer functions, all memories, knowledge and consciousness is lost at the moment of death. As far as the dead person is concerned they have now left their friends behind. However, from a live person's perspective the memories of the dead person live on in their minds.

There are many questions relating to death that must have changed as we evolved as a species. It is not clear how our minds worked in the past in relation to death. It is fascinating to consider whether certain animals know whether they were to die or not.

Does a dog know that it is going to die?

Similarly, if a human being were raised on a desert island on their own, would they instinctively know that one day they would die?

When a man died a million years ago did they bury him?

Did people a million years ago do anything with dead bodies?

Did people a million years ago die 'on-the-job' like a pigeon does today?

When was the first man buried?

What was the motive of burying the first man?

When was sadness first felt on planet Earth?

When was the first tear shed for a dead man and why?

When you die, do you know you are dead?

Do you only know you are dying up to the point when you die and then it is like your television having its aerial pulled out?

Do you only know you are dying up to the point when you die and then it is like your television having its aerial and electricity pulled out?

When you die is there still a signal, but no electricity?

When you die is there still electricity, but no signal?

When you die is there still electricity and a signal, but you are dead nevertheless?

If you die in your sleep, do you know you have died?

Do we die as a consequence of all other actions throughout the Universe and there is no way of preventing it – following the view of Determinism?

Death goes hand in hand with a lack of blood circulating within the brain. When the brain decides to give up the ghost, this is when death kicks in. However, most people think of death as a moment – you are either dead or alive. Could there be varying degrees of death?

There are many unanswered questions about death – and no one will ever know what it is like until it happens to them. Even then, it is doubtful they will ever know.

Death is unfortunately one of the only things in life that is guaranteed. The only other two things that are guaranteed are tax and a free annual trip round the Sun!

Later within the chapter 'Does Infinity Exist', infinity reveals itself as a potential property of the Universe, which is quite possible. What this then means is that every human being will be reassembled just as they are now, an infinite number of times. This is the nature of infinity.

What do people go around the world thinking?

- *The world is modelled in someone's mind by their thoughts – the world is then dealt with according to their inner aspirations.*
- *To be told pleasant things about yourself provides you with a sense of comfort and happiness that makes you feel of value.*
- *Elation is a short-lived sensation; as time passes, normality becomes prevalent and elation is sought once again.*

It is impossible to comprehend what someone is thinking just by looking at them. Thinking allows people to model the world around them and deal with it according to their wishes, objectives, plans and desires. Individual's thought processes are influenced by people around them and the environment they are within.

People within different environments such as watching a football match, riding a horse or digging for coal will have very different thought experiences. Having said this, when there is something pressing on someone's mind such as divorce, financial difficulties, bereavement or illness, this may well embed itself firmly in the forefront of their thought, often to the detriment of their normal worldly judgment.

Human beings in general are quite fragile and very often in need of reassurance. To be told that you are useful, glamorous, intelligent or precious, conjures a sense of enormous wellbeing, gives you a skip in your step and makes you feel treasured. This is an important factor in making life worth living, and so it is important that, if you are wishing to influence people, use the correct language – you may find it works extremely well.

Try saying to someone extremely positive comments and see how their demeanour changes for the better. Just as saying negative comments will result in ill thoughts and lead on to falling-out and arguments. An imbalance in a person's wellbeing can lead to disastrous feelings, depression, anxiety and in worst cases suicide.

Sentience within consciousness describes the ability to have sensations or experiences. Some may know this also as a quale. The plural is qualia. These are direct experiences that have to be witnessed to understand them, such as becoming a parent, passing your driving test or tasting great food. Trying to describe a quale does not relay the first-hand experience that anyone would truly feel in reality. If you really want to witness what a bungee jump is like then do one.

The word 'quale' comes for the Latin for 'what sort' or 'what kind', and is used as a philosophical term to describe the subjective quality of conscious experiences.

People's thoughts are effectively qualia as it is impossible to explain to someone what a thought is like, especially as it involves accompanying experiences and emotions. Imagine trying to explain to someone about your feelings and emotions when winning a race or being presented an award, these sorts of feelings and emotions can only really be experienced first-hand.

What follows are four scenes. They are all the same scene, all the same actions, but show how people could observe or think different things in the same situations. The qualia in each of the following four scenes are different although the actions are the same.

Scene one: I walked out of the hairdresser's, glanced around, looked towards the pavement, immediately looked up and caught sight of a bus passing by. I walked down the road just literally thirty paces or so and called off at the corner shop for some groceries. As I paid I saw Gordon, who told me he had seen me earlier that day. Apparently, he had parped his horn but I had not heard him.

Scene two: I walked out of the hairdresser's, I had a quick look to see if there was anyone I knew outside as I thought the hairdresser had overdone the gel on my hair, not my true style really, but who cares! I noticed a sweet wrapper on the pavement in front of me – funny I had eaten one earlier that day but would never think of throwing the wrapper down like that.

I saw a bus whiz by and tried to see the silhouette of anyone I knew; no one as far as I could make out. Remembered the short list of things I needed from the corner shop and walked in, trying to recite the list in my head. It was only when I saw Gordon near the checkout I remembered the cauliflower. I had heard a car peep its horn earlier that day but I had not thought it anything to do with me. It all made sense as Gordon told me it was him.

Scene three: As I walked out of the hairdresser's I was feeling terrible because the hairdresser had told me that a dog had been run over along my street – I was just hoping it was not my friend John's. As I glanced out of the shop door I saw the back of Jason walking off in the distance but decided that to get back home quickly was better, to check all was well with my friend John.

I glanced at some rubbish on the pavement and a bus zoomed by. I called off quickly at the corner shop and nearly forgot to buy the cauliflower. I saw my friend Gordon in there who muttered something about a horn peeping or something. I had not caught what he had said at all, but acknowledged him anyway and smiled and gesticulated that I was in a bit of a hurry. Without hurrying too much I took my change and made off for John's.

Scene four: I could not believe my luck from the night before, having met this new girlfriend. She was gorgeous, and all the time the hairdresser was cutting my hair I had to listen really hard to what he was saying because my mind kept drifting off into this fantasy world that she had created in my mind. I left the hairdresser's in a daze, thinking how she might take my new haircut. I could have walked out of the shop in front of the bus for all I knew as my mind was totally elsewhere.

All I can remember was thinking "cauliflower", "cauliflower", "cauliflower", as I had promised I would get one for tea for my mother as she had a fetish for cauliflower cheese. When I saw Gordon in the shop, I had just located the cauliflowers but had carefully started queuing as I was in a hurry, and then pretended to have just forgotten the cauliflower just as the cashier was scanning the other items, otherwise I would have been much further back in the queue.

Gordon mentioned that he had parped me in his car earlier in the day; I had heard it but had my head in the clouds thinking about my new girlfriend. I was going to tell him about her but then remembered I was seeing her again in about half an hour, but needed to wash my hair in a hurry to get the gel out, which I had accidently agreed for the hairdresser to put in whilst in my daze.

Now, what is the difference between the four scenes?

Answer: absolutely nothing whatsoever, the only difference between all of the scenes were to do with what I was thinking. The exact same motions were made in every scene. All passersby, everyone on the bus, the hairdresser, Gordon, the shopkeeper and everyone in the shop would have noticed no difference in any one of the four scenes.

Even if I was acting this out on a stage with an audience of hundreds, they would have noticed no difference whatsoever. It was just the thoughts in my mind that were different. How was anyone to know what I was thinking?

Would my thoughts have any impact within the Universe whatsoever?

Did it matter to the people around me what I was thinking?

Was I influenced by other peoples' thoughts during my actions?

Without a means to touch on someone's thought without enquiring, there appears to be no way anyone can ever know what someone is thinking. People have tried reading people's minds but it has never been scientifically proven.

Now put yourself in the position at the hairdresser's and walk out of the door yet again. Go through the motions you now know so well. In your present frame of mind, what sorts of thoughts would be racing through you as your body goes through those unchangeable actions?

Perhaps there is an ill relation about whom you constantly worry as you go about these actions, perhaps there is an ache in your leg, perhaps there is a headache you have had for couple of weeks and wonder if a trip to the doctor's would help.

I can guarantee that each time anyone walks out a hairdresser's their thoughts are different. Every person contains their thoughts inside their minds that act like their 'own known universe'. Your own specific qualia have kicked in.

The only way we can attempt to determine specific thoughts that someone may have occupying their minds is by making a judgement about them. Looking at their hair, their facial expression, their eyes, their clothing and their stance, we think we can tell quite a bit about them. However, we all know that looks can be very deceptive.

Traits that constitute happiness

- *Sad people cannot become happy simply by someone telling them to become happy – there are complex personal triggers and events that do this.*
- *When a major positive event hits someone's life then the feeling of elation can last for a few weeks – each one of those days feels like their birthday.*
- *Relationships, sex, eating, exercise, music and success all bring happiness – wealth is not on this list!*

One notable human characteristic is that when people are in a sad frame of mind, it is nigh on impossible for them to be happy by pretending to be happy. It is impossible for a person to be confident in a situation where they feel unconfident. When someone is nervous, it is impossible for them to be calm. There seems to be an inbuilt source of feelings within people that spark them into life or knock them over with a sledgehammer.

This is one reason why people say, "Money cannot buy you happiness," quite simply because it is not possible to wave a magic wand and instantly make a sad person happy. Lottery winners may be happy for a short period of time immediately after a win, but as time goes by the feeling of joy will wear off as the person returns to routine and normality.

I experienced a tremendously uplifting feeling of happiness upon publishing my first book. It was very similar to the feeling I get when it is my birthday. Every single day for about four weeks I experienced a heightened level of happiness – just as if it were my eighteenth birthday. I bounced around, telling everyone about my new book.

Then suddenly it became normal. I would not talk about my book so much and people would have to prompt me to mention it.

I remember the wonderful feeling; you cannot relay this type of experience to anyone, they have to witness it for themselves. When people ask me to sign a book I do it calmly now. Initially I was so excited, it was indescribable.

Research has highlighted a number of factors that correlate with happiness. Relationships and social interaction, extraversion, marital status, employment, health, democratic freedom, optimism, religious involvement, income, proximity to other happy people and our good old friend, eccentricity – in positive doses, they all relate to happiness.

Happiness is very much associated with other people's happiness too.

Pleasing other people seems to be just as important an aspect of attaining happiness as pleasing yourself. Making your boss happy, making your friends happy and making your family happy, all contribute towards your own happiness.

Happiness is a state of mind personified by contentment, love, satisfaction, pleasure and joy. Defining happiness and identifying its sources is challenging.

Michael Argyle, 1925 – 2002, one of the best known English social psychologists of the twentieth century, studied happiness in tremendous detail. In 1987 he published his work, 'The Psychology of Happiness', within which he listed and discussed empirical findings on happiness, including that happiness is promoted by relationships, sex, eating, exercise, music and success, but probably not by wealth!

Michael Argyle's work identified the underlying elements to happiness which can be gauged, monitored and improved.

Happiness includes thoughts about the way you feel about yourself. Whether you feel pleased with the way you are, whether you feel life is rewarding, whether you feel warm towards almost everyone, whether you feel life is good, whether you feel you are in control of your life, whether you feel you could take anything on, whether you feel healthy, and whether you feel mentally alert.

Happiness is also related to physical elements of your life. Whether you feel rested after sleep, whether you commit and get involved, whether there is a gap between what you would like to do and what you have done, whether you can find time to do everything you want to do, whether you have energy, and whether you have an influence on events.

Happiness is also impacted by your outlook on life. Whether you are interested in other people, whether you feel optimistic about the future, whether you think the world is a good place, whether you think you look attractive, whether you see beauty in things, whether you find it easy to make decisions, and whether you have a sense of purpose in life.

Happiness is also affected by your level of pleasure and satisfaction. Whether you find things amusing, whether you laugh sufficiently, whether you are satisfied about everything in your life, whether you have a cheerful effect on others, whether you experience joy and elation, whether you have fun with other people, and whether you have happy memories of the past.

Combine all these factors together and you will get an appreciation of your overall level of happiness. At various times in your life you can reflect upon the changing elements of your degree of happiness.

Professor David Lykken, 1928 – 2006, was a behavioural geneticist at the University of Minnesota. His research studying identical twins found that approximately fifty per cent of people's happiness is inherent within their genes. He found that when twins were brought up in different houses, fifty per cent of their happiness still correlated. Approximately ten per cent of their happiness was as a result of measureable life circumstances, such as socioeconomic status, marital status, health, income and sex. The remaining forty per cent is a combination of factors including actions that people deliberately engage in to become happier. These actions may vary from person to person, but some examples are human interaction, exercise, reading, general hobbies and charitable work.

Perhaps more people should be confident within their lives and take happiness more seriously. People should perhaps spend time enjoying life for the moment rather than working towards being happy – it may never come!

It is the moment in hand that really matters, there is little point reflecting upon past happiness. Reflecting upon past happiness is just one tiny aspect of your overall current happiness.

Aristotle stated that happiness is the only thing that humans desire for its own sake, unlike riches, honour, health or friendship. He observed that men sought riches, honour, or health not only for their own sake but also in order to be happy.

Happiness is characteristic of a quality life, a life in which someone fulfils human nature in a first-rate fashion. Everyone has a set of purposes which are typically human and belong to our nature, personality, temperament and disposition.

A happy person has outstanding abilities and emotional tendencies which fulfil their broad-spectrum of human desires.

Why are we unable to speak with other animals?

- *If animals could speak to us we would perhaps view them as a major threat – some people would certainly take a major dislike to them.*
- *Do similar species acknowledge one another just as people in the same club acknowledge one another?*
- *If we discover creatures living on other planets, it is still only a remote possibility that we would ever be able to communicate with them.*

There are fundamental differences between the vast numbers of diverse mammal species. They differ enormously in what they eat, how they move, where they live, how big they are, how they breed, what climate they can tolerate and how they communicate. Another major difference is the time it takes for their young to reach adulthood. Amazingly, it takes almost twelve months before human babies can confidently walk, whereas the majority of other land mammals seem able to stand up and walk immediately.

Perhaps there is a good reason for this!

Survival is one obvious reason. Being capable of walking on the day you are born definitely reduces the odds of you being attacked by predators. Perhaps becoming mobile on the day you are born comes at a price. Maybe intellect is compromised for early physical capabilities – could the substitution of brains for 'early brawn' perhaps be a simple way of looking at it?

If you need to walk and run immediately after birth, then perhaps this is to the detriment of overall reasoning and general intelligence.

Humans take much longer to learn their actions and abilities. However, human actions and abilities that are ultimately achieved prove far more skilful and versatile than other mammals. A human baby has no chance whatsoever of defending itself from predators for at least two years, which implies we have evolved within an environment where either predators did not exist, or we have been able to protect our young within organised tribes.

Humans also take a long time to learn language. However, it seems well worth it in the long run as our speech is ultimately much more complex than a few grunts and squeaks!

If animals had existed with which we could converse, it is quite possible that we would have set out to kill them – we would have viewed them as a threat. Not too dissimilar to how nations with different cultural backgrounds fight with each other today!

If we were born into a world where we could speak with animals, we would not find it odd. But I am sure the expletives exchanged by badgers, foxes and rats would be awful.

When human beings from different cultural backgrounds meet, there are a number of differences to overcome. The fundamental differences that exist are much more than just speech; there are manners, behaviour, conduct, goals, objectives, activities, outlook and demeanour. All these differences play a major part in making communication extremely difficult.

These very same differences are further intensified when humans meet with other creatures. But added to these differences are that we move differently, we look different, we smell different, we eat different things, we live differently and we survive differently. Most of all we communicate totally differently – perhaps the secret to surviving also involves having a communications method that is unique, cannot be understood by any other species and stands the test of time.

The differences in general outlook will be similar when different species of the planet other than humans meet. What is it like for a giraffe to look into the eyes of a camel?

However, for purposes of positive collusion let us take a moment to consider two different species of animal which can communicate well together. There are some odd couples of this world that have what are termed 'symbiotic relationships'. These are when different species have a close relationship with one another. They are undoubtedly survival related and require a good understanding between the two.

Crocodiles and plovers have such a relationship – the crocodile realises that having a healthy mouth is a bonus, so opens its mouth and quite happily allows the plover to hop in and peck away at parasites. The crocodile resists the temptation of having a snack on the plover – perhaps just like you and I refrain from taking out revenge on our dentists when they hurt us. Just like the crocodile, we know that a positive relationship is good for us in the long run.

It would be a major breakthrough if we discovered that squirrels could communicate with mice and that cows could talk to giraffes. But the world does not seem to work like this. It has been scientifically proven that even if animals such as dogs could learn our language, they would be unable to speak because the physical composition of their mouth parts and throat would not permit the sounds to be generated. Perhaps we should try crossing a dog with a parrot to see what may happen!

For each of the ten million different species of creature on the planet, there must be ten million totally different and unique ways of communicating. Our ability to communicate with all animals is extremely poor, even with the more advanced species such as monkeys, apes, dolphins and octopus. Perhaps they are not too intelligent, or we are overlooking something – I am inclined to think we are missing something in their method of communication!

We have no means of determining how the world looks through the eyes of another species, we seem unable to have empathy or relate. We know how humans should greet each other, but we have not got a clue how we should greet each other if we were a frog, bat, slug or leech!

Do we excrete scent, do we urinate on each other, do we waggle our ears or do we jump about?

When a frog of one species meets a frog of a totally different species, do they still acknowledge one another?

Maybe the frogs somehow acknowledge each other just like two people driving an old Volkswagen Beetle wave as they pass in opposite directions along a road!

It is thought there are roughly ten million different species here on planet Earth. It is also thought there have been roughly thirty or so extinct species for each one that survives to this day. If the first species appeared here on Earth about a billion years ago, then this means that a brand new species has evolved on planet Earth approximately once every three years or so.

Species that have just evolved cannot be too dissimilar from a close relative from which they have just evolved. Perhaps the differences are so slight initially that it would be impossible to tell. Major changes perhaps only occur when a number of the species get marooned on islands in total isolation for years.

Surely, sometime in the past, there have been two different species of creature that have been able to communicate with one another – whether by coincidence or design. I would not suggest this was a frog and an armadillo, but more likely something like a kangaroo and a wallaby. The kangaroo and wallaby communicating seems a little more plausible, but we humans do not even know to what extent this may be the case!

As far as us humans are concerned, animals tend to act and move in a rather unfamiliar fashion. Our minds seem unable to assimilate their true desires and aspirations, their activities look alien to us and we get confused about their general conduct as a consequence. As far as the noises they make, apart from a few mating calls and territorial cries, we really do not have a clue!

People can dedicate their whole lives to trying to understand the actions and noises of animals and only end up with a rudimentary understanding of a few sounds that they make. As for striking up any meaningful conversation, I very much doubt we will ever have any joy.

If we were truly intelligent, we should be able to understand the innuendos and content of a lion's roar! The only real understanding we ever get from a lion roaring in a jungle is interpreted by us as, "Watch out I am going to eat you now!"

Who knows, perhaps the underlying resonance within a lion's roar actually contains extremely sophisticated information and instruction. They must be communicating to each other relatively intelligently else they would not be the sophisticated pack hunters that they are.

Picture cards and peanut dispensing machines have been used to encourage monkeys to speak to us. The most advanced we have ever got with them is to determine that they would like another peanut!

We cannot discuss the latest blockbuster film with them – this would be a breakthrough.

Scientists have recently conducted thorough research into animal communication that has revealed some truly ingenious methods and extremes.

Animal communication has developed into ways to send messages to one another for miles. Foot stomping and low frequency rumbling created by elephants that can travel up to twenty miles – elephants use this method of communication to signal other herds or members. It is thought that elephants are able to sense vibrations through their feet and interpret them as warning signals of distant danger. These waves travel from special receptors in their feet via their bones to their ear.

Some animals can sing in frequencies we cannot hear, detect colours that our eyes cannot see and send messages in worlds of scent, electricity and polarised light that we cannot begin to interpret.

Water presents a unique set of challenges to animals. Communication can be achieved from using brightly-coloured messages on bodies or even, as with the male cuttlefish, impersonating the female to find a mate. It has been speculated from recent study that the bottlenose dolphins may even use names to keep in contact with each other!

Electroreception is the ability for a creature to detect and recognise electrical impulses. This is particularly common as a trait within salt water creatures as the salt water is quite an efficient conductor. The shark is the most electrically-sensitive creature known and can detect extremely low electric currents. It is even thought that sharks may be able to navigate the oceans using the Earth's magnetic field.

Polarised light can be perceived by many insects including bees. Bees use polarised light as the basis of their means of communicating information through dances, which can direct other bees towards sources of food. Interestingly, octopus, squid, cuttlefish, mantis shrimp and even humans have sensitivity to polarised light. The human eye is weakly sensitive to polarised light, creating a very faint pattern near the centre of the visual field called the Haidinger's brush. Although the pattern is very difficult to see, with practice people can learn to detect it better.

If you wish to observe polarised light, it may be seen as a yellowish horizontal bow-tie shape which is visible in the centre of the visual field against a blue sky whilst facing away from the sun, or away from a bright background. It will appear the same size as the tip of your thumb when held at arm's length. The small yellow bow-tie shape will have fuzzy ends which is why it is called Haidinger's brush. The effect can also be observed when looking at a white area on an LCD flat panel computer screen. Wilhelm Karl Ritter von Haidinger, 1795 – 1871, was the first to observe this slight polarization dependence of the human eye.

Some animals can communicate in enormous crowds where every individual looks and sounds the same. For instance, a mother seal can find her offspring among millions of other babies.

We do not give creatures the communication credit they truly deserve. We tend to assume their communications techniques are substandard compared to ours, but this could not be further from the truth. There are many ways in which they communicate; several types of which are totally unknown to us yet. From the long distance echoes of the blue whale, the familiar and sometimes annoying click of a cricket's back legs to the chemical-scented communication displayed by ants and many other types of insect.

How would we begin to understand the mind of an ant or the mating communication of a millipede?

It is much more likely that our lack of understanding leads us to believe that animal communication techniques are substandard compared to ours!

People have tried to understand the singing of the blue whale for years. Despite all this detailed study, we still cannot chat with them!

Their lives are remarkably different to ours, their environment is extremely dissimilar and alien too. Their situation and desires are so different from ours that we seem unable to relate to their deeper conversation.

If we could speak to whales, due to their remarkably different focus and interests, we would probably not be able to discuss anything remotely interesting with them anyway. They are probably just singing different variations upon a theme that they know very well, "Does anyone know where the fish are?"

If we were expecting them to be discussing our favourite soap opera, or nattering about the extent of political integrity in Botswana, then perhaps that is where we are going wrong!

Some people say they can communicate with their dog. Some are capable of interpreting their barking, but this is limited. This communication is not meaningful enough to have discussions and conversations. We cannot have a conversation with a dog about a good film, even if they watch it with us!

We do not know the extent of animals' intelligence and vocabulary capabilities. Does a dog actually watch the television knowing what is flashing on the screen just as we do or does it simply see a whole host of suspended lumps and speckles jumping about on the screen?

As for music and sound from a television I have often wondered what a dog may think.

Having determined that we cannot speak with all the animals of the planet, this becomes rather awkward when we consider future encounters. If Mankind were to land upon an alien planet with a detected life-form, we should not hold out too much hope of communication. Our spaceship lands, we venture out the capsule and walk over to a creature to shake hands with it. This creature just looks at us in the eye, chewing the cud and you then realise it is gazing through you just like a goat does.

How disappointing would that be?

Whatever next – speaking animals?

- *Occasionally a pet may appear to be communicating with humanlike characteristics – this can appear quite weird.*
- *Animals appear to have a different type of intelligence – their problems are solved using specific solutions.*
- *Now that we are teaching our closest animals key cognitive abilities, perhaps we will see them communicate better with us as they evolve with us.*

I was watching television with a friend whilst his dog was under the table out of sight. I asked my friend, "Have you been able to get the stereo system repaired yet?"

Precisely at this point, his dog under the table made a coherent cough-like noise that sounded uncannily like "Yes."

It was one of the funniest moments.

The timing of the dog's answer was perfect and this split-second gave us the ability to witness what it would be like to be able to speak with other animals.

As we laughed loudly the dog called Micky wagged its tail profusely and was clearly pleased. However, perhaps we misunderstood this tail wagging and the dog was really interpreting our laughter and commotion as imminent indication of 'walkies'!

Following our command of rudimentary communication within the jungles and lush savannah, we invented money to barter. The invention of money truly differentiated us from other animals and brought about the creation of great civilisations. The topic of money is covered in great detail within its own chapter later in this book.

Animals share many of the building blocks that pertain towards human thought, but paradoxically there is a great cognitive gap between humans and animals.

Whereas humans have the ability to combine and recombine different types of information and knowledge to gain new information, animals do not. This has enabled humans over the passage of time to reason and develop ideas to a point where we have experimented sufficiently to develop advanced medicines that will cure us from disease and systems to launch us into outer space.

Whereas humans have the ability to apply a rule or solution from one problem to a different and new situation, animals do not. This ability has enabled humans to become familiar with such phenomena as the force of gravity, and then develop ideas from our knowledge. Therefore, not only do we become familiar with gravity, but we are able to apply further knowledge to it, develop helicopters and spacecraft as a consequence.

Whereas humans have the ability to create and easily understand symbolic representations of computation and sensory input, animals do not. This has enabled us to develop traffic lights, road signs, brand recognition and the ability to manipulate numbers. Considering that the most intelligent animals only have a rudimentary understanding of numbers, we must give ourselves a pat on the back when it comes to calculating the trajectory of a satellite launch.

Whereas humans have the ability to detach modes of thought from raw sensory and perceptual input, animals do not. This gives us more emotional states as we battle with our surroundings, making sense of situations we have never been party to before, making decisions and acting upon these decisions. These increased emotional states make us quite fragile as far as mental wellbeing is concerned, with a significant number of the population suffering from mental health problems.

In comparison, we do not see similar signs of mental health problems in wild animals. When a wildebeest is chased by a lioness and narrowly escapes death, it very quickly re-joins the herd without any visible trauma. If this had happened to a human, they would possibly have to spend the rest of their lives in a hospital trauma unit!

Animals appear to have a type of laser beam-style intelligence in which a specific solution is used to solve a specific problem. However, these solutions cannot be applied to new situations or used to solve different kinds of problems. On the contrary, human beings have a type of floodlight cognition which permits them to use thought processes in innovative ways and apply the solution of one problem to another totally different situation. As a result of human beings possessing these key cognitive abilities, it has opened up of other avenues of evolution that other animals have not been able to benefit. This evolution of the human brain is the foundation upon which cultural advancement has been built.

We are now encouraging our closest animals such as dogs, cats and horses to follow suit and possess these key cognitive abilities. It will be an exciting development to watch over the next few million years as evolution works alongside our relationship with animals.

One day there may be an announcement in the news that reads – 'Mrs Tiggy-winkle's cat has conversation with neighbour over garden fence'.

There is no reason why animals should not begin to speak one day as it took humans only seven million years to develop our cognitive abilities to how they are today. Speaking animals may well be possible in the future – but I would have thought Mankind will need to wait a few millions years yet before an article like this appears in the New York Times!

How life's stresses and strains challenge us

- *Currently a large proportion of mental breakdowns are relationship-related – as society changes, so do the reasons for mental breakdown.*
- *Everyone has a comfort zone, inside which they can cope – venture outside this comfort zone and mental instability can ensue.*
- *Modern life comes with much more to consider and generally much more to worry about!*

Having identified the sophistication of human beings and our lengthy learning process, it becomes apparent that our mind's cognitive behaviour is really quite complex. Disruptions to our extremely balanced style of life and mental trauma can bring on the onset of quite complex mental issues.

We can only presume that mental issues are prevalent within human beings; otherwise we would have noticed animal suicides. There are a few insect suicides that have been reported – quite notably autothysis, which a soldier termite or a carpenter ant can perform when protecting its nest. The termite or ant will effectively turn itself into a bomb by detonating itself with the explosive release of gas and faeces.

Nasty!

Animal suicides, other than humans, do not occur. There is an old wives' tale that lemmings jump off cliffs, but this is just a fallacy. Lemming suicides have never happened. The tale emanates from Norwegian lakes and rivers that lemmings swim across each year. This results in a few lemmings drowning. Dead lemmings can sometimes be found floating in the water near to cliffs, hence the story that portrays them committing suicide off cliffs when the population gets too high.

This myth of lemming mass-suicide is a long-standing one and has been popularized for a number of major reasons. In 1955 a comic with the title 'The Lemming with the Locket' was released which was inspired by a 1954 National Geographic Society article showing great numbers of lemmings jumping over Norwegian cliffs. This was followed by a 1958 film entitled, 'White Wilderness' which won an academy award for best documentary.

Within the documentary it showed the mass suicide of lemmings. However, it transpired a few years later that they did not jump off the cliff but were launched off the cliff using a turntable.

How bizarre is that!

Again, I have to question how much of the information we are spoon-fed from birth is true and how much is totally fabricated?

It is enough to make you go potty!

If learning about totally fabricated facts from the past has not sent you potty yet, then there are always a number of human psychotic behaviours from which you may choose to tip you over the edge!

We will take a look at what causes mental illness and triggers someone to experience symptoms of a nervous breakdown.

There are relationship problems, divorce and marital separations which contribute towards approximately a quarter of all breakdowns. Difficulties at work and school accounts for almost twenty per cent, with financial problems accounting for just over ten per cent and strangely health problems only account for roughly five per cent.

In the 1950's health issues were known to cause nearly thirty per cent of breakdowns. Health issues would appear to be less problematic to us within our fast, technological world. No longer are our lives simple, we have now got to force ourselves to work harder to cover bills for all and sundry. Mobile phones, computers, children's toys, utility bills, mortgages, rent, travel and seasonal pressures such as Christmas, all play on the minds of individuals, causing anxiety, stress and depression.

It would appear that some people develop a better tolerance to stress and strain than others. Some people suddenly find themselves subject to triggers that place them outside their comfort zone – they discover they are unable to cope. The trigger will then manifest itself within an individual as panic attacks, mood swings, outbursts, self-harming, periods of mania and general uncontrolled delirium. Time seems to be a good healer for this type of illness – the gradual passage of time ultimately restores normality as the initial trigger becomes more and more of a distant memory.

Everyone has their own individual pressures in life and no two lives are the same – we are all individuals. Due to our 'non-laser-beam-style' intelligence, our floodlight cognition, our ability to detach modes of thought from raw sensory and perceptual input, we are destined to malfunction when put under the pressures of our now 'abnormal' everyday lives.

Studies of the causes and processes involved in the development of mental illness have found there are physical, social, environmental and psychological causes.

The physical causes are those which are biological in nature. They include our individual genetic make-up and the way this might put us at more or less risk than others. It has also been found that those who have suffered head injuries can also experience changes to their personality. In some cases they may begin to experience schizophrenia and psychotic-type symptoms. The misuse of substances or illness of mothers during pregnancy can also lead to changes in their baby's development which may ultimately affect their mental health. Recent reports have also suggested that vitamin and mineral deficiencies such as Vitamin D, zinc and certain fatty acids may also be related to our mental health and the development of neurological symptoms.

The social and environmental causes are those factors around us such as where we live, whether we have strong support networks, family and friends who make us feel safe and on whom we can rely, our place of work and how and where we can relax. Physical environments such as the neighbourhood where you live can be very stressful, particularly when there are problems with neighbours, or where there are high crime rates. Whether you enjoy your work, or feel you are under too much pressure, are unable to find employment or hold down a job, can all put pressure on your mental well-being.

These kinds of problems will increase the amount of stress people are under, and can cause depression and anxiety, especially in situations where individuals are unable to make changes to alleviate the stressors. When we face difficult times our support networks become very important. Those who do not have close friends or families, or those who do not live near the people who support them may find it increasingly difficult to cope alone.

Psychological factors influence your mental and emotional state, particularly if you are coping with a traumatic and abusive past. Significant life events like bereavement or divorce or self-destructive thought patterns and perceptions, will not help anyone's psychological state. If someone dotes on this for too long it might adversely influence their mental health.

Compared with two hundred years ago we have so much more to consider, think about, remember, live up to, discuss and generally worry about!

Perhaps if it were not for our desire to keep up with the Joneses, this would not be such a problem!

How to keep a positive mind

- *Studies have been made to see whether genius and intellect can be spread positively.*
- *There are a number of human traits that seem to enable someone to become whom they want to be.*
- *Generosity, creativity, forgiveness, sincere emotion, honour and the ability to be peaceful – provide a positive mind.*

Staying on an even keel and keeping a positive mind is not something that anyone is normally taught at school. It is learnt during a lifelong voyage of discovery. The pressures of life can often build-up which may result in negative thoughts and bad vibes. The art of keeping these negative aspects of life at bay is now being studied within a recently developed branch of psychology called positive psychology.

Positive psychology studies the strengths and virtues that enable individuals and communities to thrive. It seeks to find and nurture genius and talent with a view to determining what may encourage it to spread positively.

Other areas of specific interest are the study of what constitutes an enjoyable life. This aspect of positive psychology investigates such topics as relationships, hobbies, interests and entertainment. There is also the study of immersion and absorption which looks at an individual's match to their primary activities; the more confident they feel, the better they can accomplish their tasks. Then there is the study of an individual's meaning within life known as life affiliation, which looks at how individuals derive a sense of well-being, belonging, meaning and purpose.

A very interesting aspect of the study is how positively-minded people broaden their horizons and build on these experiences.

Examples of this are when curiosity about a landscape can become a navigational aid, positive interactions where strangers can turn into supportive friends, aimless physical exercise can become physical excellence and hobbies such as computing can positively influence someone's career path.

Perhaps children should be subject to more life studies from an early age to promote a more confident approach to the rigours of life by studying financial matters, life's happiness indicators, life issues and positive pleasures. This may give children a more substantial start in life and establish a more positive attitude.

Study shows that if you wish to keep a positive mind you must live your dreams, matter, count, achieve, stand for something, make a difference, expand your potential, become who you want to be, flourish, be authentic, put everything into what you do, and create your own destiny. This is helped by expanding your perception of the world, learning, seeking wisdom, avoiding suffering, seeking happiness, facing fears, accepting lessons, finding meaning, finding purpose, understanding the nature of reality, and knowing your reason to live.

With the correct approach you may benefit others, give more than you take, prevent suffering, promote equality, challenge oppression, distribute wealth, be generous, be creative, be forgiving, be emotionally sincere, be honourable, be peaceful, contribute to the well-being and spirit of others, be helpful, be innovative, make the world a better place, and leave things better than you found them.

We learnt how our minds can play tricks on us with delusional disorders like intermetamorphosis. We learnt about some deep issues that lead to the realisation of whether you were destined to be here or were placed here by pure chance. We learnt how the Universe is entering a new era where future change is being brought about by 'intelligent' human activity. We learnt a little about death and why no one will ever know how it truly feels until it happens to them. We learnt that a happy person has outstanding abilities and emotional tendencies which fulfil their broad spectrum of human desires. We learnt how, despite being intelligent, we have a great deal to learn about communication with animals – and how disappointing it may be when we discover life on other planets. We learnt what makes us get worried, have nervous breakdowns, what keeps us on an even keel and keeps a positive mind.

We now need to look at where your journey here all started in relation to your surroundings – hold onto your hats, it is going to a bit of a rollercoaster ride!

THE UNIMAGINABLE ODDS OF EXISTENCE

Where we learn about our odds of existence and how some things are more probable than others. We look at how the four forces of nature play their part in governing the evolution of everything and how all that exists has to comply with their rules. We take a look at how life could have originated and whether there could be other conscious entities throughout the Universe.

The unimaginable odds of your existence

- *Each person's journey into this world is extremely remote, very few realise how truly remote it is.*
- *Of all the permutations of how all the atoms in the Universe could be arranged, we just so happen to have a permutation arranged with us in it!*
- *We could have been the result of a massive coincidence or have been carefully calculated from the birth of the Universe.*

One of life's major secrets is undoubtedly how we evolved. Once the human race materialised we subsequently colonised the continents, the oceans and Earth's orbit.

Incidentally there has always been someone in Earth's orbit since October 31st 2000 aboard the International Space Station.

How did the human race evolve over the millennia and ultimately produce the uncanny presence of 'you', in the specific guise that you appear today?

Few people truly consider the unimaginable odds of their existence, their luck, their ancestors' ordeals, their inconceivable journey into life, or their general presence on the planet. We often take this fact too much for granted!

If we were to calculate the odds of you being here reading this book based on the probability from the Big Bang, we have totally unimaginable odds. For you to be here, as you are today, each and every one of your predecessors must have existed precisely as they did going back to the beginning of time – these precise people and organisms only – other people and other organisms would not do. Your specific line of ancestry goes back to the very beginning of Mankind and beyond to the primordial soup, the amoebas, the single-celled organisms and all the matter scattered throughout the Universe before this.

With everything bouncing off everything else, just one molecule of matter bouncing in the wrong direction throughout the history of the Universe would have had catastrophic consequences – no you!

Every single one of your ancestors was completely successful at breeding – not one of them failed!

This goes all the way back to the first creature that ever evolved – your oldest ancestor.

With this in mind, effectively this Universe has evolved to create you. Whether this was deliberate, an accident, design, random or inevitable – here you are today. Well done!

It is interesting to consider whether just one tiny alteration throughout the history of the Universe would have resulted in you not being here.

Sir Isaac Newton, 1643 – 1727, stated, "To every action there is an equal and opposite reaction."

This would imply that from the very beginning of the Universe there has been a steady progression of actions and reactions that ultimately resulted in you being born. Does this imply that your presence here in the Universe was inevitable from the very beginning?

Just as you throw a dice thousands of times, you know that at some stage you will be throwing a six. Taking the same view, this may well mean that your presence in the Universe was an inevitable consequence – you were bound to materialise at some time.

As events transpired – you have appeared now!

If every action does has an equal and opposite reaction, then this implies that from the very early origin of the Universe your presence here today was predestined. It is very similar to having two stars on an inevitable collision course for billions of years with nothing to prevent the collision taking place. It was inevitable from the start of the Universe that the collision was to occur!

When two marbles collide we can fairly accurately predict the outcome. Anything larger than this becomes rather tricky – but let us image for a moment we can do this. If we had the most sophisticated supercomputer that could calculate every action and reaction between every particle throughout the Universe, then so long as the properties of everything are known from the outset, it should be possible to calculate how precisely everything will materialise in the future.

Maybe there is such knowledge hidden deep within the Universe – interesting concept!

Either you believe that the presence of yourself on this planet was a complete coincidence or it was inevitable from the outset. This mystery is open to personal opinion and interpretation.

Perhaps, just as everything that has happened previously cannot be changed – as a consequence the future cannot be changed either.

If the future cannot be changed then perhaps everything has already been determined and there is no future as such – there is only 'the life of the Universe', which is fixed and reveals itself over time.

In the very early days of the Universe it is very doubtful that there were any intelligent 'thinking' species – it therefore stands to reason that the only changes that ever occurred at this time would have been due to collisions, chemical reactions and the interaction of celestial bodies. These collisions and interactions would have been inevitable from the outset of the Universe and nothing but the bumping of matter could ever change anything.

Only when intelligent 'thinking' species arrived could this situation be altered. Their powers of thought would suddenly have the ability to alter future events within the Universe by creating change in a totally different way. No longer would the Universe be left to evolve in a purely mathematical matter-bashing fashion. Suddenly we have decisions being made within intellectual minds that then move things around and bash things about differently.

If this is true, then the outcome of this will mean that the Universe's predestined future path will change as a result of the alterations made by intelligent species. Prior to this, only particles of matter bashing into each other could create change, but now we have seemingly random and unpredictable changes altering future events.

It is interesting to contemplate whether our imaginary supercomputer which monitors all movement throughout the Universe would be able to calculate the motion of every atom through to the formation of the first intelligent 'thinking' species. It seems far-fetched to think that all this movement of planets, stars, moons, cosmic dust, gases and comets, suddenly brought about human beings – but it obviously did!

If it was not just the bumping of the cosmic objects that randomly created life, then perhaps there could be a natural force within the Universe which encourages intelligence in a natural way – an invisible force of nature that is similar to how gravity and electromagnetism exist. Maybe there is no way of detecting this natural force and we can only observe its results on a physical level by observing the life-forms as we see them.

Writing the actual odds of your existence down on a piece of paper is impossible as it would involve calculating the number of atoms in the Universe, extracting those which make up you and factoring this against all other possible permutations of the configuration of the Universe.

This number is huge and believe me when written down it may have to span perhaps more than the width of the Universe!

That is not the end of the calculation, as the environment into which we appear needs to have evolved precisely too. One part of me is concerned that the calculation would involve factoring all the atoms in the Universe against all those that make you and your habitable environment exist, alongside every moment in time prior to this.

Quite a large number!

To put into perspective the truly unlikely and incredible way in which your body has been assembled, just imagine a tiny part of your body consisting of a small number of atoms which need to be assembled correctly. There are many, many ways this small number of atoms can be arranged, but nature seems to know precisely where everything goes to make us who we are – it certainly is not random or else we would never have evolved, as the following illustration highlights.

To expose the unbelievable raw talent that nature possesses we can derive an analogy for arranging atoms by simply comparing the number of ways that twenty-seven small white cubes can be positioned

We can take twenty-seven small white cubes, and from them construct one larger cube made up of three layers containing nine cubes each – thus resembling a Rubik cube.

Imagine we now paint the outside of this larger white cube with black paint, then disassemble the twenty-seven cubes and place them randomly on a table.

Next we get a blindfolded person to reassemble the small white cubes into the larger cube again.

Now just give some thought to how many attempts the blindfolded person needs to make before they construct the larger cube from the smaller one such that its surface is totally black – just as when we first painted it. Assuming they cannot feel the paint!

With a relatively small number of cubes you would think that there are not too many combinations into which they could be arranged – but how wrong you would be!

Most people's initial rough guesses seem to be anywhere between one hundred and a few million.

I point out that the blindfolded person needs to pick each piece correctly and then place it correctly too. For example, each of the eight corner pieces needs to be correctly picked from the twenty-seven available and then each placed precisely with the three black sides facing outward.

Each cube may be placed in twenty-four different ways, but only one of these is correct for each of the eight. The twelve side pieces of the cube with two black squares may be placed in two different ways and the six middle cubes with one black square can assume four different positions. There is of course the centre cube which is the only one without any black paint on it, so long as this is chosen for the centre it may be placed in any of twenty-four possible positions.

We are dealing here with just twenty-seven small cubes, but the average number of attempts the blindfolded person will need to make before reconstructing the larger totally black cube will be roughly five and a half billion, billion, billion, billion times!

5,465,062,811,999,459,151,238,583,897,240,371,200 is the precise number of times!

So exactly how our bodies manage to blindly assemble trillions and trillions of atoms in the correct way, not just twenty-seven as described in our analogy, is so truly incomprehensible that it is off the scale of our ability to grasp.

The conclusion I have reached is that there must be an invisible universal force which encourages life – very similar to other invisible forces we know are present like gravity. However, unlike the force of gravity which we can witness affecting matter within our physical Universe, we cannot witness the force that creates life – unless the pure observation of ourselves is the raw means of observing this force!

The invisible forces in the Universe become known to us when we see things like electricity flowing along a wire to power a light, a ball falling when dropped, a magnet attracting a piece of metal, a waterwheel creating electricity, petrol fuelling an engine, our planet orbiting the Sun and time passing.

So perhaps we should now view life as a type of force – a seed growing when given water, animals reproducing, trees growing tall, butterflies developing, algae spreading throughout a stagnant pond or brambles producing new shoots.

Without doubt, something very interesting has been going on in the Universe that has made creatures come alive as they have. These creatures could not request that they became alive and conscious – but the very fact that they have, implies that there is a force that encourages life.

Discovering that there is a force that encourages life would be no more far-fetched than the forces of electromagnetism and gravity – in fact it seems quite inevitable that one exists. The force of life may even be the result of a number of forces working together.

The force of life may tug molecules of various types together, and once gathered they may then form a cell – the basic unit of life. All the forces of the whole Universe must be working together to make this happen, pushing, pulling, moulding, connecting, evolving and shebang – life.

How else could something so amazing come about?

If life is not as a result of a force, or a collection of forces, then I will be most surprised!

Later within the chapter 'The four forces of the Universe', the odds of a universe developing with the physical forces corresponding to the ones we are familiar with are debated and calculated. It transpires that only one Universe in ten quindecillion, vigintillion, vigintillion, vigintillion which materialises, that is a one with two hundred and thirty-eight zeroes, has the correct values attributed to the four forces to permit life precisely as we know it to form.

This makes your odds of existence tremendously more improbable!

The incalculable odds of existence

- *People rarely seem to consider the truly remarkable odds of being alive here on planet Earth.*
- *Everyone's chance of existence is made even more remote when you consider that the odds of your mother meeting your father were roughly four billion to one.*
- *Your mother meeting your father is just the start of the remote odds – you then have to win a spectacular and highly-contested swimming race.*

As people go about their normal day-to-day activities, the average person is totally oblivious to the incalculable odds of their existence. When you realise how improbable your existence transpires to be, weighing up the odds of your presence here makes you wonder whether it came about as a coincidence of enormous proportion, the nature of infinity, some type of miracle happening or whether it could be imaginary.

When calculating the odds of your existence, you need a starting point.

At which point within the cosmological evolution are you going to commence your calculation?

Are you going to calculate the odds of your existence from the moment of the Big Bang or from the point your parents were born?

Vastly different odds, but either way we need to calculate the latter in order to calculate the former. However, the odds of your existence are remarkably more probable once the calculation starting point is from the birth of your parents.

We shall imagine a time in the past when your parents were alive but yet to meet. There are approximately seven billion people on the planet, roughly half men and half women; so the chance that your mum meets your dad as a loving partner is roughly just under four billion to one. There are, however, geographical factors, but this simply means that if your parents are from different countries, then your odds of existence are substantially greater. The average then equates back to roughly four billion to one.

You commenced your potential life as one of your father's sperm and were one of approximately one hundred and eighty million sperms that were dispatched each time he made love. Let us say an average man dispenses with his sperm one thousand times before having a child; this is based on dispensing with his sperm one hundred times a year for ten years before having a child. This means that your chances of being within the successful batch of one hundred and eighty million sperms with the chance of hitting home, is now compounded by the fact that you also have to be within the correct batch of sperms. This multiplies your already low odds of existence by another thousand!

Once you have discovered you are within the successful batch of sperm by some remote stroke of luck, you then have to win your extremely highly challenged mini-Olympics – swimming up the passageways to be first to fertilise the female egg, beating all the other one hundred and eighty million competitors.

You have heard people on stalls at fairgrounds shouting, "Everyone's a winner."

I think it is to this which they are referring!

Your existence makes winning the lottery look like a breeze

- *Your parents had to meet for you to be here.*
- *Triumphing in the realms of incalculable, unimaginable odds.*
- *Supporting all this life are numerous, seemingly improbable events that make existence a truly remote possibility.*

When analysing the incredible factors that brought you into this world, we find that when your parents were born your chances of potential existence were roughly one thousand, billion, billion to one. This is a one with twenty-two noughts on the end. This is without incorporating the additional complications that transpire when taking into consideration your mother's reproductive organs, then before this the complete history of the Universe with all its past chronological events.

If your parents never met, you would not be here. With this in mind I do not think anyone has ever thanked them enough. It was only in 1910 when Sonora Dodd arranged the first-ever celebration of Father's Day there seemed to be any type of recognition. This is now celebrated in almost every country of the world.

As for Mother's Day, this is equally worthy of note; a recognised day campaigned for by Anna Jarvis in 1907. Just as your father's reproductive anomalies made your existence improbable, your mother's attempt at making your arrival improbable was fairly influential too. Women produce over five thousand eggs on average in their lifetimes; if we now factor this into our calculation then your odds of existing reduce a further five thousand times more than our previous calculation. Straight away it is clear that the first number was ridiculous enough, so now we are getting into the realms of incalculable, unimaginable odds, and we have only been investigating the odds of our existence since our parents were already alive.

Behind the scenes, there are countless other factors making a mockery of our potential existence, the atoms and molecules jiggling around and bouncing off each other. These ultimately organise themselves, having originated from distant stars in various parts of the Universe, and in due course find their way into the male body as food. They do not waste much time being made into roughly ten thousand sperm a second. That is an incredible amount of sperm being made inside a man every second, all day, and every day. Truly mind-boggling!

This really makes your existence make winning the lottery look like a breeze!

Life could be imaginary

- *In many ways our odds of existence appear extremely remote, to such an extent we have to consider other options.*
- *There are illogical and bizarre happenings all around us that appear strange and weird.*
- *If life is due to probability alone, then it was an extremely remote chance that we struck upon that became victorious.*

The odds of our existence is so improbable, it brings into question whether our lives are actually real. This is not as daft a suggestion as you may first think – because the true odds of our existence seem so ridiculously remote.

Perhaps something else is going on!

Due to these gigantic improbable odds, it may be possible to convince ourselves mathematically that it is more likely that life is purely imaginary. We have already seen how our brain interprets the world in an imaginary way – so why cannot the whole of our lives be imaginary?

We could be just fooled into thinking that what we are sensing and experiencing is real.

We could have materialised here in a virtual state upon a virtual earth to experience the offerings of a virtual planet. We could even be some kind of Douglas Adams 'style' experiment. Perhaps we are within another advanced civilisation's pseudo three-dimensional world. Having been provided with the sensations of touch, smell, sight, hearing and taste they then set us on our way to see what happens.

Periodically, we may be thrown difficult global conundrums to tackle to see how we cope, such as global warming – all in the name of research or a hobbyist, extraterrestrial experiment of sorts.

The problem with being part of an advanced civilisation's experiment implies that another advanced civilisation successfully evolved to create us. The chances of this may not be too likely – however, it only needs to happen once in the Universe, then the same formula can be used millions of times over throughout the Universe.

Human beings could be heading towards becoming one of these advanced civilisations too – we already have the ability to manipulate atoms on a computer screen in two-dimensions. All we now need is to develop this into a three-dimensional hologram format and we will have a new world in which to live!

Once we have mastered quantum computing we can place beings with rudimentary behavioural patterns into this world, and over time they may well evolve into intelligent beings.

This brings our off-the-wall theory a step closer to reality.

Perhaps we live inside a three-dimensional quantum device. Quite an interesting concept as the whole Universe is quantum by nature – so there is a strong possibility that we are living within the most perfect quantum world anyone or anything could ever envisage. It is interesting to think that human beings may be striving to invent something within which we are already living!

Our lives could be monitored and influenced from elsewhere, perhaps within a dimension not yet known to us. If we are not real, this would explain a great deal about the nonsensical world we have discovered – as we probe deeper for answers we often see physical experiments that exhibit phenomena that possess no logical comprehension. Superposition, quantum weirdness, Heisenberg's Uncertainty Principle, non-locality and a lack of a grand unified theory, are just a few non-logical phenomena which have surfaced.

Rather than getting bogged down with the intricate details of each of these bizarre phenomena here, just remember that these are as weird as weird can be. In a nutshell, these phenomena can be likened to having massively energetic items in everyday life indiscriminately darting about, duplicating themselves, lying to us, disappearing, and reappearing somewhere else.

It is when realising events like these, that we realise the world makes little sense and that the odds of our existence are so massively improbable, that it is only sensible that we question our intuition, gut feeling and common sense. We have to look at alternative views regarding how we came into existence. The chances that we came into existence through the channel of pure probability are extremely doubtful when we consider all the remote events that needed to fall into place to bring us here. Somehow it does not make sense – we need a radical push in the right direction to understand this further.

No wonder that making theoretical headway at the quantum subatomic level is such a slow process.

All manner of outcomes are possible

- *When you consider something remote, it will either exist or not exist – the Universe is so large that this may be more likely than you expected.*
- *A large eternal creature which is self-repairing would have no need to possess a reproductive mechanism.*
- *Nature is expert at disguising how things have transpired – there is no history of events or an evolutionary manual.*

With so many permutations and outcomes possible throughout the Universe, it makes you wonder about the mind-blowing coincidences that occurred to allow you to occupy the space that you do!

How many perceivable outcomes and possibilities could there have been during the evolution of the Universe?

The Universe is so large that there must be creatures and happenings that surpass our wildest imagination.

Surely with so many permutations possible throughout the seemingly infinite Universe, any type of creature may have evolved and exist somewhere. So the likelihood that somewhere within the depths of the Universe lurks an eternal super-creature is relatively plausible.

Is your existence more or less likely than the existence of this eternal super-creature?

Your arrival within this world was an extremely remote possibility, even with your parents already alive!

With an open mind you can imagine the existence of an eternal super-creature. The evolution of this eternal super-creature seems fairly easy to imagine as it does not need ancestors, it does not require a complex reproductive system and its inevitability is so much simpler to contemplate!

We could almost reduce its possibility to the flip of a coin.

It either exists, or it does not. It is purely fifty-fifty!

Our existence is shrouded in mystery as nature is excellent at hiding everything it has done in the past. The chemistry, the physics and the biology of the world seem impossible to reverse engineer – totally unlike the movements of a clock.

There are so many intricate details of the Universe and human structure that go unnoticed, that when thinking about them, again it makes our odds of existence even less probable, wildly so!

There are the natural forces of the Universe that are so finely tuned to absolute perfection that they uncannily form a universe which can sustain itself – rather than imploding in on itself or blowing itself apart. No one knows why the four known forces exist, where the energy comes from to maintain them, how they were originally generated or even whether they need fuelling to keep them alive and active. All that scientists can do is attempt to understand what they do and get to grips with their general effects. As for why life popped into existence – scientists have less of a clue than Harry Potter.

Trying to determine how everything came together as a catalyst to define the existence of the four forces, we are still in the complete dark. We are no further ahead than the Greeks' vision that Zeus created thunder!

What Mankind is striving for ultimately is the 'Theory of Everything', which is the understanding of how electromagnetism, gravity and the nuclear strong and weak forces interact. There is also something called the 'Unified Theory' which is again the understanding of how electromagnetism, gravity and the nuclear strong and weak forces interact, but also includes the interaction and understanding of nature and its associated forces.

Plenty of work to be done before we will get the answers we need!

No point fooling ourselves that it is going to be done quickly.

How did life originate?

- *Cells – the basic building blocks of all life on earth, were not discovered until the nineteenth century.*
- *Early Earth had a very hostile atmosphere and certainly would not have been able to support the creatures and plants here today.*
- *Life could possibly be universe-wide – a natural phenomenon as abundant as the stars themselves.*

This truly eccentric Universe never ceases to hurl bizarre and baffling conundrums at us; the harder we seek the answers to where we came from and how we evolved, the more bizarre our findings.

Cells were first discovered by Robert Hooke in 1665, but it was not until 1839 that Theodor Schwann and Matthias Schleiden described the principle that plants and animals are made up of cells. It was only at this point in history that the cell was discovered as the basic building block of all life. Since this time researchers have been attempting to understand everything about them.

What makes cells exist?

What makes cells function?

What makes cells proliferate?

What instructs cells?

What is the ultimate purpose of cells?

Cells have now been studied and manipulated to such an extent that it is now possible to explain their components and behaviour in precise detail – but not why the first one ever came into existence.

No one knows the exact conditions and circumstances that were required which created the first cell. However, whatever the conditions and circumstances were, they must have been accompanied with either a coincidence of enormous proportion, a miracle of some sort or the incredible effort of nature itself.

Who knows?

But one thing is for sure – it did happen!

Whether the cells which make up creatures first evolved here on Earth or out in space is really not known. However, there are certain findings and discoveries that we can use to make an educated guess.

We customarily envisage life to have evolved by some miracle within a type of primordial soup – incredibly producing the first signs of primitive life, with all other life-forms being derived from here. The early Earth possessed a climate of zero or little oxygen with hydrogen, methane and ammonia making up a pretty nasty primordial atmospheric soup. Add to this an extremely electrically active weather system and you have all the ingredients to produce the amino acids we require as the basis of life.

Whether life originated here on Earth or it drifted here in the form of a spore from another planet, who knows, but it had obviously to develop somewhere initially. Life would be more logical as a phenomenon if the Universe had existed for an infinite period of time, this would have provided ample time for life to have evolved. However, as the Universe appears to have had a limited time span, this implies that the formation of life is perhaps an inbuilt natural wonder – which means life could be sprouting up all over the Universe.

In 1952 Stanley Miller decided to replicate the conditions of the early atmosphere within a laboratory test, mixing an amount of water with hydrogen, methane and ammonia. Incredibly, after only two days there were signs of amino acids in the water, and after one week there was a thick brown soup of them. At the time he identified the creation of five amino acids, but since his death in 2007 there have been a further twenty or so identified within the original samples he collected in 1952. They were found within a samples box he left, which makes his first experiment even more successful than originally thought. However, Stanley Miller was not able to create life from his amino acids, he could not get them arranged into something that could feed, reproduce and survive.

Although a primordial soup-style environment is possibly the origin of the first signs of life on Earth, it is now thought that some of life's components developed externally from Earth and were already floating about the Universe!

A truly groundbreaking discovery came in an area of study called astrobiology, when in 2003 Doctor Zeta Martins of Imperial

College London was able to reveal an awesome link to life within a meteorite that is understood to be close to five billion years old.

The meteorite fell to Earth at Murchison in Australia on 29th September 1969, but the equipment to analyse it correctly did not exist at the time. The meteorite broke up upon entry, and fell over an area approximately six miles square. The villagers of Murchison managed to collect enough to accumulate over one hundred kilograms, the largest piece of which weighed seven kilograms.

Later, when specialist analysis equipment had been developed, it was found that inside the meteorite that fell at Murchison there were amino acids which could have acted as the basis of a genetic code, the signature for one of the letters that makes up the genetic code for all life.

This discovery is something I personally believe is one of the most profound of our era. Although it is relatively well-known, it has still not received the recognition it deserves.

There were found to be roughly ninety different amino acids within this meteorite. Bearing in mind that amino acids are the building blocks of life and only twenty have ever been found on Earth, this is truly staggering.

Consider for a moment – how there could have been amino acids, the building blocks of life, within a meteor that hit Earth. This discovery implies that there is an extremely strong possibility that life is a universal-wide phenomenon.

Quite a spectacular revelation!

Making cells from scratch

- *Each cell has many different protein types operating within it to perform a particular task – thus creating life.*
- *The function of DNA is to manufacture proteins by assembling amino acids in a precise order.*
- *DNA performs the ultimate magic trick – the likes of which, no magician could ever match.*

Professor George Church of Harvard University has taken biology one step further and is now making new cell types from scratch. The understanding of cells has progressed to a point where the constituent parts may be assembled and manipulated to create useful bacteria and organisms to produce oil or even eradicate cancer cells.

Although we know that a cell is comprised of a multitude of lifeless chemicals, there must be something within it that gives it a burst of life. Perhaps it is purely the formation of the cells in the way the chemicals combine together that give it life – unless there is something fundamental like a 'force of life' that we are failing to notice.

Amino acids are the essential molecules that form the basis of all life. They join together to form more complex molecules called proteins for the support of life. There are thousands of different proteins, all working within cells with specific functions facilitating life within the cell. All these proteins can be created by the combination of just twenty different amino acids.

In 1953 James Watson and Francis Crick discovered that Deoxyribonucleic acid, commonly known as DNA, within cells allow the assembly of amino acids in a specific order to create proteins. DNA stores information about the construction in the form of four chemicals known as bases, and we interpret these bases as being the letters A, T, C and G. The information inside DNA is a precise set of instructions explaining the exact order in which to join up the amino acids to make proteins.

In 1958 Francis Crick discovered that when a cell wishes to create a particular protein, it copies part of the DNA strand from inside the cell's nucleus to form Ribonucleic acid, known as RNA. This RNA then penetrates through the nucleus wall of the cell to join forces with a ribosome to manufacture the protein. The ribosome reads the RNA strand and as it does so assembles the required protein by fitting the various amino acids together in the correct order.

The production of proteins within an individual cell, utilising this mechanism, is performed thousands of times a minute to keep the cell alive and functional.

It is amazing how DNA can store the design for building a human being, with its miraculous series of chemical bases that map out exactly what should go where.

In May 2010 came the announcement that Craig Venter and his team had successfully synthesized a modified version of the

Mycoplasma mycoides genome and implanted it into a DNA-free bacterial shell of Mycoplasma capriolum. The organism named 'Synthia' is miraculously self-replicating, a feat of ingenuity that has seriously progressed Mankind's capability to manipulate and control the natural world.

Craig Venter was also the biologist who led the effort to map the human genome. What he had produced in 'Synthia' was the first cell controlled entirely by man-made genetic instructions. This breakthrough paved the way for the creation of organisms with specialised functions that could never evolve through nature. There is no end to the technologies ability to create microbes that make vaccines, disease-fighting organisms and algae that convert carbon dioxide into hydrocarbon biofuels.

The cell starts its life stored on a computer. Four bottles of chemicals are used to effectively boot up the DNA within a living organism. In Craig Venter's words, "Though this is a baby step, it enables a change in philosophy, a change in thinking, a change in the tools we have. This cell we've made is not a miracle cell that's useful for anything; it is a proof of concept. But the proof of concept was key, otherwise it is just speculation and science fiction. This takes us across that border, into a new world."

The scientists headed by Venter, made a synthetic copy of the genome of a bacterium, Mycoplasma mycoides. They then added a number of inert DNA watermarks to distinguish the synthetic genome from the natural one. The man-made genome was then transplanted into a related bacterium, Mycoplasma capricolum. This rebooted the cell so that it was controlled by the synthetic genome, transforming it into another species!

The cell is then free to divide billons of times!

The applications of this technology are enormous.

Pat Mooney, a campaigner against biotechnology, said, "This is a Pandora's box moment. We'll all have to deal with the fallout from this alarming experiment."

In the wrong hands this technology could lead to all sorts of problems!

Let us hope that the worst does not happen.

The programming of DNA now takes the helm at the cutting edge of computing. I wonder how Charles Babbage and Alan Turing would feel about that!

Charles Babbage, 1791 – 1871, thought he was only inventing an analytical engine when he became the first person to conceptualise a fully-programmable mechanical computer. Alan Turing, 1912 – 1954, known as the founder of modern computing, thought he was only developing a machine to crack wartime code. Little did he know, his code-cracking technology would one day develop further to be used to create new DNA code for a new species.

The way DNA encodes information about structure is best relayed through the description of an ingenious card trick.

A magician shuffles a standard pack of fifty-two cards and asks someone in the audience to pick five at random. The magician then looks at the five cards and chooses one, then hands it back for safekeeping with the audience. He then discards the other four unwanted cards face up on the table. This is just the start of a very interesting code-cracking style card trick which is fascinating for all occasions.

The magician then brings in his accomplice from outside who has not witnessed these proceedings and invites them to state which card from the pack is with the audience. The accomplice immediately states the correct card upon looking at the discarded four cards on the table; he does this without any other signals or indicators. How can this be possible?

It is quite straightforward really. If you think that of any five cards taken from a pack, at least two will be of the same suit. Therefore you choose one that is of the same suit and place the other on the leftmost side on the table to denote the suit.

Two cards of the same suit are never more than six cards apart; the furthest apart would be say the two of spades and the eight of spades with a difference of six. The two of spades and the nine of spades is only five apart counting upwards from nine through to two. As the magician is capable of choosing the card, they choose the one which has the shortest distance upwards from the other card making it always a number between one and six.

The other three cards can now be positioned to the right of the suit card to represent this number. There are six different combinations to organise the cards to represent this number. The order of high, medium then low value cards could represent perhaps the number one and six could be represented by the order low, medium, high. So long as it is agreed in advance with your accomplice which orders represent which number from one to six, it does not matter. Suits need an order too, so if you have the two of hearts and the two of spades then you both need to agree that the two of hearts is lower alphabetically than the other as hearts is alphabetically lower than spades.

In this way your accomplice can simply add the number to the card positioned on the left of the discarded four and surprise the audience!

Nature beat us to this clever way of encoding. It has exceeded all expectations, such that the whole of the construction of a human being is encoded into a strand of DNA.

It puts our card trick to shame!

Everything serves a purpose towards life

- *Even nature gets rid of unwanted eyes in cave-dwelling creatures.*
- *Everything that nature develops seems to have a purpose – it is difficult to name something within nature that does not serve a purpose.*
- *Even the defunct features of animals, plants and humans have a historical reason for their existence.*

Is it possible to determine whether everything within the Universe is hierarchical and leads towards the support of life?

It would be fascinating to learn that throughout the history of the Universe everything has been organising itself to contribute towards life. This would be a revelation. Given the relatively short period of time the Universe has been in existence, this could possibly explain why life has been able to form quickly and develop into so many diverse species.

There are certain things in this world that we know are right and there are those we know are wrong. If you were to see heavy rocks float in the air, time go backwards, the earth spin in the opposite direction, socks growing on trees or squirrels playing football, you would no doubt be witnessing something that you instinctively know is wrong!

However, there is a little known and extremely strange phenomenon, which provides an insight into the type of force which may be required to encourage life. The force affects matter in a non-intuitive fashion – in a way you may expect a force to operate if it were to be encouraging life-like components to congeal together.

If you sat down to enjoy a cool drink on a summer's day, when suddenly the drink climbed up the inside of the glass and dropped down the outside spilling out on the table – you would think there was something remarkably strange going on. But this is exactly what happens to helium at cold temperatures.

Scientists have been thoroughly bamboozled after having observed super-cooled liquid helium at close to absolute zero. When placed in a cup it will astonishingly climb up the wall of the cup in a thin layer and spill out over the rim.

Thoroughly bizarre!

Could this phenomenon be one of the forces which encourages the formation of life?

Super-cooled liquid helium is also bizarrely affected by light. If an open ended tube is placed partially into the liquid helium, a light-source shone towards the tube will make the helium move upwards like a fountain.

Totally and utterly bizarre!

It contravenes all the forces of gravity and seems to have almost paranormal qualities.

The conclusion we may draw from this most peculiar liquid is that there are many forces we still do not understand; under different temperatures and pressures matter will behave in very strange ways. Scientists will be studying matter under various conditions for centuries to come – especially chemicals at super-cool and super-hot temperatures. As helium is the second most prolific substance in the Universe after hydrogen, it is an awfully large amount of matter to know very little about.

If everything in the Universe does move stealthily towards life, then this would explain how it originated so successfully on Earth. If we look at nature, we see that everything has a purpose. Nature has been pure genius at developing legs, arms, wings, tentacles, antenna, flippers, whiskers, feelers and fins. All that we observe in nature appears to serve a purpose. When trying to think of anything in nature which does not serve a meaningful purpose, it is impossible to think of anything.

Within our bodies we have the odd superfluous organ like the appendix which is a blind-ended tube connected to the cecum. The most widely accepted view is that the appendix was originally some type of supplementary digestive system that has now lost its function. It is thought that our ancestors may have relied upon it when they lived upon a diet of rich foliage. Devout herbivores such as the koala utilise it to host bacteria specifically for cellulose breakdown. Other theories regarding the appendix include that it may harbour bacteria valuable in the function of the human colon, the production of endocrine cells, acting as a lymphatic organ, infection-fighting for the immune system, training the immune system and the production of hormones in foetal development.

It sounds to me that it may well still serve a purpose!

Doctors used to remove the appendix as a precautionary measure during other surgical procedures. However, they no longer practise this as it can be successfully transplanted into the urinary tract to rebuild the sphincter muscle to reconstruct the bladder.

So it definitely has a purpose after all!

It was grown as a spare part!

There are whimsical man-made, non-purposeful things like surreal art forms, figureheads on ships and ornate numbers on clocks, but even these things have a decorative purpose – similar to some of the elaborate colours we see in nature.

If nothing evolves in a counterproductive fashion in nature, then could the same apply to the Universe at large?

Within the Universe we have asteroids, meteorites and comets which seem to be superfluous. However, planets and moons supposedly make use of these celestial bodies to build themselves up. Very large amounts of dust and debris form to become new stars.

If everything in the Universe serves a purpose then there must be a clear objective that it is trying to achieve. Just like the old saying 'All roads lead to Rome', perhaps we can say 'All the things in the Universe lead towards life'. In essence it is virtually impossible for us to determine how much of the Universe serves a purpose and how much is a non-purposeful white elephant.

If the creation of life is a coincidental random event, then the more matter there is in the Universe, the more likely it is that life will transpire.

Nature would seem to discard defunct organs such as the unwanted eyes in cave-dwelling creatures, the unwanted legs on crawling lizards and birds that lose their ability to fly when there are no predators. Nipples on male humans are as they were at the age of fourteen weeks in the womb, which is the exact time that the foetus's sex is determined – perhaps this is a sign of nature hedging its bets!

There does not seem to be anything we can identify on Earth that nature has developed whimsically. All animals with horns seem to utilise them for combat; birds of paradise use their elongated and elaborate feathers extending from the beak, wings or head for mating, the long neck of a giraffe, the trunk of an elephant and even ostrich wings had a historical purpose.

If everything we observe in the Universe serves a purpose, then can we safely deduce that stars shine for a good reason?

Why would stars be there in the first instance?

Well, interestingly, stars are a precursor to life because much of what human beings are made from has been created inside previously active stars. Carbon, which makes up roughly twenty per cent of a human being, formed as a simultaneous triple collision of helium nuclei alpha particles within the core of a giant or supergiant star. This could not happen at the time of the Big Bang as the high temperatures and the concentration of helium was not sufficient.

Therefore carbon-based life can only exist when this star debris is scattered into space as dust within a supernova explosion, and then becomes material within a second or third-generation star system. The space dust must settle successfully upon the planets around these new stars to become carbon-based life such as ourselves.

So did stars develop deliberately, specifically for the creation of life, or are they just random events that just so happened to contribute towards life as a by-product?

Does the fact that stars shine mean they communicate?

Do stars have survival tactics?

Are stars self-aware and know they exist?

Do stars shine to let everything know they are there?

Could the stars be one giant universal chat room?

Some stars have binary partners, comets fly through solar systems in elliptical orbits and many star systems have planets – there are moons around these planets and perhaps some of these lovely planets have little green men on them!

One day we may know a great deal more, but for the moment we shall just have to keep guessing and use our imaginations as best we can.

If I were to have to place money on whether everything serves a purpose towards life – then I would be quite happy to put a large amount of my life's savings on it.

We learnt how unlikely it is that your body has become assembled in the way that it has – not only is your life here extremely remote, but you also had to win a heavily contested swimming race against one hundred and eighty million other competitors. We learnt that the odds of your existence makes winning the lottery look like a breeze! We learnt how our lives could be imaginary as our existence is so improbable. We learnt how there are so many permutations and outcomes possible throughout the Universe, that it can only have been as a result of many mind-blowing coincidences that allowed you to exist. We discovered that there is an extremely strong possibility that life is a universal-wide phenomenon. We learnt how DNA can store the design for building a human being, with its chemical bases that map out exactly what should go where. We also learnt how at least twenty per cent of a human being is made up of star dust.

So we have an extremely remote chance of existence – on top of this our environment must be perfect and our conscious minds must be perfect too. We will look at the development of our conscious minds next.

Relax back into your chair – you need to be in an almost trancelike state to absorb this, after all, you are to be learning about consciousness through your conscious mind!

WHAT IS CONSCIOUSNESS?

Where we investigate how consciousness could materialise, how it helps us interpret everything around us and exactly what it is monitoring. We look at what consciousness could be and to what degree it functions. We look at how other creatures' consciousness compares to ours and the way that a bluebottle is capable of outwitting us.

Defining consciousness

- *The brain of a fly connects in the same way as ours to the fundamental building blocks of the Universe.*
- *Your consciousness is aroused as your taste buds zing with delight at the sight of food.*
- *If we shape the matter in the Universe to our advantage, it will bring about a more pleasant experience for everyone – one of the ultimate aims in life.*

Does size matter where consciousness is concerned?

The North American ruby-throated hummingbird has a brain weighing less than a gram, and a blue whale has a six kilogram brain. Yet both show a marvellous variety of behaviours. Both sing, defend territories, attract mates, raise young and migrate seasonally for long distances. The tiny-brained, ruby-throated hummingbird also has an elaborate courtship dance, builds nests and solves some interesting pattern-recognition problems, and has no difficulty finding the correct flowers.

It is extremely difficult to judge whether the consciousness of a creature is related to its brain size. We have no idea how our consciousness compares with a hamster, dog or cat.

If you take the smallest human dwarves and put them head to head in an intelligence test with the largest humans, I doubt very much there would be any significant difference in results. There is no reason why a fly could not be more intelligent, in its own way, than a human. The brain of a fly may be just as conscious as we are. After all, a fly's brain is connected to the fundamental building blocks of the Universe in the same way that our brains are.

Perhaps consciousness comes in a 'one-size-fits-all' package just like life itself. A human may be no more biologically alive and conscious than a worm. How can a human be alive any magnitude greater than a worm?

A human is either dead or alive, just like a worm is either dead or alive.

The magnitude of 'aliveness' is of the same value within a worm as a human.

Perhaps this is the same for consciousness, which would mean that all animals would have the same level of consciousness attributed to them – they are either conscious to the exact same degree as we are, or dead.

Just because we cannot assimilate to the function of a fly's brain, we cannot see its intelligence and ignore its ability to be conscious. As a consequence we are quite happy to splat them!

If we try to compare consciousness to how a computer operates, our senses could be viewed as having values attributed to them; they could each take values from a reading of zero to a million. Therefore the smell of a gorgeous rose equals a million and the smell of disgusting dog dirt equals zero. The sound of your favourite music equals a million and the sound of ear-piercing chalk scraping upon the blackboard equals zero. The taste of your wonderfully-prepared favourite food dish equals a million and the accidental ingestion of bleach equals zero. The touch of silk equals a million and the scorch of a hot flame equates to a painful zero. The sight of a stunningly beautiful view equals a million and the sight of an atrocity equates to a stomach-churning zero.

The interesting thing is that the senses all interact in the 'now moment', such that the worst sight imaginable, coupled with your favourite music, will create mixed feelings that are difficult to comprehend within your consciousness; although listening to your favourite music with your eyes closed may bring back memories of a beautiful view where you once listened to the music on one of your happiest days.

Our senses must be instrumental in the operation of our conscience because our brains interact quickly with another sense's experience when a particular sense is tantalised to an extreme. For example, when experiencing a fantastic meal, and your taste buds are zinging with excitement, your consciousness is aroused. You can visualise in your mind's eye the place you were the last time you had such a delightful dish.

Based upon the one million states of each of your senses, your consciousness can assume a combination of one million, trillion, trillion different states of consciousness, making each and every moment of your life different. Perhaps when we get similar values registered that we remember from earlier in our lives we get such delusions as déjà vu!

Humans are fully aware of what they experience in their everyday lives, but have enormous difficulty putting it into words. Philosophers have struggled to attribute a definition to consciousness and scientists have more than struggled with what it is.

The German mathematician and philosopher, Gottfried Leibniz, 1646 – 1716, takes a courageous place in history for both his mathematical and philosophical innovations. He invented the binary system, which is the foundation of all computing. He also developed infinitesimal calculus independently of Sir Isaac Newton. Within philosophy he was famous for having highlighted that the Universe is quite possibly the best that could have been made. However, his greatest philosophical thought has to be his analogy with consciousness.

Leibniz drew inspiration from his mathematical idea and offered a theory of the mind in 1686 called 'Discourse on Metaphysics'. Within the work he explained there are 'infinitely many degrees of consciousness' and suggested some thoughts were unconscious. He called these unconscious thoughts 'petites perceptions'. Leibniz was first to distinguish explicitly between perception and apperception, which equates roughly to awareness and self-awareness.

In his famous work called 'The Monadology' published in 1714, he proposes his analogy to a mill building to express his belief that consciousness could not arise from mere matter. He suggests a thought experiment where he asks the reader to imagine someone walking through an expanded brain as someone might walk through a mill. Leibniz invites the reader to walk around examining the physical nature of all the mechanical operations and physical items. Nowhere, he asserts, would such an observer see any conscious thoughts.

What type of device or organ would you imagine could exude consciousness?

Is the source of consciousness inside our brains, external to our brains or a combination of both?

Louis Armstrong was once asked to define jazz. He replied, "If you've gotta ask, you ain't never gonna know."

If you do not know what something is, it is very difficult to help out by putting into words what it is. Consciousness is very similar in this respect, and a great deal more of a challenge to define than jazz.

Consciousness is both blindingly obvious and extremely elusive at the same time. It is constantly before us and yet hardly ever noticed throughout our waking hours. It seems impossible to pin down in words what consciousness is. It is not part of the wider world that we already understand, and there is nothing similar that we can compare it to.

Let us try to define what consciousness is; perhaps we can then expand upon it further. An almost impossible task, but here goes.

Consciousness happens in the 'now' moment. It is an immediate experience that moves in synchronization with time.

You know you are conscious.

You do not have to be told you are conscious.

You instinctively know you are conscious of all that is surrounding you.

Consciousness is a private experience that only allows your conscious thoughts to be experienced by you alone. No one else experiences the sensations that your conscious generates.

Consciousness relays to you the degrees of pleasure or discomfort associated with your senses. Who you are and what you are experiencing is all held within real-time in your conscious mind. Whatever feeling is held in your conscious mind is acted upon by your brain, both consciously and sub-consciously.

If your conscious mind reports that you are trapped by fire then you will make an immediate conscious attempt to get out of danger. If you are enjoying life's pleasures, it will make a conscious attempt to keep within this environment. You will be consciously polite to those that enhance your comfort and possibly unpleasant to those people that obstruct it.

Your conscious will be interrupted by important messages of pain, whether this is a mild intermittent ache or excruciating agony. Your conscious is a central reporting place for all sensations of pleasure and pain. Your body's senses relay all input from the world and your conscious mind analyses this – conjuring up feelings recognised to you as anger, frustration, grief, hunger, happiness, sadness, terror, anxiety and delight. Your conscious mind also has the power to deduce what actions should be taken as a consequence of your feelings. There seems to be a hierarchy associated with the sensations felt; the feeling of happiness will be interrupted by acute pain and the onset of grief will disrupt any feeling of delight.

Reflecting upon this, our conscious minds act as a hub of input, feeling, thought and deduced output. The special feature of this conscious hub allows you to imagine, remember, think, calculate, have ideas, speculate, mediate, ponder, love, doubt and question.

It would appear that someone 'is' who their conscious mind tells them they are. This may have come about from all their experiences and how they managed to react to these – making each of us individual. It would be logical to suggest that the conscious minds of all humans work in exactly the same way.

Consciousness appears to be unlike a lump of matter such as a rock; it appears to have no obvious dimensions, no measurable volume, no weighable mass, and no specific location in space. It is not possible to touch a lump of fear, to measure the length of a feeling, to weigh a sound, or determine the width of happiness. It appears visually invisible to us, but at the same time it is the only thing that is visible to us in our minds – without it we would clearly not exist.

An interesting fact which I have not been able to verify is that the Internet is getting so enormous now that the information whizzing around it weighs as much as a strawberry; this is absolutely incredible if it is true!

The reason why this is incredible is quite simply because the type of substance that constitutes the information-bearing aspect of the Internet weighs so little. In fact, it is all the electricity and the data combined available to the Internet that weighs the same as a strawberry; if we were to just count the data alone, it would weigh much, much less. It is interesting at this juncture to look at the collective weight of the Internet over time – if the progressive weight of the Internet were to grow at an exponential rate, then it would eventually take over the composition of the whole Universe.

What are the implications of this?

Is the human race just witnessing the creation of the Internet within an already existing advanced Internet?

This type of thought is a little like thinking that atoms are small stars and that our whole Universe is just one small atom in a larger scale universe.

We had better move on!

Let us take a look at some deeper aspects of our conscious minds which we can all relate to. There are distinct traits that our conscious minds have that a lump of rock fails to exhibit.

Firstly, we have the direct knowledge that something is happening, such as a bang on the head, a tap on the shoulder or toothache. We are directly wired into these sensations and cannot mistake them ourselves for anything else. However, our conscious is totally unaware of other peoples' feelings, pain or discomfort – we have to be informed of these events directly, else we are not aware of them.

Being the object of direct knowledge is a quality of consciousness.

Next, there is the privileged access that your consciousness provides to you that is denied to others around you. You may have no physical signs of injury, but be in tremendous pain that can only be relayed through speech to your doctor. A rock may be cracked in half or someone may have a physical rash on their arm; these are obvious to everyone and are not considered privileged access. Privileged access is considered to be a characteristic of consciousness.

Next, there is the infallibility of consciousness. This trait of consciousness is all about what you feel; if you feel pain then you cannot be mistaken. If you feel pain, then you are definitely experiencing pain. It is unlike the mistake of hearing a phone on the television and thinking it is your phone ringing. The infallibility of consciousness means that you are never wrong about the direct sensations you feel.

Next, there is the lack of shape, space, length or dimension to a conscious thought or experience. We cannot compare the ache of a broken limb to the exhilarating experience of a rollercoaster ride. Although our conscious minds are responsible for both sensations, there is no spatial type of link between the two types of experience.

Finally, there is intentionality, which is the distinction between the mind and the physical. The philosopher Franz Brentano, 1838 – 1917, said, "To claim that mental states are intentional is to say that mental states unlike physical objects have the property of being about something, that they have a content of some kind."

What this means is that when we are in an angry mental state, our anger is about something, for example, the car not starting. When we are in the mental state of happiness, our happiness is about something, for example, we have just passed an exam. Memories, fears, anxieties, ideas, plans, loves, hates, desires and hopes, all have the property of being about something. However, physical objects lack this property. It is difficult to make any sense of the idea that a mountain, a lump of rock or a table, are about something.

Many people are instinctively drawn to a dualist way of thinking, holding the belief that they are non-physical conscious minds somehow housed in physical bodies. However, dualism raises so many difficulties and contradictions that it has been challenged time after time by recent thinkers. The philosopher Colin McGinn stated, "Expecting human beings to solve the mind-body problem is like expecting armadillos to understand algebra: like them we lack the necessary intellectual capacity and apparatus."

So if Colin McGinn is correct, what can we do to overcome our shortfall in knowledge to pin it down?

Perhaps we shall just have to wait a few thousand years!

Consciousness of a fly and absolute perfection

- *A bluebottle is an excellent example of a creature with fast thought and consciousness – able to evaluate what to do in the event of pending danger.*
- *There must be a very complex and powerful set of rules governing a conscious mind.*
- *Absolute perfection seems to be at the heart of everything that exists.*

Quite incredibly, whenever you have tried to swat a bluebottle you are witnessing something that is an evolutionary winner at super-fast consciousness and thought. By the time you have begun to swish your arm towards the fly, it has noticed it coming, decided to move, decided in which direction to move to avoid impact, flapped its wings a few hundred times and embarked upon some totally different event in its life. Your hand then thumps the table, leaving you frustrated, and not knowing how the Dickens something as small and as fidgety as a bluebottle has outsmarted something as big and as smart as you.

The simple reason is that the same universal force that looks after the life within you also looks after the life within the bluebottle too. We are on equal terms!

The bluebottle's life is more than likely being saved by quantum-level processing events. For example, the fast quantum processing of thought at the tiny subatomic level will make itself visible by immediately transitioning to the scale of the very large within death-defying situations. It is clear that super-rapid thought is essential for physical survival. An adrenalin rush may promote more simultaneous, instant, entangled and super-positional thoughts that assist with defying death. Entangled quantum particles and superposition quantum-level phenomena may well contribute, but are very difficult to understand within the world of the large. We will look in more detail at these bizarre effects later.

Within the quantum world, these types of events are totally normal, and a constant means of interaction with the all-pervading, unified field.

Superposition may facilitate consciousness choice, allowing waves of potential subatomic particles to be anywhere they wish to be, effectively allowing particles to pop into and out of existence at will to finalise a decision.

These subatomic particles which make up everything in the Universe, including our thoughts, may communicate instantly, and are capable of being in many places at the same time – hence covering options that are available to thought and consciousness!

Scientists have speculated that other parts of the human anatomy also use quantum effects to relay messages to our brain – specifically our ears.

No doubt these effects include the instant interaction and conformity with the universal, unified field for the fly.

Whatever consciousness is, it must abide by an extremely powerful set of rules which will undoubtedly utilise quantum physics as its underlying mechanism. It would seem rather improbable that something as miraculous as the world of quantum physics could be overlooked when developing consciousness, especially when it is available for use. It would be a little like someone deciding to hop home on one leg when there is a Ferrari available to drive them!

It would be impossible for human beings to exist without a supporting Universe. The unified field of the Universe, coupled with our DNA instructions, moulds us into what we are. Our DNA instructions must be in complete harmony with the Universe, and fully aware of all its forces. The unified field moulds us into what we are, based upon our DNA instructions.

On a more simplistic level, it is like blowing up a Mickey Mouse-shaped balloon with air. You instinctively know pretty much what it will look like when you have finished, even before you have started. But if you were to blow up the Mickey Mouse-shaped balloon at the bottom of the ocean or inside the crevice of a wall, it would take on a totally different shape. Similarly, we need the unified force to shape us physically and mentally into what we are. The unified field takes your skeleton outline defined within your DNA and fills it with you.

The whole Universe is extremely active and alive at the quantum level – the whole planet, the whole galaxy, the void between the galaxies, everything. Every atom has electrons whizzing around its nucleus at millions of times a second. This is no dormant, inert Universe.

So, "What?" You may ask, "Was it that brought about this pure, life-providing, conscious Universe?"

We can only presume that if anything is going to exist, then it can only be built based upon absolute perfection. Perhaps there are billions and billions of other universes that are defunct and have disappeared because their unified field was imperfect and died a stagnant death. Within the infinity of the structure of everything, appeared a universe that held the perfect formula for intelligence, consciousness and life. This type of Universe and its all-pervading, omnipresent intelligence would no doubt survive amongst other irregular or imperfect universes.

The mixture of perceived sensations

- *Humans reach a state of consciousness as their sensations are intermingled together inside an all-pervading omnipresent brain.*
- *Are all the senses essential to conjure a conscious mind? Would a conscious mind evolve without sight, hearing, touch, taste and smell?*
- *A bat's hearing seems to be much better than human hearing, despite human ears being much larger than those of bats!*

Imagine a creature which senses objects and events as they truly are. The creature's vision consists of seeing jiggling atoms upon a hard substance; its hearing consists of the pure detection of changes in air pressure and vibrating airwaves. Touch is now the difference between the ends of the fingertips and the substance being touched; taste and smell are pure acknowledgements of the vibrations of food and floating particles.

This type of creature may not need a brain to process this information, just external peripherals with a central processing area to link these sensations together to make decisions about what to do next. Sounds very much like a computer to me!

Whereas a computer cannot reach a state of consciousness, human beings can. This appears to be because their perceived sensations are being mixed together within an all-pervading, omnipresent brain.

Could it be that our brains are in cahoots with universal forces not yet known to us?

Is it possible that we have some type of conduit through to the Universe's unified field, and hence witness the Universe's intelligence in our minds?

As the unified field theory is a means of representing all the forces of nature as one, it would make sense that consciousness is related to it as the brain is certainly governed by them. The unified field combines the forces of electromagnetism, gravity and the strong and weak nuclear forces – the brain utilises these forces within its construction, so any lack of knowledge of these seems impossible.

Whatever creates our consciousness is then mixed with our senses within our brain and provides us with a state of self-awareness. Although consciousness could have developed as a survival tactic, it is more than this; it creates our internal identity and magically enables us to relate to ourselves.

Without consciousness, where would this leave us?

Perhaps it would leave us as inept as a computer!

There could well be some relation and interaction between human beings' conscious minds and the Universe's unified field. Perhaps the Universe's unified field is conscious and aware also; living organisms are perhaps a means of the Universe expressing its consciousness.

This may not be as daft as it sounds.

When venturing into unknown territory if often pays to think differently – who knows what the answer may be?

Does a person without sight, hearing, touch, taste and smell have the ability to create a conscious mind?

Perhaps it is the sensations of the five senses that provide input for a conscious mind, so that it can react, deduce, predict, experience, live and die. Perhaps, without these inputs, our minds would merely be in a state of meditation of sorts.

If we look upon the input to our consciousness as purely the current state of interaction between our five main senses and the billions and billions of states these can take, then an example of this could be as follows. We take the vision of the beautiful sea;

the gorgeous smell of fresh air; the taste of a 'just-eaten', scrumptious meal; the gentle touch of a warm light breeze passing your face and fingers; and the lovely sound of crashing waves. This input to our conscious equates to contentment and happiness.

Compare this with the vision of a dead cat on the road; the smell of rotting flesh; the sound of a spade scraping the road to remove the cat; the taste of acid in your mouth just prior to vomiting; and the touch of the freezing air numbing your fingertips. This equates within our conscious to an unpleasant and disturbing experience.

It appears that the input or stimulus for consciousness may be a constant assessment of negative and positive sensory values, whilst our minds compare the sensation with experiences from our past.

Human consciousness may be advanced, but how well-developed is this mixture of perceived sensations within other creatures of the planet?

Many questions spring to mind when considering the consciousness of other animals and plants.

Does an ant taste and smell anything like we do?

Does a tree feel pain when it is felled?

Do hamsters see images like we do?

Do birds smell freshly-mown grass just as we do?

Can slugs taste lettuce just as we do?

Do cats hear the radio as just a crackling or hissing sound?

Do dogs hear noises like we do?

Can dogs view television exactly as we do?

We know that dogs have a much greater spectrum of hearing than we have, so with this in mind we should assume that a slug has a different spectrum of taste to ourselves. Perhaps lettuce to a slug tastes like our favourite dish. Unless your favourite dish is a lettuce!

Animals have senses that look and appear to function similarly to ours, but are often tuned somewhat differently. The ability for a bat to hear seems to be immeasurably better than ours, even though our ears are bigger; the ability for a cat to see in the dark is immeasurably better than ours; the ability for a dog to smell is immeasurably better than ours.

Animals seem to have a number of other senses too, such as a homing pigeon which has the ability to sense the magnetic fields of the Earth; a rattlesnake which has the ability to detect heat; bats and dolphins which have the ability to echolocate.

All indications are that animals have a similar experience of consciousness as we do. Although some people think that plants plug straight into the Universe's unified field as their source of consciousness and decision-making capabilities. A particularly interesting concept, which would explain their lack of a brain!

Now – no sleepless nights thinking about this!

String theory and consciousness

- *Strings are thought to be what everything throughout the whole Universe is made from.*
- *String theory is popular because it describes a 'theory of everything' which has never been achieved before.*
- *Everything within the Universe is thought to be made up of energy strings that are connected at both ends to the Universe's membrane.*

Once you realise that light can only travel one metre every three nanoseconds, you can see that there is a great deal of thought to be accomplished by a conscious mind within a split instant, utilising this limited speed. With a nanosecond equating to a billionth of a second, you may consider this speed sufficient, but with the human mind being so complex this may not be enough. Thinking at this speed does not seem to be right. With all the complex things going on inside our bodies, something faster than this speed must be happening. I do not know about you, but I feel that my mind performs parallel thought rather than sequential thought.

Can the mind work faster than the speed of light?

Can the brain work faster than the speed of light?

In November 2011, scientists at CERN began to speculate that neutrinos may travel fractionally faster than the speed of light. If this ultimately proves to be correct, then it would make sense that our brains work at neutrino speed. This would better allow us to visualise and process events that operate at the speed of light. To think that something operating at the speed of light can monitor something else accurately at the speed of light is comparable to expecting a ball to bounce forever, or produce an engine which has 100% efficiency.

How is it that we can react so quickly to our surroundings?

How can our bodies organise themselves so quickly and efficiently?

Our mind has so many things to manage and control, so how could it operate at a speed of thought equating to just one metre each three nanoseconds?

Could consciousness and fast thought be achieved with parallel thought?

Interestingly, there is a solution potentially available to us buried deep within the very fundamental makeup of the Universe. Over the last few decades, people have been studying matter within our Universe from very large objects to very small. Now that nature's behaviour and make-up has been analysed from galaxies, planets, rocks, molecules, atoms, protons and neutrons, to quarks, we have a much better understanding of what the fabric of the Universe consists.

Many scientists believe that everything is made up from tiny, tiny little circles of energy called strings. These strings are speculated to be the underlying building blocks of the whole Universe, and make up everything we are and see. They are so tiny that it has been estimated that if you expand an atom to the size of the solar system, then one of these tiny, tiny circles of energy called strings would be the size of a tree.

At these tiny levels, all manner of bizarre happenings occur. Particles of matter can disappear and reappear somewhere else, matter can be in two or more places at once and effects can be relayed instantly across great distances.

Could this Universe-wide set of features be something that our conscious minds have latched onto and utilised?

The unified field theory amalgamates all the forces there are between particles within the Universe, and represents them as one principal force. This one force dictates how everything within the Universe should behave, including nature. Through all this probing at the subatomic level, we have been getting to understand the unified theory better.

Einstein dedicated half his working life to the potential discovery of the unified field theory, but was unfortunately unable to complete his findings. As the years have passed we have developed a better insight into the theory, but it is by no means perfect.

String theory describes a purely theoretical view of the fundamental building blocks of the Universe and demonstrates the characteristics expected of a unified theory. Many scientists believe that string theory contains the correct fundamental description of nature as it also correctly describes the standard model of particle physics. This model describes electromagnetism plus the nuclear strong and weak force, including their fundamental particles which facilitate their interactions.

Incidentally, the reason for gravity is still not yet known. However, people speculate gravitons are the mediators that trigger gravity – similar to the massless elementary particles that facilitate all the other forces.

String theory has been developed slowly over time since the late 1960's and early 1970's by a number of scientists. It is popular because it is the first time that anything close to a 'theory of everything' has been achieved to describe the whole Universe. Although Albert Einstein had spent two decades attempting to discover the 'theory of everything' and failed, at least he highlighted its importance.

It is not really clear what benefits understanding the 'theory of everything' will truly bring. However, we may get an insight into the workings of consciousness, the whole Universe, nature, matter and life itself.

The string theory is heavily criticised for never having provided truly measureable, experimental predictions. Testing string theory to unequivocally prove its accuracy is well beyond human engineering capabilities. Without going into any mathematical detail, what has already been ascertained is that string theory can never be proven.

This does not mean that our knowledge of physics at the very small subatomic level is approaching a brick wall. We can still continue to develop new theories and ideologies until we feel comfortable with our speculations. We may then test these speculative theories with intricate physical experiments, such as those conducted within particle accelerators.

String theory requires new mathematical and physical ideas to be incorporated together, creating a truly alternative approach to any previously-proposed solution. One of these new approaches within string theory is the eleven-dimensional M-theory, which, as the name would suggest, requires time and space to have eleven dimensions rather than our normal four of time and space.

We may well indeed live within a universe that has more dimensions than originally meet the eye. People who first started announcing there may be extra dimensions within the Universe were often referred to as eccentric crack-pots. But it transpires that we may be trapped upon a slice of a higher-dimensional universe!

If this is true, then it could be very interesting to investigate with regards to the conscious mind. It might explain how our minds could tap into extra dimensions within this higher-dimensional universe. If this is not true ... at least it sounds convincing.

Careful who you tell, as you may get some strange looks!

String theory gives us a new understanding and a new perspective of the inner-workings of our Universe. The theory starts with the premise that everything within the Universe, whether it be matter, gravity, electricity or a human being, is made up of these incredibly tiny vibrating strands of energy called strings. If the theory is to be believed, it changes everything we ever thought about the Universe. Particularly, strings change our understanding of what we thought we knew about space.

Although we may perceive our neighbourhood out in space as flat grids that are static and unchanging, Albert Einstein explained that space can warp and stretch. Shortcuts can be made from one location in space to another through wormholes. These wormholes can theoretically be created by making a rip in the fabric of space, therefore enabling travel through to another place within the Universe. String theory incorporates this effect to the extent that shortcuts are made by strings creating these rips in space. However, the precise mechanism proposed contravenes some of Einstein's beliefs.

But no one is right all the time!

Our Universe could be one small part of something a great deal bigger. Taking ideas to the extreme, some scientists have suggested that we may be living within a three-dimensional membrane that floats within higher-dimensional space. There could be worlds right next to ours that are completely invisible because of the true make-up of space. It is this notion that gives rise to the possibility of there being parallel universes. Many scientists now feel this is not a particularly strange idea.

Perhaps we will be a step closer to understanding our conscious minds when we know a little more about this phenomenon. After all, we would stand no chance of understanding a goldfish if we did not know it lived in water.

Reality is a strange thing, especially as we live in a world which we do not understand. Within this scary, unfathomable world, our minds play tricks on us to such an extent that we go to the lengths of creating items that are not real; for example, a Dalek. Having done this we then conjure a whole host of facts which go with it, until Daleks are normal and accepted and are viewed as an actual reality within the World. We have got to the point where there have been so many books written on Daleks, nearly one thousand listings, that we know more about these 'made up' fanciful creatures than we do about real ones.

If you do not know what a Dalek is, then you will need to know that they are an extraterrestrial race of mutants from the British science fiction television series Doctor Who. They were of course created by the scientist Davros during the final years of a thousand-year war against the Thals.

People seem more comfortable with the Daleks than they are with understanding such things as atoms, molecules, gauge bosons and nuclear forces; which are much more interesting.

If you wish to take the popularity of fictional characters one step further, then my research on the Amazon website shows further startling results. There are roughly twenty-five thousand listings for books that specialise in Stars Wars and the famous R2-D2 and C-3PO characters. Although they are quaint, they have nothing to do with our 'normal' lives or our survival, unlike quantum physics. Compared with our fictional Star Wars friends, quantum physics as a topic, in comparison, has five thousand fewer books written about it. Such is the popularity of fiction to factual books in this world!

I personally find this extremely intriguing and feel it may have something to do with a human requiring comfort in a world they cannot understand. This would further explain the creation of gods in years gone by. Perhaps R2-D2 is a modern-day god.

Christmas, Easter, elves, pixies, fairies, Santa, werewolves, Satan, football rules, official working hours, time, money, soap operas and ghosts are all other examples of things we have made up in our heads. Perhaps we ought to get shot of these Dalek-esk nonsensical things that cloud our creative thought and concentrate on purely the 'real' things in this world which will help us make progress. Electricity, magnetism, water, air, elements, nuclear forces, animals, plants, electrons, protons, neutrons and quarks; strangely this is it – this short list is pretty much it for real things. Therefore, upon reflection, most things are delusional interpretations of our creative minds. Even a cup of tea is water mixed with a few other chemical elements.

String theory became increasingly popular as scientists realised that electrons, protons, neutrons and quarks could also be made up of tiny vibrating strings of energy. This groundbreaking theory simplified everything within the whole Universe. Everything is made from just one fundamental building block; strings.

This simplification appeals as quite an attractive thought.

Everything has been made up from strings!

These strings all vibrate at different speeds, rhythms, frequencies and in different patterns to make up everything that exists throughout the Universe. When all these tiny, dancing energy loops are positioned together, we get the beauty of the Universe that we witness.

If we could understand the rhythm of these strings we would be in a strong position to master the secrets and make-up of matter, forces and nature itself. We could get an understanding of all that is going on within the Universe, and possibly develop the ability to change its destiny.

Rather confusingly, scientists ended up developing five different string theories. This had the effect of casting doubt about the theory in some peoples' minds – how could it be a serious contender for the theory of everything?

Then in 1995, Edward Witten, an extremely well-respected scientist specialising in string theory, announced his discovery that all five string theories were in actual fact the very same one. He made it clear that the five string theories were purely being looked at from different angles, and was able to show how they all came together to form one theory, which he called M-theory. Edward Witten, following his deep study and creation of M-theory, subsequently said that the 'M' stands for magic, mystery or matrix according to your particular taste!

The mathematics that Edward Witten used to demonstrate M-theory revealed that the original ten dimensions of the Universe which string theory required did not quite weigh up. He proposed that within M-theory there were in actual fact eleven dimensions.

An extremely difficult concept for us mere mortals to wrap our heads around!

The more dimensions or degrees of freedom something within the Universe has, the more it can do. So if strings are within eleven dimensions they will be able to do a great deal more than when trapped within four dimensions. When people started to delve further into this world of eleven dimensions, they began to realise that other objects could live within this type of world that were not just strings. These objects resembled membranes or surfaces. The extra dimension that Edward Witten had introduced to M-theory allowed strings to stretch to become something like a three-dimensional membrane.

Coupled with a sufficient amount of energy, these membranes can grow to an enormous size. They are purported to be able to grow to the size of a universe!

The concept of these giant membranes and the extra dimensions are the reason why scientists believe our Universe could be upon a giant membrane within a much larger higher-dimensional space. All these extra dimensions could be revealing all types of illusionary phenomena. It may be an illusion that everything appears so far apart. It could be that everything is much more tightly packed, but the eleven dimensions give so many more routes from A to B that everything seems that much more distant.

Incidentally, with everything being much more tightly packed together, this would very much help our understanding of consciousness, and explain why it is so immediately reactive!

Within such a complex number of dimensions, all manner of possibilities would arise. M-theory could now explain the phenomenon of superposition, which is also known as the non-local property of particles. This is where particles are observed to be in two or more places at once.

The membrane upon which we live may be one within an immensely larger array of membranes that scientists call the bulk. The other universes or other membranes might be right alongside us, but as atoms cannot escape from one membrane to another then we will never witness their presence.

Where string theory begins to reveal its true worth is when considering the force of gravity, which is immensely weaker than electromagnetism. As we all know, the power of a small magnet is extremely strong, and can easily combat the force of gravity created by the Earth. A small magnet will hold a piece of metal away from the force exerted by the gravitational pull of the whole Earth.

Electromagnetism has been calculated by scientists to be one thousand billion, billion, billion, billion times stronger than gravity. That is a one with thirty-nine noughts on the end.

It is a good job that this is the case. It is for this reason we are able to pick things up against the force of gravity with our weak arms.

String theory reveals that gravity may well be as strong as electromagnetism, but appears weak because we just cannot feel its strength. The majority of gravity's strength could seep through to other areas or dimensions of the Universe.

It could even be used to create time!

All matter in the universe is thought to be made up of single lengths of energy strings that are tethered down on either end to our Universe's membrane. However, there are other types of energy strings of which gravity strings are thought to be just one. These are quite possibly free-floating, closed circles of energy which are not tied down to our three-dimensional membrane.

The free-floating, closed circles of energy that make up gravity do not tether themselves to our membrane, and are free to float throughout other membranes. These closed circles of energy associated with gravity have been termed 'gravitons', and as they are free to float within other dimensions and through to other membranes we only witness a tiny amount of their strength.

All of this explains how we could be living very close to other membranes that are sharing gravitons between us and other membranes. It has been suggested that one day we could perhaps harness energy to create a source of gravitons to produce gravity waves. This could ultimately allow us to communicate using gravitons through to other membrane universes that could perhaps be harbouring intelligent life.

Another breakthrough that string theory brings is in the study of the Big Bang that supposedly created the Universe. Scientists have perpetually been confused about our expanding Universe that seems to have originated from nothing. All the theories that have been contemplated in the past have never been able to suggest how the Universe came into existence. What we do know is that it came into existence through an almighty blast of energy. This energy could well have been as a result of two membranes colliding with each other.

String theory can therefore explain where the Universe came from!

However, we have to be careful not to get too excited, as there is still so much to be discovered, especially where the membranes came from initially.

We do not know whether particle accelerators such as the Large Hadron Collider will ever be able to detect gravitons. There is a remote possibility that when protons are collided at close to twice the speed of light, the subatomic particle shower that results may well momentarily reveal a graviton before it disappears off into the other invisible dimensions. It has also been said that gravitons could even be detected by their absence from the other particles produced.

We wait to see!

The eleven dimensions of the Universe

- *Mathematics can cope with many dimensions – however, special problems occur when we try to envisage multiple dimensions within our minds.*
- *Imagining eleven dimensions is difficult – where does someone begin when there are only the three dimensions of space and one of time for us to relate to?*
- *Our minds are not particularly good at conjuring up multidimensional worlds – perhaps we need a decent analogy.*

Let us try to imagine a world with a different number of dimensions to those with which we are accustomed.

If we close our eyes, we could be fooled into thinking we are within a zero-dimensional world! Nothing goes up, down, left or right and you cannot see the clock to tell what time it is!

It has always been extremely difficult to imagine different dimensional worlds within our minds; we seem to only be able to relate to up, down, left, right, forwards, backwards and the passage of time.

Perhaps if we introduce words like in, out, visible, invisible, bright and dark, we can perhaps get our minds to think that these are other dimensions.

So imagine a visible light bulb that is bright that travels into a dark box upwards and forwards just over to your right. Thinking like this may help us visualise a world with more dimensions – we could be onto something here!

The 'disappearing into a dark box' aspect of the image in your mind can act as a type of wormhole illustration, something that our minds have had difficulty perceiving. The box could represent a 'hidden' or 'not hidden' status. The bright-dark or on-off aspect of the light bulb could represent an active or non-active dimension within a dimension.

It is difficult to imagine other ways of visualising up to eleven dimensions. This way of thinking helps to satisfy our lack of understanding about how to visualise multiple dimensions over and above our usual three, plus the one of time.

Dimensions can easily be handled within mathematics, but when trying to relate to extra dimensions in our minds, within our perceived physical world, we hit problems. The problem is that human beings are not suited to visualising multiple dimensions in their minds. It is like trying to get a fish to describe what it is like on Mars. If humans are unable to visualise a concept clearly they are unlikely to relate to the concept. They will be more inclined to disbelieve and seek an alternative view.

We have seen enough drivel of that nature before, when trying to explain how 'things' came into being. One great example is the explanation from the ancient Greeks, who thought Thor was the creator of thunder. Now we know a little more about the Earth we can explain the evaporation of water, formation of clouds and the precipitation as a consequence.

No longer do we reference Thor on our weather forecasts!

Perhaps we could bravely submerge ourselves into the following thoughts as an approach to visualising the eleven dimensions that exist within string theory; this is of course if you are willing to buy into the concept of string theory!

Remember that string theory is purely a mathematical theory that works to create a unified theory of everything. However, it requires patience, understanding and the realisation that it can never be verified in the real world. The only conciliation you get for being a believer in string theory is that it is as promising a theory we have at the moment to explain everything.

So now on to our visualisation.

To attempt to visualise the eleven dimensions we shall use the following states: up-down, left-right, forward-backward, in-out, bright-dark, visible-invisible, hot-cold, soft-hard, sweet-sour, spinning-stationary plus the passage of time; that makes eleven.

Now we can visualise a world with eleven dimensions.

Picture this.

I walked around the corner in the sweets manufacturing area and I could not believe my eyes. I saw a hot, visible, bright-red, luminescent, soft, and hard to chew humbug spinning forwards on my right side, travelling upwards into a gift box that will no doubt be opened at Christmas.

There are eleven aspects to the description of the humbug; giving it motion up, motion along, motion right, motion in, temperature, visibility, brightness, hardness, sweetness, spin movement and a timescale. This is somewhat easier to visualise rather than attempting to imagine parallel universes, wormholes, space-time and dimensions that curl up around everything. These are far too many paraphernalia for our minds to cope with.

Let us experiment with the logic again using an everyday event such as travelling along a road in a car just as you enter a tunnel. The visualisation of a world with eleven dimensions could be a little like this, "I was in my robust hot and very visible bright orange, extremely sweet car, staying on the horizontal plain, moving forwards along the road not veering left or right or spinning as I entered into the tunnel, becoming invisible to my friends who were waving goodbye to me until I see them again next week."

Rather a mouthful, but at least if we look at each of the descriptive words as dimensions, we have something that makes sense and can be treated as a reasonably decent analogy. After all, I am sure if we were to try visualising the true eleven dimensions, it would be totally incomprehensible to our minds with light bending, solid objects disengaging and swashbuckling all over the place.

Now – sleep easy tonight, do not think too hard about this!

The Universe has extreme intelligence at its core

- *There is an unknown universal mechanism that unites gravity, light, electromagnetism, radioactivity and nuclear forces.*
- *Consciousness may be created by something external to the brain – it may be disconnected with the brain's molecular structure and chemical processes.*
- *Everything within the Universe is a wave of vibration connected to the underlying unified field.*

The Universe has an enormous amount of extreme intelligence at its core. Without the human race, the Universe would undoubtedly continue to exist and would not even notice us missing. It is a relentlessly vast void of unimaginable beauty, which harbours inconceivable wonders.

As the Universe evolved it created the elements that make us. These elements came together in a miraculous fashion to form each one of us. It could only have been the Universe at large that was responsible for bringing us into existence. In a similar way, we have been responsible for bringing television sets into existence. As a comparison, if we were to think we are the most intelligent entities within the Universe, it is like a television set thinking that it is the most intelligent thing in the Universe!

The Universe is a single universal field of intelligence, a field which unites gravity, light, electromagnetism, radioactivity and nuclear forces together. Scientists are convinced that all the forces of nature and everything that makes up matter are somehow interrelated. As we know, the race is on to produce a formula to reveal that they are one.

Everything that makes up our entire Universe, atoms, quarks, leptons, absolutely everything, are just different ripples within a single ocean of existence. This is what is called the unified field, sometimes known as the superstring field.

All the disciplines of quantum physics, molecular biology and neuroscience say we create our own realities, including the way we experience the world and the way the world behaves towards us. Quite a remarkable claim!

Once we have a better grasp of the unified field we may have an opportunity to discover the secrets of the physics of consciousness. We will be in a very strong position to decipher exactly what consciousness is, and perhaps able to determine from where it came.

It is quite possible that our understanding of the unified field could lead us to discover that life at its core is fundamentally one. It may transpire that at the centre of all life's diversity there is unity. It may be that we can describe all living creatures and the Universe as one entity.

This unity at the centre of mind and matter could well be what is responsible for bringing about consciousness – the perpetual universal consciousness we witness. Knowing there is this unified field, we know that there is a deep unity within the Universe; as a result it could be that consciousness is not created by the brain, neither might it be as a result of the molecular structure or chemical processes within the brain.

Consciousness appears to be fundamental within the natural world, and manifests itself at the very core of nature; this knowledge comes as a result of understanding the consequence of the unified field. Our understanding of the unified field will enable us to see how consciousness bubbles up through our physiology to become the consciousness we experience.

Exciting times are afoot, as science is now on the verge of being able to understand the link between neuroscience and quantum physics.

Everything within the Universe, people, trees, planets, rocks and stars are just simply waves of vibration within the underlying unified field. Everything is united at its core. We quite possibly individualise our consciousness from the unified field through the existence of our nervous system and brain.

It makes sense – in very much the same way as a television set miraculously gets its life from electromagnetism!

I rest my case.

Your thoughts on this matter will be much appreciated.

Nothing in the Universe is inactive

- *At a quick glance, the Universe could be thought of as a fairly inactive place – this is most inaccurate.*
- *The most innocuous object such as a billiard ball is made up of billions and billions of tremendously energetic particles.*
- *Could our thoughts be made up of the same waves that ripple within the unified field?*

If we were to think that the majority of matter within the Universe is uninteresting, inert, lifeless and decomposing, we could not be further from the truth. We are living firmly within a universe of enormous activity. All matter is alive, every spec of it is whizzing around at a rate of knots.

Nothing within the Universe is inactive, even if we are looking at a lump of rock that appears not to have budged for billions of years; it is full of atoms that are as full of life as those that make up you.

Could the Universe be so active throughout that it generates an almighty invisible field, making it fundamentally conscious at its core? This is very dissimilar to the way we may instinctively imagine it to be.

A little like an electric eel surrounds itself with an invisible force of electricity and a magnet reaches out with invisible fields to affect certain matter in its reach.

Some people's natural instinct steers them to believe that the Universe is full of dead and decaying matter of little relevance. This could not be further from the truth. Every spec of matter is active, and could well be contributing towards the consciousness of the Universe as a whole.

Before the discovery of the quantum world, we had been able to study the interaction of only seemingly-lifeless large chunks of matter, often referred to as billiard ball mechanics. Scientists were able to observe what happened when a billiard ball bumped into another billiard ball. They were totally oblivious to the makeup of the billiard ball, with its billions and billions of extremely active subatomic particles that combine all the forces to make the billiard ball. How wrong we were to see everything as being lifeless and static.

Amazingly, within the quantum world, the concept of a particle is replaced by the wave function. A particle may exist as a wave until it decides to transition to a particle. This means that the waves that ripple within the unified field, which we know make everything in the Universe, are actually the same thing of which thoughts are made. Thoughts, consciousness and matter are made up of the same thing!

We are living within a 'Thought Universe', a conceptual Universe. Thought is all around us in every direction we look – even inside our heads!

The deeper we delve into the structure of natural law, the less dead it appears. The more alive it becomes, the more conscious the Universe becomes.

When we take another look at the unified field of the Universe, it could be considered as the foundation of the Universe. It can be seen as a field of pure 'being' and pure 'intelligence'.

The unified field can be viewed as intelligence as it is quite simply the fountainhead of all the laws of nature, all the fundamental forces, all the fundamental particles, all the laws governing life at all levels within the Universe. The laws that govern life have their unified source within this single unified field. This is what makes the unified field such a concentrated, intelligent and self-aware field throughout the whole Universe, and is responsible for the generation of our consciousness.

Quite a concept really!

Consciousness as an inherent part of the Universe

- *Consciousness within the Universe is all around us – consciousness can only have been derived from the fundamental building blocks of the Universe.*
- *It is quite possible that consciousness streams through the Universe in a quantum-style state.*
- *Consciousness gives us the ability to reason with all the information streaming into our brains – we are very dissimilar to a computer.*

There is a strong possibility that consciousness could have transpired through sheer Universal necessity. Alternatively, it could have materialised by design, accident, chance, inevitability or trial and error.

But my research has led me to believe that the consciousness we experience is an inherent part of the Universe. Perhaps its existence is inseparable from the Universe itself, and has been developed via the mysterious intelligence of the four forces of the Universe.

Would our conscious minds still work outside the Universe?

It is as if the consciousness within the Universe is all around us, having been originated from the fundamental building blocks of the Universe itself.

If we look closely at the evolution of all creatures through to the human being, it is doubtful that consciousness evaded early creatures. But it is interesting to imagine the extremes of consciousness between primitive single-celled creatures to the likes of ourselves. It is also interesting to consider how the first conscious thought came about, what it would feel like and what it would comprise.

To what extent would the first conscious thought have reached?

How vivid would the first conscious thought have been?

What would the first conscious thought have been about?

Perhaps the first-ever conscious thought would have been nothing more than a tiny blip – perhaps a moment of pleasant surprise in reference to a little warmth or comfort. Imagining those initial conscious thoughts are decidedly difficult to picture, but intriguing all the same.

Acknowledging a flash of light; acknowledging heat; acknowledging another of your kind next to you; acknowledging yourself; acknowledging you have ingested some nutrition.

Interestingly, scientists have never been able to pinpoint any part of the brain which is responsible for consciousness; they can only presume all of it makes the conscious feeling. Could it be coming from the DNA at the core of every cell?

Perhaps it is the quantum behaviour of the DNA and the intricacies of the synaptic gaps?

It could well be a combination of these, plus a multitude of other of nature's secrets.

Our brain's active mind seems unlike a serial processing computer, it certainly does not feel like our thoughts are attributing degrees of intensity or values to represent feelings and experiences as a computer does. We seem to have the ability to witness something real and natural, wholly instantaneous, complete, non-hazy and extremely rounded.

The phenomena that are more than likely at work within our minds, which could assist enormously with consciousness, may well be entanglement and superposition. These would provide tremendous speed of thought. Interestingly, this is precisely how the leading quantum-computing experts are trying to get computers to operate in the future. Within the brain's composition, a combination of entanglement and superposition would facilitate instant communication across the brain's synapses. A very useful property when trying to perform such complex processing within the mind, and facilitate super-fast thought.

There is a possibility that advanced quantum computing could begin to develop artificial, or even actual, consciousness.

Albert Einstein said, "A human being is part of a whole, called by us the Universe, a part limited in time and space. He experiences himself, his thoughts and feelings, as something separated from the rest ... a kind of optical delusion of his consciousness. This delusion is a kind of prison for us, restricting us to our personal desires and to affection for a few persons nearest us. Our task must be to free ourselves from this prison by widening our circles of compassion to embrace all living creatures and the whole of nature in its beauty."

The evolution of consciousness is truly fascinating.

Where do you think it comes from?

We learnt how direct knowledge is a quality of consciousness, and how the privileged access your consciousness provides is denied to others around you. We learnt how the same universal force that looks after the consciousness within you also looks after the life within other creatures. We learnt that there could well be some relation and interaction between human beings' conscious minds and the Universe's unified field. We learnt how everything could be made up from tiny, tiny little circles of energy called strings, and how understanding this may give us insight into the workings of consciousness. We discovered that visualising eleven dimensions is totally incomprehensible within our minds. We learnt that we cannot consider ourselves the most intelligent thing in the Universe. We learnt how there are exciting times afoot, as science is begins to understand the link between neuroscience and quantum physics. We learnt how life, and thus consciousness, originates from the Universe's unified field. We also learnt, of course how Einstein saw a human being experience himself, his thoughts and feelings, as something separated from the rest – a kind of optical delusion of his consciousness.

And ... if you do not understand this type of stuff – read Enid Blyton!

FASCINATING ASSORTED UNIVERSAL MATTERS

Where we look at a whole array of mysterious Universal phenomena and what may cause them. We look at some of the Universe's greatest secrets and question how they came about. We investigate the size of the Universe and how explosive the visible Universe is. We thoroughly investigate infinity and see how it cannot hide from us in the real world.

The eccentric Universe

- *What a strange place we live in! Strings, quarks, leptons, protons, neutrons, atoms, molecules, photons and entangled particle-pairs.*
- *The Universe is fascinating – it is thirteen point seven billion years old and its rate of expansion is still increasing.*
- *Some interesting creatures have evolved in the Universe's lifetime – the potential for future evolution is astounding.*

Our mental vision of the seemingly boundless Universe does not match any model or image we can conjure in our minds. With its vast array of fascinating celestial eccentricities, we seem incapable of picturing the entire Universe within one single coherent thought.

The composition of the Universe's cosmic structure appears unconventional and odd. We have to understand that the bizarre cosmic 'goings-on' are 'normal behaviour' to the Universe and that it may view humans' as the diverse, weird ones. The eccentricities of the Universe!

The term 'eccentric', to describe unconventional and odd behaviour originated from the Greek 'ekkentros', meaning 'out of the centre'. This led to the Medieval Latin 'eccentricus' which in turn led to the English word 'eccentric'. It first appeared in 1551 as an astronomical term, meaning 'a circle in which an orbiting body deviates from its centre'. It was only in 1685 that the word 'eccentric' evolved to mean 'unconventional or odd', and it was not until 1832 that the term applied to people. Moreover, eccentricity is often associated with genius, intellect and creativity.

The Universe is overwhelmed with unconventional and odd cosmic structures which possess genius, intellect and creativity. There are particles, planets, moons, stars, black holes, quasars, brown dwarfs, red giants, super clusters, dark matter, dark energy, comets, asteroids, nebulae, supernova, galaxies and gaseous clouds. It is a vast formation that contains trillions of diverse but interconnected celestial objects.

Similar to eccentric people, the Universe can be considered whimsical or quirky, although it can also be strange and disturbing. Thinking excessively about the expanse of the Universe can leave one's head spinning, and create an extraordinary feeling of insignificance.

Attention seeking and extravagance are eccentric traits. The Universe certainly grabs our attention and is particularly extravagant, especially in terms of its abundance.

The Universe is the entirety of all space, time, matter and energy. The Earth is merely a tiny island of matter and energy within this vast ocean of space and time. It is often said that 'truth is stranger than fiction', and this applies very fittingly to our Universe. The things that have been discovered in the vast realms of the Universe are unbelievable, and very often beyond imagination.

Asteroids, comets and meteorites are lumps of rock and ice debris left over from the formation of planets and stars that get caught in eccentric orbits around celestial bodies.

Galaxies are spiralling whirlpools of matter and energy made up of billions of stars and clusters of stars with orbiting planets. Galaxies are thought to have central black holes which devour all matter that comes close. Our Milky Way galaxy is one hundred thousand light years in diameter and one thousand light years thick. It consists of approximately two hundred billion stars such as our Sun. There are thought to be at least one hundred billion such galaxies in our visible Universe!

Stars are categorised into various types according to age and size. A protostar is the youngest of stars and a supergiant star is the largest. Black dwarfs and brown dwarfs are failed stars that never heated up enough to become fully-fledged stars. Some scientists think that these failed stars contribute towards the invisible dark matter that is lurking throughout the Universe. If it were not for dark matter then galaxies would be unable to bond together as they do – planets and stars would be hurled outwards for all eternity.

When the hydrogen fuel at the core of a star is exhausted, it will expand to become a red giant. Sometimes a star meets its end in an enormous explosion called a supernova. The remaining core from a supernova is called a neutron star. These possess huge magnetic fields.

When a neutron star spins at speed, emitting radiation, it is known as a pulsar. These pulsars can be detected rotating at phenomenal speeds of a few milliseconds to a few seconds. When pulsars were first detected, their lighthouse-style bursts of radiation were mistaken for a signal from an alien civilisation!

In fact, the first pulsar source identified was referred to as 'LGM', which stood for 'Little Green Men'!

Quasars were first mistaken for stars, but were later discovered to be galaxies with very high red-shift, due to their enormous speed that is very close to the speed of light. Quasars are the most distant and energetic galaxies, with an energy output unsurpassed by any other object in the Universe. The energy from a quasar is equivalent to about one trillion Suns. The enormous energy is thought to be created by the heat of a super-massive black hole at its centre as it swallows up matter, gas and dust whilst it hurtles through space.

Black holes form when large stars run out of fuel and collapse under their own gravity, forming a hole in the fabric of space-time. Nothing can escape the phenomenal gravitational pull of a black hole, not even light.

The earth began life approximately four billion years ago as a ball of swirling dust and gas. This swirling dust and gas formed our wonderful Earth as we know it today, which currently orbits around the Sun travelling at over a million and a half miles per day!

Following a complete analysis of the visible galaxies, astronomers realised that space was even weirder than they had originally thought. They discovered that the Universe is subject to a phenomenon termed 'inflation', which has the effect of forcing the Universe to expand. They discovered that the total mass of the Universe, including the hidden dark matter, made up only about a third of the critical density needed to satisfy the effect of inflation. Cosmologists thought that inflation must be wrong. However, measurements of the cosmic microwave background left over from the Big Bang showed that the total density of the Universe did add up to this special critical density, so inflation was correct.

This all meant that there is something else even stranger than dark matter. It must be invisible and possess a repulsive force, otherwise it would merge to be part of the galaxies and affect their movement. This mysteriously strange matter acts like antigravity that counteracts the attractive force of gravity. Cosmologists term this 'dark energy' – but they have not a clue what it is!

Instead of the Universe slowing down as one might think it should, it is speeding up with increasing pace.

Amazingly, Albert Einstein in 1917 calculated that the Universe was expanding. However, just like everyone else at the time, he nevertheless was convinced that the Universe was static. To counteract what he considered initially to be a flaw in his mathematical formulae, he invented a number to correct his equations so that the Universe remained static. He called this the 'cosmological constant'.

Amazingly, Edwin Hubble later discovered that the Universe was indeed expanding. Einstein had therefore been originally correct when he had calculated that the Universe was expanding. Upon hearing the news from Hubble, Einstein hastily retracted his 'cosmological constant', calling it his 'biggest blunder'.

Ironically, eighty years later, Einstein's cosmological constant has been resurrected to account for the mysterious force of dark energy.

How extraordinary all this is, especially when you think that everything is supposedly made up of the same tiny substance called quarks – so why should everything on the large scale be so different?

The Universe does not present its particulars very well at all, leading the human race to develop some extremely botched interpretations over the years. The Aztecs believed that the Sun died every night and needed human blood to give it strength to rise the next day. They sacrificed fifteen thousand men a year to appease their Sun god, Huitzilopochtli.

Nice!

The Ptolemaic view of the Universe saw the Sun and planets orbit the stationary Earth with a sphere of stars just beyond them. We have lately developed what we perceive to be a more accurate depiction of the Universe, but it still has scope for significant improvement.

Thus far the Universe has existed for thirteen point seven billion years and is expanding. Amazingly the rate of this expansion is increasing. This implies that the Universe will eventually meet with a long-drawn-out and cold end – but who are we to speculate the fate of such a genius of an object?

Perhaps other forces will become prevalent over time which will alter the outcome.

There are so many things we do not know about our Universe. There are also many human-derived influences embedded within our view of it. Our narrow vision within the light spectrum, our inability to detect its vibration, the lack of any sound, our inability to touch it and our self-centred view of time, all contribute towards a distorted view of our Universe.

There are unseen aspects of the Universe within the infrared and x-ray spectrums which are just as real as those we see. The Universe is resonating and stretching the fabric of space-time throughout, but we are unable to detect this.

We developed the day and the year based on our observation of the Earth's orbit around the Sun, but there could well be a more appropriate timescale lurking somewhere. One timescale I thought quite appropriate for our solar system is connected with the solar maximum. The solar maximum is the Sun's period of greatest activity in the solar cycle. Sunspots appear during a solar maximum which take an average of eleven years to go from one cycle to the next.

Upon this timescale I am only four-year-old at the moment; I could quite take to this view of time!

Perhaps all the components of the Universe are connected in some way to form one enormous biological creature of sorts. After all, if the Universe can create something as magical and as intricate as a biological cell, then it must be thoroughly intelligent and well-organised. Alternatively, you could believe that the Universe is in a state of chaos, and out of the chaos came order and structure.

If the Universe has managed to develop humans inside thirteen point seven billion years, then just imagine what it could produce after existing for one hundred trillion years!

After such a period of lengthy evolution we may perhaps witness humans that are born with inherited knowledge – relating to pretty much everything Mankind knows. Perhaps there will be a day when we are born with instinctive knowledge that may be accessed through our DNA!

Our analysis has undoubtedly highlighted that the Universe has unconventional and odd traits of eccentricity.

As we get to learn more and more about the Universe, we get to realise that its genius is far superior than we could have ever imagined.

The size of the Universe

- *The light from the first moments of the Universe have been travelling endlessly since its creation.*
- *The size of the Universe is particularly difficult to imagine, as we are accustomed to such small distances here on Earth.*
- *Attempting to mull over the enormity of the Universe leaves people giddy.*

When Claudius Ptolemaeus, in the second century AD, proposed his Ptolemaic view of the Universe with the Earth at the centre, he pictured the Sun, Moon and planets orbiting the Earth with a sphere of stars just beyond this. With the Universe described like this, it is easy to imagine the stars fixed at a distance. With such a simple explanation, it is easy to see why it was accepted and no one raised any eyebrows. No awkward questions were asked and no one tried to dig deeper for over a thousand years!

But as soon as Copernicus and Galileo challenged these views, we were faced with the tricky and somewhat daunting prospect of coming up with a new answer. This was compounded at the start of the twentieth century, when it was discovered that the Sun is only one of billions of stars in the Milky Way, and is located far from the centre of the galaxy. Everything became even more daunting when it became clear that there were billions of other galaxies scattered throughout vast distances of the Universe.

Estimates of the size of the Universe have been bettered as methods of measuring galactic and intergalactic distances have improved. Close stellar distances were at first found by measuring a star's trigonometric parallax. A more powerful, contemporary method is to analyse the light reaching the earth from an object by means of a spectroscope. A very faint, distant object can be estimated by comparing its apparent brightness to those of similar objects at known distances.

Another ingenious method of measuring distance relies on the fact that the Universe appears to be expanding – as indicated by the red-shift in the light emitted from distant galaxies. A formula known as Hubble's Law then allows the distance to be calculated, based upon the speed they are seen to be rushing away from Earth.

If we are to believe in the Big Bang, then we believe that the Universe arose and expanded from a single point in space approximately thirteen point seven billion years ago. Within this model of the Universe there is what is known as the observable Universe, which consists of the galaxies and other matter we can see from Earth today. The Universe is thought to be of relatively uniform consistency throughout, with galaxies evenly spread. The edges of the observable Universe are defined by the distance to the far reaches of light, and other observable signals from the most distant objects. The observable Universe is therefore a spherical shape centred on the observer, irrespective of the shape of the entire Universe.

So how wide is the Universe?

If we multiply the speed of light, which is six hundred and seventy million miles per hour, by the number of years the Universe has existed, then multiply this by the number of hours in a day and then by the number of days in the year, we arrive at the distance that light has travelled since the birth of the Universe. However, we must also multiply this number by two as the light has been travelling in two directions.

The Universe has not only expanded outwards from the Big Bang, but the space in between objects has also expanded due to dark energy. The effect of this has meant that much of the Universe is not visible to us. The light from these distant objects has not had time to reach us, and what is more, the light may never reach us as we are moving away from these objects faster than the speed of light.

Einstein said that it is not possible for objects to travel faster than the speed of light. However, the Universe appears to have contravened this rule!

Perhaps neutrinos that travel faster than light might have had a part to play the Universe's 'faster-than-light expansion' – the next few years will reveal much more on this topic.

This apparent contravention is just an illusion. When objects accelerate away from each other, due to the space expanding in between these objects, the net result provides an impression that merely looks like the more distant objects have expanded faster than the speed of light. However, nothing in reality has actually travelled faster than the speed of light.

To visualise how this works, imagine a deflated balloon uniformly covered in dots. When you inflate the balloon, the dots that are close together move apart slowly, but the dots that are positioned further away from each other move apart much faster.

As the Universe is roughly thirteen point seven billion years old, one might assume that the diameter of the Universe is double this at twenty-seven point four light years, or 161,074,335,225,230,851,825,340 miles wide. However, this is incorrect as the expansion effect of dark energy has not been included.

To visualise the expansion effect of the Universe, imagine a photon of light that travels for one year whilst the Universe is just one million years old. At this time the Universe was a thousand times smaller than it is today. Therefore, the light that travelled for the one year all that time ago has now stretched to become a thousand light years. When we add up all the expansion that has taken place since the birth of the Universe in this way, the width gets staggering. A photon of light that has been travelling since the early Universe is now seventy-eight billion light years away. Twice this distance will therefore be the actual width of the Universe. This means that the width of the Universe equates to be one hundred and fifty-six billion light years, or 917,065,558,216,642,806,012,884 miles wide.

Of course I may be out by a few miles – if you spot any discrepancy of any sort, please let me know!

To some, this number may seem large, but personally I find it quite incredibly small in comparison with other numbers we find within the Universe; such as the number of atoms in a sugar cube, the number of revolutions an electron has made around its nucleus and the odds of our existence.

Irrespective of your view, the Universe is still very large and leaves you dizzy when trying to think about it in its entirety. There are statistics that can put the size of the Universe into perspective. One statistic I like is relating to the occurrence of supernova within our Milky Way galaxy. It has been estimated that on average three stars blow up to become supernova every century inside our Milky Way galaxy. Now, if every visible galaxy behaves in this way, then one star will blow up somewhere within the visible Universe creating a supernova explosion every second.

Who would need fireworks night if we could see all that was going on!

Another way to visualise the vast expanse of space is by looking at a very short distance. The distance from the Earth to the nearest star, Proxima Centauri, is 24,943,007,458,418 miles or approximately 24.9 trillion miles. One trillion miles is a million, million. As it takes roughly two weeks to count out aloud to a million, scaling this up, if you were to count out aloud to the number of miles to Proxima Centauri, it would take roughly one million years to do this – that is without any sleep!

Why matter exists

- *If we are to investigate why matter exists, we also need to investigate where the laws of physics came from to govern the matter once it materialises.*
- *The creation of matter is clearly achievable within the Universe – perhaps just because it is possible for matter to exist, it will come into existence.*
- *If there were nothing in the Universe, there would still be a place where nothing existed.*

There are more tiny particles of matter buzzing around than we can care to imagine. Matter naturally forms shapes, organises itself into clusters and can combine together to create enormous celestial structures. But, what exactly is matter?

Scientists have discovered that there are only three different fundamental particles of matter at the tiniest level, called quarks, leptons and gauge bosons. These three fundamental building blocks of matter make up everything in the Universe. Electricity, energy, plasma, light, sparks, fire, solids, liquids and gases are all made up of the same fundamental particles.

How strange that these three particles have managed to arrange themselves to become so many different elements.

So from where did matter originally come?

Perhaps our understanding of where matter came from is as advanced now as our understanding of supersonic aircraft was in the tenth century!

But if we have to speculate where matter came from, let us touch for a moment on string theory. As we know from earlier in the book, string theory and M-theory predict that our whole Universe could be living upon one giant membrane, within a much larger and higher-dimensional space.

So to imagine this, picture many Universe-sized slices of enormous energy layered one on top of another. Two of these enormous energy membranes collide into each other to trigger what we refer to as the Big Bang. The collision of these energy membranes allows the energy to convert to matter that we see in galaxies.

It is known that energy cannot be created or destroyed. However, interestingly this only remains the case so long as the laws of physics do not change!

At this point I think we need a serious high-level 'ingredients recap' to visualise what is required to make a universe. Firstly, we need a set of laws to which energy must conform. The energy must then materialise. Once the energy exists, then so long as the laws that the energy can follow are appropriate, the energy may convert into matter.

Most of us learnt that matter is made of atoms and that atoms are made of protons, neutrons and electrons. What we do not generally learn is that for each particle of matter, there is another particle called an anti-particle of matter. The anti-particle is identical in every way, but with an opposite electric charge.

It is thought that the Big Bang saw the creation of matter and antimatter from a vast amount of energy. Matter and antimatter would normally cancel each other out, leaving nothing but light, but if matter was produced in slightly more quantity than antimatter, some would be left over.

Matter was initially thought to be symmetrical, but scientists found that a particle called a B meson is unsymmetrical and undergoes what they call charge-parity. With a small amount of B mesons undergoing charge-parity, and because the Big Bang created such an enormous amount of matter and antimatter, eventually all the matter we see came into existence.

This is all well and good, but the matter and antimatter had to materialise into somewhere that has rules. These rules must be there to be complied with by the particles. In the absence of rules it would be like expecting nothing to go missing when opening Fort Knox up to thieves!

According to our 'understanding' of the laws of particle physics, matter and antimatter should be present within the Universe in equal amounts. However, our observations report that there is much more matter than anti-matter. One of our greatest challenges is to try to discover what has happened to all the antimatter.

As everything so far sounds rather illogical, let us take one step back and ask the question – from where have the laws of physics come from?

Perhaps the laws of physics were set at the moment of the Big Bang – or shortly afterwards. Alternatively, they may have always been set the way they are and there is no alternative – it is just the way that things are!

It is very difficult to imagine matter existing in an area of space where there are no laws of physics. Does this mean that the laws of physics must have come first before matter appeared within space?

Alternatively the laws of physics could have materialised at the same time as matter appeared.

Could it be that the laws of physics change over time?

Let us imagine for a moment that the laws of physics alter. This altering of the laws of physics within a large area could have the most amazing effects. Perhaps a small change in the laws of electromagnetism may cause massive amounts of matter to spontaneously materialise. A little like a change in pressure causes moisture to fall as rain from a cloud.

Who knows?

There is just one additional approach which may suggest a mechanism through which matter originated – perhaps it originated from nothing!

If you imagine everything started off in a huge empty void of nothingness, then somehow this nothingness developed a shape. This nothingness in the form of a shape organised the void of nothing to become evenly distributed throughout – a little like how the atmosphere of the Earth creates wind to even out pressure differences. Perhaps all this progressed successfully until a lump appeared within the void of nothing; this lump then insisted that something was borrowed from the future to even out the present moment.

Imagine that the act of borrowing from the future is permitted within these early laws of physics, as perhaps time is not a concept at this point, which incidentally would make borrowing from the future much easier. Once the present borrows from the future, suddenly there is an almighty tug on the borrowed particles in the 'now' moment and the Universe is incapable of giving back to the future what it borrowed from the present.

Hey presto – matter!

What we are trying to picture here is an uneven balance which will jostle with the laws of physics, the past and the future to eventually force something to materialise which is similar to borrowing. It could be this uneven distribution between past and future that forces the charge-parity changes in B mesons to create matter.

String theory says that everything could be much more tightly packed, and the eleven dimensions provide many more routes from A to B such that everything appears much more distant, when in reality this is not the case. In a world like this, borrowing from the future seems no more bizarre.

Just an added thought – if matter were to be borrowed from the future near a black hole, then upon being created, the antimatter could be sucked into the black hole leaving the remaining matter free to enter the physical Universe.

There are some big gaping questions.

Why is the Universe crammed full of physical matter?

Why does energy choose to become matter?

Why is there energy in the Universe at all?

Could energy become other things, other than matter?

Would it be more logical if there was absolutely nothing at all?

Perhaps just because matter can exist, it will. This may be a simple statement, but it is just like why people climb mountains – because they are there to be climbed. But just for a moment imagine the Universe never allowed matter to ever exist. There would never be anything at all and we would never know there was nothing, quite simply because there would be nobody or anything to know there was nothing.

If you try to imagine nothing whatsoever in the Universe, you will find your thoughts keep failing.

I have tried on many occasions to think such a thought, and it becomes alien to your mind when trying to fixate upon it. I came to the conclusion that trying to concentrate on imagining there to be 'nothing in the Universe whatsoever' turns out to be an impossible thought.

Let me know if you have any success with this thought – and how you went about it.

Perhaps the Universe could be in a state of ultra-superposition – this would mean that it is cohabitated by particles that interfere with themselves. Perhaps these particles came about by the presence of observation, simply brought about by the inability for nothing to exist. After all, for nothing to exist, this still requires something to think and bring about the concept of nothingness.

If you follow this logic carefully, it implies that the existence of matter is one hundred per cent probable. For nothing to be a reality requires something to acknowledge it. Therefore matter in the Universe has to exist.

Every single thing around us has to be considered as an absolute sensation – every particle is an outright winner in this bizarre Universal pandemonium. Everything that exists has survived years and years of tugging by incredible universal forces, which therefore gives each and every particle in the Universe an individual label of complete and utter success.

We can only imagine that all these particles spewed out from a central point of singularity. Events prior to the Big Bang were obviously such that this was forced to happen, but it is not appropriate to address that just yet.

So what was the Big Bang?

For some reason, thirteen point seven billion years ago, an almighty flurry of energy burst into existence and formed our Universe. Perhaps a reasonable explanation of why this happened is associated with a philosophical argument that it is impossible for nothing to exist.

For there to be 'somewhere' for nothing to exist, there needs to be 'somewhere' for it not to exist within. This then inadvertently creates 'somewhere' for nothing to exist. At this point there is the advantage that we know there is 'somewhere' as we witness being on the inside of it, otherwise we would not be here. By knowing there is a 'somewhere', then within this 'somewhere', 'something' blasts into existence to define the nature of the 'somewhere'. It is almost as if reason, instinct and intuition is suggesting it is impossible to have nowhere with nothing within it. This leads us to a situation where there is always somewhere with something in it.

How does this 'something' or matter blast into existence?

Let us imagine prior to the Big Bang there were huge swathes of nothingness within an enormous number of vast areas we call membranes. These enormous areas, devoid of content, organised themselves into gigantic, empty spaces. Remember that M-theory states there are eleven dimensions, so these vast swathes of eleven dimensions that contain nothing actually have structure, but no content. It is likely that the eleven dimensions of the membranes organised the empty space into tiny, curled-up areas in some ratio of a Planck distance apart. A Planck distance is the smallest size of quanta known to Mankind, as defined by Max Planck.

The vast membranes of nothing would be a little like a massive pile of empty boxes. Smash a pile of empty boxes hard enough together and you will find that some of the boxes will end up containing shards of box inside them although previously you would have said there was nothing to put inside the boxes.

Some boxes are now full of vibrating particles from the impact.

Could this cardboard box theory, or C-theory, be how everything came into existence?

All it takes now is a small bump from one of these vast swathes of 'nothing' to bump into another, to set off a chain of events that produces vibrations in eleven dimensions. Suddenly there are tiny, independent shards of space that splintered off the vast swathe of nothing at the moment of impact. Upon impact there would be enormous ripple effects travelling at incomprehensible speeds outwards – vibrating space as it goes. The vibrating ripple effects would work in a similar way to how ice crystals form within liquid water as it freezes.

This would explain how matter arose from nothing. Subatomic particles could be tiny, vibrating elements of empty space. At the moment of the collision between the two membrane structures, enormous vibrations caused tiny shards of empty space to define themselves differently to their surroundings. Matter exists because vibrating space is different to non-vibrating space – this is how we can differentiate what we see as a particle from empty space.

Remember there are plenty of possibilities within an area that has eleven dimensions. Imagine your empty space is a length of metal. When this metal is clamped in a vice and the tail end twanged, the end clamped in the vice remains perfectly stationary. The observer looking down over the end of the vice will observe the metal as being stationary. If the observer views the twanged metal in the vice form the side, especially with a strobe light, they will observe the very same effect. The metal also becomes hotter as it twangs away, this will be one of its properties. The vibrating metal, or the empty vibrating space, becomes defined as the body of it pulsates and takes on a form of its own. This vibrating motion could bring about the existence of what we interpret as a particle. The existence of a 'somewhere' or a potential 'somewhere', which will always be in existence, is the reason why these vibrating eleven dimensions flourished.

So in summary, vibrating the envelopes of empty space that exist due to the eleven dimensions of empty space, is how matter comes into existence.

If matter is actually vibrating 'nothing', then perhaps our understanding of matter is flawed. Maybe our image of matter that is depicted within text books, with electrons orbiting a nucleus, is based purely upon the way it behaves – it may not be that way in reality. What we have witnessed from matter's behaviour has given us the impression that its make-up is how it has been documented within text books – this is what we have been led to believe, when in reality it actually just acts like that – its true shape and make-up could be wildly different!

As an analogy, it could be similar to a police force thinking they have tracked down a burglar from his actions and movements, only to find that it was not a human after all but a dog. The assumption all along was that a person had been responsible for the incident, but it was only when delving deeper and questioning the matter further it transpired the culprit was a dog.

In an identical way, observations and experiments could have led scientists to believe that atoms are made up of electrons orbiting a central nucleus. In reality, however, perhaps an atom is just 'nothing' vibrating, which just gives the impression that electrons are orbiting around a central nucleus.

Our initial interpretations of many phenomena are often very unlike the interpretation we ultimately settle upon. One classic example is our understanding of the Universe. Our initial understanding was hideous, but we have steadily improved our understanding over the passage of centuries.

It is getting clearer now that matter does not actually exist, it is purely the Universe's interpretation of what forms when nothing vibrates.

Around each atom is an electromagnetic field, which is actually energy rather than matter. The electron which orbits the atom is probably another electromagnetic field, and the neutrons and protons are possibly another electromagnetic field of sorts. Therefore, it is a possibility that an atom is made up purely of energy and has no physical matter associated with it.

If matter is made up of electromagnetic force fields then it is possible to say that matter does not really exist. At least not in the way we think of matter. We think of matter as something solid and energy as something ghostly, like a radio wave. This may mean that matter does not really exist and only appears real, in the same way that we can interpret a scene from an image on a postcard.

The card is solid and the ink on the card is solid. However the scenery you see is just implied and therefore does not really exist. With this very same thought in mind I have come up with what I term the 'Implied Image Theory'. This is where we think that something is solid, but in reality it is just an implied three-dimensional image made up of energy; with energy made up of the vibrations of nothing.

Scientists are part way to replicating or mimicking how the Universe creates matter now that we can project holograms. It is possible to wirelessly transmit a three-dimensional image of anything from one side of the planet to the other; it just needs a bright individual to next work out how to project the mass of the image along with it.

To someone who lived one hundred years ago, this would be nothing short of a miracle. The funny thing is that I know how this works in intricate detail and it is not that complicated. Perhaps this goes to show that the ultimate solution regarding how matter is created is not that complicated either. Now that Aaron O'Connell's has discovered a simple large scale experiment which shows trillions of atoms simultaneously residing in two places at once, scientists must be just a stone's throw away from the ultimate objective of being able to recreate matter.

Later within the chapter 'Wave particle duality of light', Aaron O'Connell's groundbreaking discovery is described in further detail.

The four forces of the Universe

- *If anyone had the opportunity to choose the properties of four universal forces – they could not have chosen any weirder than those we have!*
- *The force of electromagnetism is inordinately stronger than that of gravity.*
- *The proportions of the various forces are inconceivably different in magnitude.*

Everything that happens within the Universe has to comply with the four fundamental forces. Everything within the Universe has been shaped and moulded by these four fundamental forces; electromagnetism, gravity, the strong nuclear force and the weak nuclear force.

The strong and weak nuclear forces operate over very tiny distances and govern the interaction between components of an atom. When you study these forces, you come to the conclusion that you could not dream of a more bizarre set of properties.

According to scientific calculations, the nuclear strong force is one hundred trillion, trillion, trillion times stronger than gravity. Whereas the nuclear strong force only acts within a miniscule distance, gravity acts within what appears to be an infinite distance. What gravity loses in strength, it would appear to gain in distance. Gravity has an infinite range compared with the nuclear strong force that only acts over one thousand trillionth of a metre!

As gravity has such an infinitely massive range, this is why it is so weak. It is the underlying force that shapes the Universe, bringing everything together in the shapes we see. As gravity has an infinite reach, it also has the ability to carry information about all objects to all other objects – therefore, every conceivable part of space is influenced and aware of all other parts of space. Gravity moulds everything in relation to everything else.

The electromagnetic force acts between electric charges and is also hugely strong compared with gravity, but also seems to act within an infinite distance just like gravity.

The electromagnetic force is ten trillion, trillion, trillion times stronger than gravity – and this can be proved quite simply by comparing the Earth's gravitational pull with a magnet and a piece of metal.

Not only does a small magnet counteract the complete influence of the gravitational pull of the whole planet, but when trying to force the metal off the magnet in line with gravity it is often an effort, or even impossible with stronger magnets. Electromagnetism is so strong that it renders any gravity local to it irrelevant to such an extent that gravity can be omitted from calculations.

The weak nuclear force is responsible for the emission of electrons from an atom and again, despite its name, is some ten trillion, trillion times stronger than gravity. The weak nuclear force acts over the shortest distance of all the four forces. It only acts over one million trillionths of a metre!

It has been suggested that gravity may lose most of its strength to create time. If gravity puts almost all of its energy into the creation of time, then this could explain why gravity is so weak.

No one has ever been able to explain how time is maintained, or touch on precisely what time is, or how it was initiated. Scientists have never determined why gravity is so weak. It could therefore make sense to propose that the passage of time is created by gravity.

The four forces are a true mystery – why four?

Why not one hundred and four?

Or even none?

Gravity is thought to have a messenger particle called a graviton that is without mass. The electromagnetic force has a messenger particle called a photon that is without mass. The strong nuclear force has a messenger particle called a gluon that is without mass. However, the weak nuclear force has a relatively heavy messenger particle called a weak gauge boson.

It is totally baffling why three of the messenger particles are without mass and the fourth, the weak gauge boson, is one of the heaviest known particles. Similarly, why do the ranges and strengths of the forces differ so drastically?

Irrespective of why these forces are as they are, it cannot be disputed that the universe would be strikingly different if the force properties were anything different.

The formation of stable nuclei is dependent upon the ratio of the strong and electromagnetic forces.

The protons in a nucleus repel each other. However, the strong force overcomes this repulsion. A small change in their relative strengths could allow the electromagnetic force to overcome the strong force, and atoms would not exist.

If electrons were any more massive, then electrons and protons would bond to form neutrons, thus disrupting the formation of heavy elements.

If gravity were any stronger, stellar matter would bind more strongly, and stars would use their nuclear fuel much faster.

If gravity were any weaker, atoms and molecules would not attract each other as they do to form larger structures. This would prevent the formation of stars and planets.

If a universe popping into existence is provided with a random figure for its forces, then a universe conforming to our Universe's exact forces would be very, very rare indeed. For the purposes of calculating this, we will take the strong nuclear force as a maximum force value on our scale of values for each of the four forces. If each of the forces could assume a value zero through to that of the strong nuclear force – and of course this value could be much higher – the number of different settings for a particular force could be one of one hundred trillion, trillion, trillion.

If we wish to calculate the odds of our Universe being created exactly in the way that it was, then we need to look at all the permutations that this range of values could possibly take. We will take the tiny weak nuclear force as the minimum lower boundary value that can be allocated, and take the strong nuclear force to be the upper boundary. When we do this we find that each force may take one value from any of one hundred million trillion, trillion, trillion, trillion different settings.

So we can see that there are an inordinate number of possible settings for the forces within a universe. The odds of our Universe getting the exact value we experience for its forces are one with two hundred and sixteen zeroes following.

This is a massive number, one of the largest numbers I have seen calculated, so this brings into question whether the Universe was random or designed?

Certainly, if this is the only way things can be, to allow stable matter within the Universe, then it is a massive, massive coincidence of mega-proportion that we exist within such a remote improbability.

Factoring this in with our odds of existence calculated earlier makes our 'true' overall odds of existence now an incredible – one with two hundred and thirty eight zeroes on the end, to one.

This makes winning the lottery look pathetic!

Does infinity exist?

- *Infinity is much larger than your imagination will allow you to picture in your mind.*
- *When someone dies, imagine they remain dead for a considerable period of time – or even remain dead infinitely – can this be correct?*
- *The repeating of events is part of the properties of infinity – it is very probable that an extremely large part of infinity repeats itself.*

For a significant proportion of people, infinity is tremendously difficult to imagine, something that goes on and on forever without ever coming to an end. The further you look into it the more it goes on forever. You cannot even fathom infinity within your imagination, as whatever you think of is totally insignificant as far as infinity goes.

Infinity can exist in many different guises, and manifests itself into several different styles of connotation within these guises. We have this phenomenon called time, which some think never had a start and may never have an end. It is possible to subdivide a second into an infinite number of moments, just as it is theoretically possible to cut a piece of material in half forever, always having something left to cut in half. Mathematics has examples of infinity, such as the calculation of a bouncing ball, which will never come totally to rest as it always bounces back a proportion of the distance it fell. Space can be perceived as stretching out forever infinitely.

Once you die, you remain dead forever.

Or do you?

Some of these ideas are theoretical and can be discounted in the real world such as the bouncing ball and the cutting in half of the material, but there are the true mysteries regarding whether the Universe continues infinitely and whether you remain dead infinitely.

We can try imagining infinity as person running around a permanently present world, without ever stopping each time we return to our starting point. When we refer to infinity, we mean something like time which never ends, or the Universe that is totally everywhere and has no edges. I have always viewed infinity as something within which pretty much everything is possible – teapot-shaped planets, aliens with green heads and one big frontal eye, replica planets like ours and distant, intelligent life-forms with massive spaceships.

Infinity can exist within a number of different concepts.

We can speculate the types of things that could be considered infinite, but we have to avoid the mathematical calculation of infinity which only exists on paper.

What about the possibilities of infinity within the Universe?

If the Universe has no edges and continues forever, or if there are an infinite number of universes, then surely infinity exists.

What about time?

If time never had a beginning then time could be considered infinite.

What about death?

Is death infinite?

Do we die forever?

If we die forever could this be considered as infinite?

To appreciate the bizarre nature of infinity, imagine a straight line projects out instantaneously from the end of your finger when you point it into the air. The line projects into the deep of the Universe infinitely, until it hits a 158-legged piano being played by 457 nine-eyed frogs.

The nature of infinity says this is certain to happen!

It is impossible that this piano does not exist in the direction you are pointing within an infinite Universe.

What is more, I am now going to change the odds of this possibility for future generations by leaving a request in my will. I would like a monument in my memory which depicts a 158-legged piano, being played by 457 nine-eyed frogs. If the Universe is infinite, then I certainly will not be the first person to have requested this.

In fact the nature of infinity is even more wonderful, within an infinite Universe our straight line will actually encounter an infinite number of 158-legged pianos being played by 457 nine-eyed frogs.

Honest – this is the nature of infinity!

Another way to contemplate infinity is to imagine a man who never dies whatever the circumstances – and he starts counting. This chap will carry on counting forever but will never reach infinity. Infinity is not a number and cannot really be thought of in terms of a number. It is perhaps best thought of as an image in the mind.

Infinity is something that never comes to an end.

Therefore imagining this guy counting forever will not yield infinity – especially as he will never get to the end of his task!

Nature has come close to showing us a means of displaying infinity within a circle. When dividing the diameter of a circle by its circumference, the resulting number called pi, according to mathematicians, never repeats the same series of numbers despite carrying on infinitely.

Infinity is such a ridiculously large concept that nothing is impossible within it. This is an extremely difficult concept to grasp.

If we study pi closely we realise it has infinitesimal characteristics. If mathematics is correct and pi never repeats the same series of numbers, then mysteriously hidden or encrypted within pi will be the blueprint for DNA, the complete works of Shakespeare, all the achievements of Mankind, the complete history of the Universe and an infinite number of future outcomes for the Universe.

This may be difficult to relate to, but it is true!

As most things in the Universe seem to congregate into circle shapes, such as stars, planets, moons, atoms and water droplets, then we can be sure that nature knows about pi. In a strange way, by pulling everything naturally into a circle, the Universe is pulling everything towards a ratio that is infinite – the magical number you get when dividing the circumference of a circle by its diameter.

If infinity were a number, perhaps it would start with a three!

If the human race's research into the quantum world thus far is anything to go by, then I would be very surprised if the Universe was oblivious to pi. If the Universe is intelligent and has access to this realm of infinite possibilities, then there is no reason why it would not utilise this magical special concept of infinity.

If the Universe is also capable of working instantaneously, as has been proved in several physical experiments, then we have a recipe for an ultra-intelligent Universe that could constantly recall outcomes and possibilities. This type of Universe would immediately be able to determine the optimum settings to achieve any physical outcome. This could even explain the settings of the four forces that some believe were established a split moment after the Big Bang.

Within a split instant at the beginning of the Universe, it would be able to devise a type of decoding language to read and understand infinity. It would then instantly scan the infinite possibilities available to it, and no doubt implement what it feels is best.

If one of the Universe's major objectives at the beginning of the Universe was to create Mankind, then it did a splendid job. But, just as a master marksman will take aim at a distance and be out by perhaps one millimetre, the Universe shows signs of slight imperfection.

This is perhaps why we see imperfections within this world such as corruption, greed, dishonesty and fluff in our tummy buttons!

The Universe therefore shows signs of being able to calculate whatever it wants instantly, and has every permutation of absolutely everything at its fingertips.

Rather a powerful tool!

No wonder we came into being!

This concept is rather exciting. It means the Universe could have achieved perfection at the instant of the Big Bang. This would explain how the balance of all that we see within the Universe came about.

The fact that infinity was known to the Universe as soon as it came into being, meant that it instantly had all possible permutations for the future available to it.

Infinity is so large that nothing is impossible within it.

Infinity is so large that somewhere within it enormous sections will be repeated.

We can look upon death as a possible infinite event. So long as death continues forever then it can be treated as an example of infinity. However, it is a spooky thought that within an infinite Universe, the exact same atoms that make you as you are today will reassemble once again at some time in the future. This is bound to happen, not just once, but actually happen an infinite number of times in the future.

This is the nature of infinity!

"Many times have I been born, and many times have you been also. I remember mine, but yours you have forgotten. Birth-less am I, the everlasting self, lord of all creatures, yet I preside over nature and I manifest through my inscrutable power of illusion. O son of Bharata, when there is a failure of justice and virtue, and vice and impiety reign, I body myself forth from age to age, for the protection of good men and the removal of wickedness", Bhagavad Gita.

A classic way of understanding infinity is by referencing a funny tale I was once told. Two men, identical twins, died at the same time and went to Heaven. There at the gates, Saint Peter explained that due to a clerical error, they could not be told apart. What was worse, only one of them was to be allowed in! The other would have to go to Hell.

Peter said, "However, we've come up with a solution that we think is equitable."

"You," he said, pointing to one of the twins, "Will go to Hell for eternity. But every February 29th, you will change places with your brother for the day."

The man was angry.

He cried, "That's terrible! You mean that I might deserve to go to heaven, but my brother will go there and I'll only get out one day every four years?"

St. Peter calmly responded, "That's correct, but in the long run you'll spend the same amount of time in both places."

That is the nature of infinity!

The infinitely dormant Universe

- *Top physicists struggle with the Universe being infinite.*
- *If the Universe comes to an end, will it remain in this state infinitely?*
- *The Universe or universes could perhaps perpetually pop into and out of existence on an infinite basis.*

Let us for a moment imagine that the Universe had a beginning and will eventually come to an end. The Universe we are describing here would not be infinite. Many people would disagree with this notion, but somehow I do not think we are truly able to relate to the Universe being infinite.

If we are of the opinion that the Universe will come to an end, then let us take a look at how long the Universe will remain in this 'ended' state. This 'ended' state must be infinite, because as soon as it were to come into existence again, the Universe could possibly be considered infinite, unless it merely reappears a few times never to be seen again – which again would mean it would disappear infinitely!

What we have narrowed down here is that one of these states has presented itself as an infinite state. Either, the Universe keeps disappearing and reappearing on an infinite basis, or it disappears forever after appearing once or several times and remains vanished infinitely.

I am not sure about your spinning head at the moment, but mine is leaning towards the view that the Universe cannot realistically pop out of existence in its entirety forever.

Infinity cannot hide from us due to the fact that the opposite of infinity is again infinity. Either the Universe is infinite, or it disappears for an infinite period of time. The absence of something for an infinite period of time is surely another form of infinity!

Personally I cannot see how the Universe could materialise once then die, never to return – it just does not seem right. If it were to disappear for a long period, I cannot see why at some point it could not come back again, even if that period of absence were extremely long.

For this reason I will put my money on there being a universe that reappears on an infinite basis. This also makes my mind content that my exact atoms will someday reform as they are today, but I promise that the book I write in my next lifetime will be much better and easier to understand!

Is it possible that infinity can be attributed to time such that time never ends?

If time is infinite, would it make sense that time would need to be infinite in both directions?

Is it possible to have infinity, plus, infinity?

The outcome would either be infinity again, or you could say that it is not possible to add the two infinities together – a little like trying to divide an elephant by a tree.

Could time have a beginning and no end?

Could it be that time had no beginning?

Perhaps it has always been here and will never go away.

These kinds of questions make me also question exactly what the concept of 'now' really is.

If time has a beginning and comes to an end then the Universe will not be infinite.

If time has no beginning and no end then the Universe will be infinite in both directions.

If time has a beginning and no end the Universe would be infinite.

If time has no beginning and comes to an end then the Universe would rather interestingly be infinite into the past – but not the future. However, with an infinite past it would not be possible to reach the future.

If the past is infinite then the future must be infinite too, otherwise we would have never been able to reach our current 'now' moment!

To think that the past is infinite makes the moment that we exist within 'now' extremely improbable, due to all the possible moments that it could possibly have been. This view makes a mockery of time itself, unless time really is infinite and we just so happen to be extremely fortunate and living 'now'.

One conclusion we could draw is that time is infinite for the following reason. Imagine there was nothing in the Universe whatsoever; you may think that time could have disappeared and there be nothing for time to exist for. However, if all of a sudden a piece of matter materialised, if it knew how to behave, this would mean that the knowledge of how matter behaved had survived the period of time when no matter was there. This would imply that time is present in the complete absence of matter. If any knowledge, matter or memory of anything is to be kept, then it makes sense that we need time to keep it.

This spontaneous behaviour implies that forces are keeping an account of what is happening, in order for matter to have a type of behaviour when it materialises. This implies a behaviour-conscious Universe even in the absence of matter, which in turn implies at worst we have an infinitely behaviour-conscious Universe!

Quite a revelation really!

Before we move on I think it only fair whilst on the topic of infinity to contemplate whether we are alone in the Universe. If we live within an infinite Universe, and the Universe is constant throughout, then life elsewhere would be a certainty. However, if the Universe turns out to be vast but has a finite boundary, then the probability of other life elsewhere diminishes.

We learnt about the eccentricities of the Universe and how its genius is far superior than we normally give it credit for. We learnt that the Universe is just 917,065,558,216,642,806,012,884 miles wide and that within just the visible Universe a star blows up every second. We attempted to theorise how matter came into existence, and discovered what a conundrum it presents. We learnt about the four forces and how lucky we are that they are set so precisely. We have learnt how interesting the concept of infinity is, and how important it is to our understanding of everything. We learnt how, within an infinite Universe, our atoms that make us will materialise identically an infinite number of times in the future. We also learnt that the Universe cannot have it two ways – the Universe is either here infinitely – or when it ends, it will be gone infinitely.

It is this last point above which makes me think that the Universe is actually infinite. Just as it is impossible to think of nothing, it is equally difficult to think of the Universe disappearing infinitely.

Your thoughts on this matter would be most welcome.

If you though infinity was hard work – then have a cold shower before reading about time!

THE MYSTERY OF TIME

Where we take a look at the origins of time, and how it could manifest itself throughout the Universe. We investigate the concept of time and look at what it could be. We also look at what its implications are within the Universe and whether it is the same throughout. We touch upon what could be happening within the smallest moment of time and what 'now' is.

The difference between an invention and a discovery

- *In the future we will look back to today's technology and shriek with laughter, just as we do now when we look back at the first brick-like mobile phones.*
- *Is time an invention or a discovery?*
- *Time would have been invented and calibrated differently if our planet's orbit was different.*

What further inventions and discoveries could there be in a million years' time?

The latest wave of technological computing has only been in progress for a few decades, so people of the future are in store for a whole host of intriguing breakthroughs.

Within a generation, people will witness developments that will far exceed their wildest expectations – just as we have experienced over the last few decades. The format of music changed rapidly from vinyl to audio cassette, from audio cassette to compact disc, then from compact disc to digital. Similar advances have taken place in map reading, television, telephony, medicine, genetics, written communication and household appliances.

What inventions and discoveries may we expect from the future?

Are we to remain with current communication and Internet as it is today?

Will the written book survive the test of time?

Will language change to some kind of dolphin-style text speech?

Will our brains one day be coupled with quantum computers to enhance our thought capabilities?

Will country borders have broken down due to technology, such that their boundaries are just a distant historical memory?

No one knows!

The answers to these and many more questions will only be known with the passage of time. For the moment, we will have to be satisfied with present-day knowledge. The knowledge we currently have will act as our foundation for future discoveries and inventions.

Both an invention and a discovery contribute towards our knowledge and capabilities, but there is a fundamental difference between the two, as the following highlights.

Man discovered what types of material floated on water, and from his discoveries invented the boat. Man sailed in his boat he had invented, and discovered a large land mass in the Southern Hemisphere and invented the word Australia for it.

With the passage of time, physical inventions improve. When looking at the inventions we already have, such as toasters, computers, coffee tables, cookers, tin openers, undergarments and music players, we can visualise what these things looked like fifty or more years ago. We can obviously see them now, but to visualise them in fifty years' time is extremely difficult. Progress is always going to hammer away within the present and make us laugh as we look back in time to a point when things were not quite as they are now.

Tinned food was invented in 1810 by Nicolas Cugnot, but was originally opened precariously with a knife because the tin opener had yet to be invented.

When tin openers were finally invented forty-eight years later, thanks to Ezra Warner in 1858, they were well overdue. Over the years they have begun to look more like sophisticated Frederick Roland Emmett-style contraptions that make a meal out of everything!

Once a discovery has been made it will never disappear in today's society – certain discoveries in the past have been lost to Mankind and then rediscovered later. Now it has been discovered, there will always be water on Mars; there will always be water on the Moon; quantum physics will not go away; the structure of DNA will remain unchanged; and quasars will remain as quasars. As time seems to flow in one direction, we will not wake up one day not knowing whether there is water on Mars.

Is time an invention or a discovery?

The question is not as straightforward as you may first think. Initially, you may think that as time has been here for such a long time it could only be a discovery. However, some people have difficulty determining whether time is an invention or a discovery. Just as we invented clocks that calibrated time, time itself as a consequence could be viewed as an invention.

We have become so attached to our twenty-four hours in a day and our three hundred and sixty-five days in a year that we sometimes consider time as a discovery. Time can be viewed as an invention when we take into account that the system that calibrates its passing was invented.

Time would be apportioned very differently if we lived upon another planet with a different period of day and night – we would have invented a totally different system of time. It is also difficult to imagine how we may have developed our concept of time if we lived upon a planet in permanent daylight.

Although time itself seems to allow one moment to proceed to the next, in a world before intelligence there would be no acknowledgement of one moment to the next or one day to the next. If there was no ability for anyone or anything to acknowledge one moment to the next, did time not exist, or did it exist but go unnoticed?

Strangely, without intelligence, there be no need for a reference point to subsequently relate to other moments – perhaps under these circumstances events could take place in time that warps, bends, goes forwards, goes backwards and makes a real mess of the conformity we currently see.

Prior to the invention of the calibration of time, it was still possible to recognise the passage of one moment to another – and the objects moving within it. However, it required the invention of time to make sense of, and make reference to, the discovery of motion!

This comprehensible reference mechanism we call time became better and better as we began to understand more about what precisely it was and where we are in our solar system.

Perhaps time was invented just as a means of calibrating the discovery of movement.

I have lived life telling everyone that time is an invention, not a discovery.

You may have different thoughts?

What exactly is time?

- *Time is unlike anything else we can imagine – it is a unique aspect of our lives to which there is nothing to compare it to.*
- *Without time, nothing in the Universe could change.*
- *When change is witnessed, it is possible to attribute time to this – without change perhaps time would not exist – is it change that creates time?*

How far back and how far forward does time stretch?

When did time itself begin?

We can read the time on a clock, we can judge the time by the angle of the Sun, and we can strategically schedule events. However, no one really knows what time is – it is a complete mystery.

We get familiar with time by the way we mark its passing. It is the wonder that governs our lives most of all, but it remains one of the greatest mysteries in the Universe. Time is unique and unlike anything else – there is nothing else we can compare it to.

We know it is there, we observe it, but we cannot touch it.

So what on earth is it?

What are we measuring?

There are certain observations and philosophical views that help us begin to understand what time could be. In a world without time, we can comprehend that there would be no movement. Therefore matter requires time so that it may move. However, perhaps it could be the movement of matter that creates time.

Time is clearly not a straightforward phenomenon to understand. Looking back at how our views of time have changed over the years, it is obvious that our comprehension of it is improving.

Within ancient Babylon, people were taught about the early kings who had lived for vast ages. The earliest kings reigned for tens of thousands of years. It was said that the longest reigned for more than forty-three thousand years. Each subsequent generation of kings reigned for a shorter time. The ancient Babylon's would study how the planet gods gave birth to other gods, and when the heavens became unstable the planet gods battled each other for supremacy and as a consequence of the ensuing battles, chaos was periodically brought to Earth. Nothing orderly or uniform was taught.

Catastrophic change and continuous change were thought to be a natural occurrence. This applied to the continuous change of time itself. The ancient Babylonians thought time changed speed continually through the seasons. This theory was enforced by the water clocks that they invented to time the length of night and day – used for night-watchmen. They measured their water for these clocks in units of mana and found they needed to pour one-sixth of a mana extra to each half-month to accommodate nights in the winter months. The emptying of the cylindrical clepsydra indicated the end of the night-watch.

How unfair looking back. All those sentry guards who spent the long winter nights thinking it was just the same length as the daylight hours.

They thought that time had speeded up or slowed down!

The ancients believed that their ancestors had lived for eons and in Aristotle's day, the word 'eon' or 'aeon' originally referred to the length of a human lifetime. It was believed that everything had been better in the past. They believed that there had been a golden age when their ancestors had lived immeasurably long lives.

Eventually the Greek philosophers invented assumptions to limit what could change in time. They were then able to invent some logical, mathematical, and natural explanations for the cosmos. This was not straightforward, and they found themselves spending hundreds of years in a long-running debate, seeking principles on which to found a system of natural science.

In just a few hours we can be taught an enormous amount about time, gravity and the motions of planets. We take this information for granted now, but seldom appreciate that it took hundreds of years of debating to accomplish.

Aristotle thought that time was a constant attribute of movement and was unable to exist on its own, but is relative to the motions of other things. He defined time as 'the number of movement in respect of before and after', so it cannot exist without succession. Aristotle also said that for time to exist, it requires the presence of a soul capable of 'numbering' the movement.

Galileo observed a pendulum swinging back and forth. It was possible that he timed the swings by feeling his pulse. He was able to observe that the swings took the same amount of time, or pulses, no matter how far the pendulum moved. He subsequently rolled brass balls down a grooved inclined plank. He measured the distance they moved relative to the pendulum swings. From this he discovered that objects increased their velocity an equal amount during every interval. They also moved a distance that increases with the square of the elapsed time.

Sir Isaac Newton explained that all objects are affected by gravity. He unified astronomy and mechanics by showing that a falling object is reacting to the same force as the orbiting Moon. Newton wrote that, 'Gravity must be caused by an agent acting constantly according to certain laws.' He arrived at these laws of motion because he took 'time' as the independent variable.

Newton wrote, 'Absolute, true and mathematical time, of itself, and from its own nature, flows equably without relation to anything external.' His understanding of space and time successfully predicted the motion of planets and explained how apples fell from trees. Newton's concept of time allowed him to discover the force of gravity.

Newton was fortunate enough to have studied in Western Europe, the only place on earth where clocks ticked, and what is more, they ticked with equally sounding seconds. No other people in history had ever imagined that time always moved at the same speed.

This was revolutionary!

Before Newton's era, Claudius Ptolemaeus 90 – 168 AD, known in English as Ptolemy, had sometimes measured astronomical events in 'equal hours'. These 'equal hours' were simply angles, which just so happened to correspond to time, measured along the equator instead of the ecliptic. He did not imagine that equal hours were really equal chunks of time!

Richard of Wallingford, 1292 – 1336, is famous for having developed the astronomical clock. It was the only clock of its type with a sophisticated mechanism. This was followed over the centuries by more intricate timepieces, clocks, pocket watches and wristwatches.

Then along came Einstein!

At the start of the twentieth century, everyone accepted Newton's gravity. However, Einstein questioned the assumptions about space and time. Within the Universe that Einstein described, each observer has their own clock and their own measuring ruler, but strangely all observers measure the speed of light equally in every reference frame.

Einstein extended Galileo's theories to include relativity of space and time. In 1915, Einstein produced his General Theory of Relativity. In Einstein's thinking, gravity is not a force pulling us towards some distant mass, but rather the evidence of space-time curvature. All straight lines are bent by the presence of massive objects, because space and time are intertwined. Time is slowed down by mass, supposedly because mass warps space-time.

Einstein could explain the acceleration of gravity without the use of an attractive force. Clocks in the vicinity of massive objects slow down. The greater the mass, the more all clocks slow down. Apples fall from trees because clocks run a tiny bit faster on the branches than on the ground. Space-time is much harder to visualize than Newton's gravity. The mathematics of gravity is also much simpler than the mathematics of space-time. No one would think of using complex fields and partial derivatives to compute an orbit, when the assumption of gravity is so much simpler.

Experiments, however, showed that clocks really do slow down in the vicinity of massive objects. Einstein, like Newton, used an operational definition of time. He said, 'Time is what clocks measure.' An operational definition is not concerned with the actuality of time.

Einstein just assumed that time is real because he defined it with clocks. Einstein's system also depends on assumptions. No one has ever directly detected any space-time or even isolated any time. If we deny the existence of time, nothing in the universe would change, therefore we have a right to claim that time is a synthetic idea.

Like Newton's gravity, Einstein's space-time involves perpetual motion without the expenditure of 'energy.' No one has ever detected bends in the vacuum of space, as space-time forces the huge earth to continually accelerate – following the bend in the vacuum.

Within Einstein's way of thinking, gravity cannot be divorced from time. In every attempt to isolate space-time as an entity in its own right, gravity waves or gravitons have ended the attempt – the two cannot be separated.

Can we prove Einstein is right?

We can prove his theory is more accurate if we accept the assumptions and operational definitions upon which his gravity theory depends.

Newton and Einstein's concepts of gravity result in different predictions. Einstein's mathematics correlates more accurately with angular changes in Mercury's orbit better than Newton's gravity. The bending of starlight in the vicinity of the Sun is more accurately predicted by space-time than by gravity. Clocks carried in aircraft or onboard rockets, compared to identical clocks remaining stationary on Earth, show the blue shift and red-shift of time in clocks as Einstein predicted.

How do we know that time has not changed pace throughout history?

How do we know that the gravitational constant is really a constant?

Some scientists believe that a black hole experiences no passage of time. The atoms within a black hole are so tightly packed together under an unusually enormous pressure of gravity that the atoms are unable to vibrate. Because the atoms are static, this prevents time occurring. However, it is extremely difficult to imagine an area of space where time cannot pass; just because atoms have come to a halt does not mean that time must freeze too.

In a world without time there would be no change, as time and change are inextricably linked. We know that when something has changed, time has also passed.

Is it impossible for movement to occur without the passage of time?

An interesting thought is to try to imagine whether change in matter is permitted because time passes, or whether time passes because change happens.

What happens to time at absolute zero?

When water turns to ice at zero degrees centigrade, the atoms within the water molecules are still moving around. You need to freeze things to absolute zero to make the molecules stop. Absolute zero is the coldest temperature the Universe will allow. This is roughly minus two hundred and seventy three degrees below zero centigrade. At absolute zero molecular movement is minimal, but scientists believe that time still passes.

Absolute zero was first theorised in 1665 by Robert Boyle, 1627 – 1691, within a paper he entitled 'New Experiments and Observations Upon Cold'. But it was William Thomson, 1st Baron Kelvin or Lord Kelvin, 1824 – 1907, who developed the true basis of absolute zero, and was able to clarify its existence.

The timescales we use have been engineered for the convenience of people living on Earth. The Earth rotates roughly three hundred and sixty-five times on its axis each time it rotates once around the Sun. The calibration of time we use was invented by our ancestors in order that we may be a little more precise with each other – rather than having to arrange to meet each other by a tree when the Sun is at its highest!

In our world, we observe change and hence attribute the passage of time to this change. The passage of time seems constant even though change happens faster or slower for various events. An eclipse of the Moon only happens once in a while, whereas the Sun rises and sets every day. The second hand on a clock moves every second, but a hummingbird may flap its wings dozens of times a second. This does not mean that different events take place within different bubbles of time. Rather, the Universe possesses a mechanism that keeps time constant amongst objects upon Earth – just that some things happen more often than others.

The correlated Universe

- *Switch on a light bulb and the room will illuminate in an instant, we do not need to wait for hours or weeks.*
- *The Sun seems to have a direct correlation with the same time that we witness.*
- *The first calendars were based upon the Sun and Moon passing over the skies – hence a lunar month.*

Time is often referred to as a dimension; it is a dimension that permits change to occur within the three spatial dimensions. One simple example of this is how a ball passes through the air. The advent of time seems to only be relevant to a particular area of space. It is only when external influences of time impact upon other items, that different items correlate within the same bubble of time. For instance, when you switch on a light bulb, it makes the room bright almost instantaneously, not yesterday or tomorrow!

The instantaneous illumination of light in a room may be obvious, but it is only because the speed of light is fast on a small scale such as a room. Imagine if light travelled more slowly; let us say one metre an hour, roughly the same speed as a slug. With light travelling at this speed, we would have to switch a light on today in order to light up the room tomorrow – interesting concept. However, this is how the Universe operates on a large scale; light from the more distant stars reach us billions of years after the event. On a small scale, there is still a delay, but it goes unnoticed due to the small distances involved.

The first humans would have witnessed very little change, other than the trees swaying in the wind and the movement of the Moon, planets and the Sun in the sky. The most significant change was the difference noticed between day and night. It is more than likely that the cycle of day and night would have been the first acknowledged significant period of time.

Humans would have noticed the Sun rise high in the sky during summer months and lower in winter. Counting the days, they determined that there were three hundred and sixty-five days before the cycle started once again. This period of time was labelled a year. When noticing the different phases of the Moon following a cycle taking roughly twenty-nine days, we labelled this period of time a month.

Early humans were hunter gatherers and moved in tribes across the Earth in search of food. When agriculture settled as the preferred means of producing food, we no longer needed to move from one place to the next to survive. When humans hunted, time was not such an issue, but when humans began to plant crops, it became essential to know the timing of the seasons.

Humans developed calendars, more than likely based on the Moon as a lunar month. The wise men of the tribes became very important, as they understood the secrets of the skies. They were able to predict when the seasons would change, and hence announce when crops should be sewn.

The calendar was born.

The origin of time

- *Time became established within peoples' daily lives four thousand years ago.*
- *Four-day weeks, five-day weeks and six-day weeks have all been used instead of our familiar seven-day weeks.*
- *The seven days in the week was settled upon as a means of representing the celestial bodies we saw moving in the sky – an arbitrary number really!*

The division of time used by us today was originally developed by the Babylonians four thousand years ago. The Sun that appeared to move around the Earth had a cycle of three hundred and sixty-five days, which they divided into parts called months. Each day was divided into twenty-four equal parts called hours. They divided each hour into sixty minutes, and each minute into sixty seconds.

As intelligent beings, the Babylonians saw the practicality of base sixty, and developed the notion of sixty minutes in an hour and sixty seconds in a minute. Sixty was found to possess eleven numbers which factored into it, making it useful for dividing into smaller segments for representing portions of time.

It sort of makes sense, but I am sure it would have been better to have divided it all up into something a little more conducive such as ten hours in the day, with perhaps one hundred minutes in each hour and have one hundred seconds within each minute. This way, perhaps children would not have to do what they call, 'learn to tell the time'. They could convert their ability to count directly to telling the time. We do not often refer to 'learning to read a measuring stick'. As soon as you can count, you can measure millimetres, centimetres and metres, there is no conversion or interpreting required, but reading or telling the time is a different kettle of fish!

Weeks are peculiar things; over the years humans have always had differences of opinion about these spans of time. We currently have seven days in a week. Fairly obvious you may think, but it is not really. There is absolutely no reason for it to be seven, other than it feels right. However, I have noticed that a seven-day period is a suitable cycle for mowing the lawn in the summer months!

In different countries over the years, weeks of differing numbers of days have been used.

The Igbo of Nigeria have a calendar with a four-day week and the Javanese people of Indonesia have a five-day week known as a Pasaran cycle. Amazingly, these are still in use to this day!

Between 1929 and 1931, the USSR changed from the seven-day week to a five-day week with seventy-two weeks making up a year, and five national holidays. Then, in 1931, they changed yet again to a six-day week. Each 6th, 12th, 18th, 24th and 30th of the month was a state rest day. The five national holidays in the earlier five-day week remained the same. However, as some months have thirty-one days, the week after the state rest day of the 30th was seven days long and the extra day was made a working day. Oh, this sounds fun!

There was obviously something fundamentally wrong with this schedule of weeks, as on 26th June 1940 it was abandoned, and the normal seven-day week reintroduced the day after.

The Akan live by a forty-two day week or cycle known as an Adaduanan. It may well have been based upon an earlier implementation of a six-day week which exists to this day in some northern Guan communities, such as the Nchumuru.

The first record of using a seven-day week was in 586 BC, by the Jewish community during the Babylonian Captivity. It has established itself globally as the most widely-used definition of a week. Although I did not know there were so many deviations from a seven-day week throughout the world until researching for this book, nothing surprises me. It seems that when seeking answers to mysteries as to how and why they have established themselves as they are, it sometimes appears that in fact they haven't; there are clearly ambiguities, variations and deviations!

The ancient Etruscans invented an eight-day week, known as the nundinal cycle, and this quickly found its way into the Roman culture around the year 6 BC. As Rome expanded its boundaries, it encountered the seven-day week which. This presented a more acceptable rhythm, and the eight-day week disappeared.

The cycle of seven days simply represented a different day for each of the Sun and Moon, and the five visible planets.

Monday, or 'Moons day', was derived from the Old English 'Monedaei' from around 1000 AD.

Tuesday is named after the Old English god Tiw, who was associated with Mars. Tuesday was derived from 'Tiwesdaeg' which literally means 'Tiw's day'.

Wednesday's origin is from the Middle English 'Wednes dei'. This first came from the Old English 'Wodnesdaeg', meaning the day of the English god Wodan in Anglo-Saxon England. 'Wednesdaeg' is similar to the Old Norse 'Odinsdagr', meaning 'Odin's day', which is an early translation of the Latin 'dies Mercurii', or Mercury's day.

Now we come onto Thursday, which is from the Old English for 'Thunor's day'. Thunor and Thor are derived from the Proto-Germanic god Thunaraz, the god of thunder. The Roman god Jupiter was the god of sky and thunder. In Latin it was known as 'Iovis dies', or 'Jupiter's day'.

Friday, in most countries with a five-day working week, is the last workday before the weekend, and viewed as a celebration, and a more relaxed approach is tolerated as a consequence. Recently, casual clothing has been encouraged in an office environment, with 'dress-down Friday' accompanied by a shorter day. I am sure it was just a managerial trick, picked up on a successful management course entitled, 'Ten tricks to make your staff happier without paying them more', devised when times were hard!

In Saudi Arabia and Iran, Friday is the last day of the weekend and Saturday is the first workday. In other countries, Friday is the first day of the weekend, with Sunday being the first workday. In Bahrain, the United Arab Emirates and Kuwait, Friday was formerly the last day of the weekend, while Saturday was the first workday. In Bahrain on 1st September 2006 this changed to Friday being the first day of the weekend, and Sunday as the beginning of the workdays. Kuwait followed this change exactly one year later. Friday comes from the Old English 'Frigedaeg' meaning the day of 'Frige', a West Germanic translation of the Latin 'dies Veneris', namely 'day of the planet Venus'.

There are no prizes for guessing that Saturday is named after Saturn and Sunday is named after the Sun. The planet Saturn was named after the Roman god of agriculture. It was called 'dies Saturni' and entered Old English as 'Saeternesdaeg'. It gradually evolved into the word 'Saturday'. Sunday was originally 'Sunnandaeg' in Old English, and settled as the word 'Sunday' over the passage of time.

The ancient Balts had a nine-day week, as did the Celts. The Chinese had a ten-day week as far back as the Shang Dynasty, 1200 – 1045 BC. It changed within the Han Dynasty, 306 – 220 BC, when it became law for officials to rest every five days, called 'mu'. It was once again changed back into ten days for the Tang Dynasty, 618 – 907 BC. Ancient Egypt also had a ten-day week, three weeks making up one month, with five days left at the end of the year.

One of the most mysterious of all changes was when France began using a ten-day week for a period of twelve years from 1793 through to 1805.

Well, we can change currency; we can change pretty much anything we want, so just because we can change time, we did.

Synchronising global time

- *The hourglass was used extensively for centuries – many people used it to tell the time, but only when the Sun was shining!*
- *The world was divided into different time zones only in 1884 – until then it appeared to be a free-for-all as to precisely what time it was!*
- *When we understand time better, we may enter an era where a myriad of new technologies emerge.*

The sundial, which simply traces the movement of the Sun across the sky, was the earliest and simplest means of measuring time. The raised stick or rod blocks the sunlight, casting a shadow which marks the surface to show hours and minutes.

One of the earliest known sundials, and arguably the oldest, is found at the Knowth passage tomb in Ireland. It dates back to approximately 5000 BC.

The sundial's major drawback is that it only operates when the Sun is shining, so other ways of telling the time were desperately needed. The human race never seems content with second best, and eventually surpassed all timekeeping expectations by developing the atomic clock.

Quite a move on from a sundial!

The hourglass was used extensively for centuries throughout the shipping industry, religion and cooking. It was the job of a ship's page to keep a range of hourglasses turned throughout the day to provide the correct timings for the ship's log. Hourglasses are still sold in shorter time formats as egg timers, and as a method of distinguishing a predefined play period for various board games.

When international communications and travel increased, it became necessary to establish a common time for areas of the globe. In 1884, an international conference was organised that came to the conclusion that the world should be divided into twenty-four time zones. The astronomical observatory in Greenwich was chosen as the starting point for these time zones, with twelve to the east and twelve to the west.

Scientists have speculated all sorts of reasons for the existence of time. Some even believe that it is possible for time to flow in a backwards direction, although this takes some deep contemplation. We do not have difficulty with time flowing forwards, people are born, grow old and die. We tend to remember the events in the past, but we do not know the future whatsoever.

Nobody has been able to present a reasonable explanation of what time truly is, and how it is generated. We can only speculate at what it is caused by.

Certainly without it we would be in rather a stagnant world!

What exactly causes time may remain a mystery for many years yet. Once we understand time further, perhaps we will enter a new era, with a whole new series of advance technologies. Just like when we first understood electricity, we will yet again develop technologies without which we will have difficulty imagining our lives without.

Let us ponder for a moment whether time and matter are interrelated – they must be linked somehow. For if there is to be matter within the Universe, it must have a 'now' moment to reside or exist within to at least allow electrons to move freely, otherwise everything would come to a complete halt, possibly infinitely. Time may percolate itself purely by the fact that matter exists within the Universe. The movement of electrons throughout the Universe requires a different snapshot of the Universe in which to reside. Therefore, perhaps time originates directly from matter itself, as there does not seem to be anything else around to be oozing it.

So the properties of matter or atoms perhaps contain the essence of time. After all, matter must understand time as it conforms to it. If matter did not understand, witness or conform to time, then similar objects, when dropped, would not fall at the same rate.

We know gravity is extremely weak compared with the other forces. Perhaps gravity is decidedly weak because the same force is used to create the function of time. Maybe the force that creates gravity is also used to act upon all matter to drag it forward in time.

Could the weakness of gravity be the secret of time?

If time is created by gravity, then we should be able to observe a few flaws in time roughly equivalent to the gravity we witness. Just as if you tow a trailer with a car, we will notice the degradation in the performance of the car equal to the trailer's drag. Just as if you take an amount away from something, what you are left with is deficient by the amount you removed. If you take one bun from a pile of a million buns, you are left with nine hundred and ninety-nine thousand, nine hundred and ninety-nine buns. Although only a small difference, it is a difference all the same.

Gravity and time could operate in this way. Time may require almost all of gravity's power to operate – like taking all the buns!

When you ask someone about what makes them confident about whether time exists, they may reply with something along the lines of, "Well, if you put an apple in a fruit bowl it will get older over time and eventually decay."

Is it possible to reverse this and say that time goes backwards?

Therefore, when you observe the apple growing in the first instance from a bud, the nutrients and moisture within the tree create the apple.

On second thoughts, perhaps the decaying of the apple is the bit where time goes backwards for the apple. So perhaps we go forwards in time whilst we are alive and then go backwards in time once we have died.

It certainly appears to me that those dinosaur fossils we see in museums have gone backwards in time not forwards. But on the other hand I am pretty sure I am going forwards in time. But, if a load of river silt fell on me I would go back in time just like those dinosaurs!

Time wreaking havoc in our lives

- *The perception of time intervals between events can seem longer or shorter depending upon other events.*
- *A younger person's hour can seem to take much longer to pass than an older person's hour.*
- *Is it possible for time to go backwards as well as forwards? Could Time speed up and slow down?*

Are there examples of time wreaking havoc in our lives?

I refer to an interesting observation made some years ago within Victorian England; they had breakfast, elevenses, lunch, afternoon tea, dinner and supper with a night cap. I am not sure about you but I struggle with breakfast, lunch and dinner.

Perhaps we are trapped within a vortex of time. Perhaps where we sit within the outer spiral arm of the Milky Way Galaxy is undergoing some type of time shift!

There is interestingly something called the Kappa effect, whereby time intervals between visual events are perceived as relatively longer or shorter depending on the relative spatial positions of the events.

An example of this is when a six-month-old child subconsciously recognises an hour as being approximately the four thousandth hour of their life, whereas a forty year-old will recognise an hour as their three hundred thousandth hour. Hence the child's hour appears to take much longer to pass than the forty year-old's hour. An hour seems much longer to a young child than an old person, even though we know the measure of time is the same.

Perhaps time is perpetually going forwards for some things and backwards in time for others, speeding up for some and slowing down for others. Perhaps the whole Universe has half of its matter progressing forwards, and the other half going backwards through time to form some type of equilibrium.

It may be like shaking a cotton sheet by all four corners and at the same moment – some parts of the sheet are travelling upwards whilst other parts are travelling downwards; momentarily some parts of the sheet could be stationary.

Dissecting a second into an infinite number of moments

- *How does time move on from one moment to the next? Does time tick, or does time run smoothly like a river?*
- *Is it possible for two things to happen absolutely instantaneously, or does one thing always happen before another?*
- *In a running race, does someone always win, or can there really be a draw at the absolute microscopic level?*

Each moment appears to seamlessly move on from one to the next. Does it tick from one moment to the next in discrete steps, or does it flow smoothly like a river?

Would it be possible to dissect a second into an infinite number of separate moments, or is there a predefined fraction of time which cannot be divided any further?

Scientists have united time with an equivalent of Planck's constant. Just to remind you if you are unsure, Planck's constant defines the smallest divisible physical object, based upon observation at the quantum level. Following this discovery, it has been thought that the smallest unit of time can only be related to how long it takes light to travel this tiny distance. This makes sense if you believe that time ticks in discrete steps from one moment to the next, making the smallest moment of time equal to an extremely tiny fraction of a second – a little over zero point, forty-three zeroes, with a five at the end.

With such a tiny moment of time, and everything happening at such a miniscule level, it makes you wonder whether there are many less atoms in the Universe than there are divisible fractions of a second. As this number is so small, it may be possible for all components of all the atoms in the Universe to move independently of each other!

If this is the case, or if a second of time can be divided into an infinite number of separate moments, it is easily possible for everything within the Universe to move independently of everything else. This effectively means that just one component of an atom could be the only thing moving at any one moment of time throughout the whole Universe.

Alternately, if we were able to subdivide a second into an infinite number of separate moments, this implies that everything in the Universe moves independently of everything else. It does not make sense at first glance, but this is the nature of infinity.

If a second can be infinitely subdivided, then it may be subdivided into segments that undoubtedly allow each tiny component of an atom to move independently of everything else in the whole Universe. Once we have experienced one component of an atom's slight movement, we move on one tiny, tiny fraction of a frame, and wait for the movement of the next component of an atom.

Perhaps the speed at which everything can move governs the rate at which time can be observed to pass!

Food for thought!

Perhaps it is an atom on the wingtip of a dragonfly that is next in turn to move, along with the slight movement of an atom falling within a stone on Mars. If both the atom on the dragonfly's wing tip and the atom within the stone wish to move just one Planck's constant of a distance from where they are now to where they need to be, and if we are taking the view that a second is like a river and flows and may be subdivided infinitely, there will be one atom that moves before the other.

By subdividing the Universe into such miniscule time fractions, it may not be possible for two things to happen instantly at any one moment of time. One event will always be triggered before the other throughout the whole Universe. If it is thought that two things commenced at once, you could simply investigate at a greater fraction of a second deeper. One will always be found to have started before the other!

If this is the case, then it will always be possible at the subatomic level to differentiate the winner of a running race, however close the finish. When investigating further into the atoms involved at the winning line, it will be found that one atom moved onto the winning line slightly before the other.

So which is it to be?

Does time move like a flowing river, or does it tick from one moment to the next in discrete steps?

If time ticks from one moment to the next in discrete steps, then this implies that two simultaneous events can occur within the Universe at the same time independently of each other – and therefore ignore their particular forces that they inflict upon one another. Somehow, this does not seem correct, as ignoring the effect that forces produce implies that a component of an atom could defy a force. An atom could defy a force if it were to move at exactly the same time as another atom – after all they are both emitting influencing forces, wherever in the Universe they are.

When two magnets are moving towards each other, they will be perpetually emitting forces. Each magnet will move towards the other a tiny fraction at a time. At all times there are forces being emitted and exchanged between the magnets. It therefore stands to reason that each atom of matter must obey the effect of the currently emitted forces – so when the magnets are moving towards each other, if two atoms move towards each other at exactly the same moment, they will have defied an element of force. It is like two chess players making a move at the same time. This seems counter-intuitive – each atom must therefore move one at a time to prevent the change in forces exerted being ignored by all other atoms.

It makes a great deal of sense that everything moves independently of everything else, wherever that may be.

The analogy with chess is a reasonable one. The players must move one at a time in order to give their opponent the opportunity to consider the next move. It would be ridiculous for two chess players to have their turns simultaneously – maybe this is the same for the movement of matter.

Remember it is the nature of infinity that makes possible this tiny discrete movement of one component of an atom at a time, in fact it makes it impossible not to be the case. Perhaps Olympic officials ought to bear this in mind in future when they declare a draw in a one hundred metre race!

Every point in the Universe ticks at a different rate

- *Particles travelling at different speeds in the same area of space will witness a different time experience.*
- *Time does not pass for a particle travelling at the speed of light.*
- *Vibrations and waves from all the different events around us build up a precise picture.*

We know from Einstein's theories that time is local, relative to a specific location, and ticks or flows independently from anywhere else. This means every point in the Universe is ticking at a different rate, just as something that travels through space at speed experiences a different pace of time too.

This implies that the same piece of space can provide two totally different time experiences for two different objects, especially when they are going at different speeds. Picture the time experienced by relatively stationary particles, similar to ourselves; compare these to particles travelling at the speed of light. Although the two sets of particles may be in the same place, they will experience different rates of time.

This principle may give rise to complications if entangled particle-pairs begin to experience different rates of time. Who knows what implications this phenomenon could produce?

Entangled particle-pairs must keep in contact with each other throughout the Universe and reveal their identity only when questioned. So if each particle-pair finds itself within a different time zone – then the detection of one particle may mean that the knowledge of the other has to be passed into a future moment.

This may be the reason why some of the bizarre quantum effects that we witness exist; namely, allowing an electron to exist in two or more places at once, or an electron being able to interfere with itself.

As an electron zooms around an atom nucleus, it is moving at such pace that it is experiencing a different time within the same space as an object at the level of the large, for example an apple. Experiments have identified that an electron is able to traverse many locations instantaneously before deciding within the same moment to become visible in one of its possible destinations.

It has become well known that electrons can interfere with themselves when in a state of superposition; this implies that the particles really do have the ability to exist in two places at once. The quantum world has proved to be really, really strange!

When travelling at the speed of light, time is not thought to pass!

Coupled with this fact, anything travelling at the speed of light has infinite mass, unless it has no mass initially, which means it will not pick up any extra mass as zero plus zero equals zero – unsurprisingly!

Light is thought to have no mass, and will therefore immediately travel at the speed of light – it does not have anything to influence or interfere with it. When it lands on something, it gives itself up to lighting the object! Its job is now done.

I am not sure about you, but I am forever witnessing the moment I think is 'now' and never seem to witness this for long before the next 'now' is moved into. What can we deduce from this perpetual witnessing of 'now'?

Perhaps this means that all previous events have disappeared, and can never be referenced other than within the context of the aftermath of the event. Whether that is the breeze created by a butterfly's wing flapping in a forest, or the impact of one galaxy colliding with another, the Universe has been shaped by all the events that took place throughout its history. The result of its history is held within the vibrations of all particles positioned throughout the whole Universe. We only have to look at the fossils we find on a beach to see that we can observe the relics of the past still vibrating alongside us. In a way, the whole Universe is one giant fossil, representing all of past time.

The famous physicist Richard Feynman once referred to a fictitious creature that could sit within a swimming pool and analyse the vibrations and waves created, such that it could deduce a picture of all that was going on in the pool.

This is quite likely the case for the Universe, whereby no event of the past is ever lost, as its vibrations and visual effects are carried forward for all time. Albeit these movements may well be integrated in with other objects, but they there all the same. It is not beyond the realm of possibility for the most intelligent of creatures to reverse engineer and untangle this web of events!

Perhaps the unified theory, or the theory of everything, is more intelligent than we originally thought!

The study of time can be relatively interesting. If anything, it may make you realise what a desperately short period of time we each have on Earth. As a consequence, perhaps it makes you realise that you should aim to achieve more in your life. Someone living to eighty years old lives for just over twenty-nine thousand days.

This does not sound too many, so we had all better make the most of what we have got left!

If all that my book does is highlight this fact, then I will at least have accomplished something.

We learnt the difference between an invention and a discovery, and realised that time could well be an invention. We learnt that time allows change to happen, and how movement could actually create time. We learnt why and how Mankind created time. We also learnt how human beings invented time approximately four thousand years ago, and how its accuracy has improved over the centuries. We learnt how some people do not abide by a seven-day week, and that some countries have experimented in recent years with different numbers of days. We learnt how sundials and hourglasses were used for many centuries, and how we ultimately developed the atomic clock. We learnt how the Kappa effect can make an hour seem longer for some people than others – depending upon their age. We learnt what would happen if a second could be divided into an infinite number of moments. We also looked at whether time flows like a river or ticks like a clock. We learnt how the Universe is like is one giant fossil, representing all of past time.

Now that we have a better grasp of time, we should now look at what matter is made of.

The Universe is made of a very peculiar substance!

THE MYSTERY OF THE ATOM

Where we contemplate the size of an atom and just how many of them make up the everyday objects we see. We look closely at how Mankind discovered that matter is made up of atoms which are far too small for us to see. Having identified the characteristics of an atom, we discover exactly how much of even the most static of objects is actually moving. We also look at what could possibly give an atom its mass.

What exactly are atoms?

- *An atom is made up of mostly empty space with an absolutely minute piece of matter.*
- *The underlying substance of all atoms is believed to be the same, but this underlying substance can assume different shapes and sizes.*
- *1661 AD saw the start of Mankind's quest to discover how matter was made up – a new science called chemistry.*

As far as scientists are concerned, atoms make up everything. They make up everything that is solid, gas or liquid which exists within the Universe. It would appear that, as far as matter goes, there is nothing else to consider. Usually, all that anyone gets regarding an insight into the understanding of atoms is a cursory glance over the topic at school. Rather a shame, as absolutely everything we see is made up of atoms which are shaped and governed by the four forces that act upon them, namely: gravity, electromagnetism, nuclear strong and nuclear weak forces.

Atoms are extraordinarily small, and it is extremely difficult for us to comprehend their size. But if you imagine a long line of them sitting side by side, for every centimetre of average matter there would be approximately two hundred million atoms in a straight line.

What is equally amazing is that the majority of an atom is almost entirely made up of empty space, such that its centre called the nucleus is approximately ten thousandth the diameter of the whole atom. This tiny nucleus of an atom is made up of a small number of protons and neutrons, with the number of protons determining the precise chemical element.

Atoms are so small that even a tiny object like a pinhead has been calculated to contain roughly one hundred billion, billion atoms!

It makes you wonder how people have been able to determine so much about something that is so small. It is not as if you can look at something and it immediately becomes clear that there are these absolutely minute atoms tightly bonded together. Everything appears to be so uniform and smooth when we look at it.

The first time that atoms were ever contemplated was within a philosophical context in ancient India and Greece. As far as ancient Greece is concerned, it was in the fifth century BC when Leucippus and his pupil Democritus proposed that all matter was composed of small, indivisible particles called atoms.

Democritus explained different substances and objects in the Universe by explaining that atoms were made of the same material, but could take on an endless amount of shapes and sizes. This was at the time when there was an argument between Heraclitus and Parmenides on the character of 'change'! Heraclitus believed that the nature of all existence was change, whereas Parmenides thought that all change was purely illusion.

Another of the earliest known theories regarding an atom-like concept was derived by an Indian philosopher called Kanada in the second century BC. Theories regarding how atoms combined into more complex objects were put forward, and ultimately further theories about how the behaviour of matter was linked to the nature of atoms.

Quite unsurprisingly, people using just their observational powers thought there were elements relating to air, earth, fire and water. Further progress was not made until philosopher Robert Boyle in 1661 argued that matter was made up of various different combinations of corpuscles or atoms. This was the beginning of the science of chemistry, and he came up with a paper called 'The Sceptical Chymist'.

Discovery of the atom

- *It was discovered that gases began to behave differently under different conditions.*
- *Electrons were being used in the creation of electricity before Joseph John Thomson discovered they existed.*
- *Ernest Rutherford discovered that the atom had a small, charged nucleus at its centre.*

Discovering what was really happening at this tiny subatomic level started with inquisitive individuals such as John Dalton, 1766 – 1844, who in 1801 realised that gases expanded when heated. He also experimented with gases under pressure and investigated colour blindness, but he was most famous for developing the atomic theory. John Dalton came to the conclusion that each chemical element is composed of atoms of a single, unique type. He identified that atoms could not be altered or destroyed by chemical reactions, but could be combined into more complex chemical compounds.

It was at this point in history that the idea of an atom was taken from the philosophical world into the scientific world. John Dalton reached his conclusions by observation, experience and experiment. Whilst studying the proportions by which various chemical elements formed different chemical compounds, he eventually determined that there were atomic weights associated with each element, and even invented chemical symbols.

A significant breakthrough had been made.

Joseph John Thomson, 1856 – 1940, came up with the first modern model of the atom, which was termed the 'plum pudding model', This portrayed a sphere of positive charges mixed together with an equal number of electrons. He has been credited with the discovery of the electron, and was subsequently awarded the Nobel Prize for his work in 1906.

Ernest Rutherford, who was one of Joseph John Thomson's students, gained the Nobel Prize in 1908 for discovering that the atom has a small, charged nucleus. He managed to achieve this by using his famous gold foil experiment, which had the effect of probing the structure of the atom. He fired a beam of alpha particles, produced by the radioactive decay of radium, onto a very thin sheet of gold foil. He then analysed the results using zinc sulphide. A sheet of zinc sulphide surrounding the gold foil acted as the detector, and would light up each time it was hit by an alpha particle.

The results showed that the current 'plum pudding model' could not be right, as all the alpha particles should have been deflected by a number of degrees. Instead it was found that a very low proportion of alpha particles were deflected through angles much larger than ninety degrees.

It had been expected that the alpha particles would have passed straight through the 'plum pudding model' without any deflection. However, what was observed was that a small proportion of the alpha particles were deflected, which was indicative of a concentrated positive charge.

The energetic electron and the pinhead phenomenon

- *Although a pinhead seems inactive and stationary, there is a great deal of energy and activity constantly within it.*
- *It is ironic that we know everything about how a car works except all the matter from which it is made.*
- *The Universe could be a tiny fragment of a much larger structure – a little like a tiny spark within a massive explosion.*

To show how amazingly energetic an electron is, we will attempt to calculate the number of kilometres an electron has spun around an atom in its entire lifetime.

Let us say that an atom has been in existence for thirteen point seven billion years, which is the approximate age of the Universe. If we now multiply this number by the number of seconds in a year and then by the number of times an electron whizzes around the nucleus of an atom in a second, which is an enormous number in itself; then the number of times an electron has spun round a typical atom turns out to be roughly an incredible ten billion, billion, billion or so revolutions.

Now, how is this possible without something driving it to do so?

What is this driving force?

Sadly, there are so many unknowns observed throughout the Universe to which we just do not know the answer. We have no means of determining how to think further about them.

As a pin head has been calculated to contain roughly one hundred billion, billion atoms, and an electron is known to revolve around the nucleus of an atom at one thousand kilometres per second, there is obviously a great deal of energy and activity occurring inside a pinhead. It just does not look like it!

Calculations show that all the electrons combined within a pinhead collectively travel roughly ten thousand billion, billion miles every second.

This fact is absolutely incredible!

When I first worked this out and told my friend we sat and looked at each other in total disbelief, whilst staring at the end of a pinhead. How can all this movement take place inside such a tiny motionless blob?

We had to agree that there is either something wrong with the understanding of an atom, or that this Universe is truly, truly incredible, holding secrets of amazing proportion that humans have difficulty fathoming.

Our total understanding of the atom can be compared with an understanding of a car. Imagine if we discovered a ready-made car in the way that we discovered the atom. Our total understanding of the atom to date would be similar to announcing that a car works because it has a dashboard, with no understanding that it has wheels, seats, axles, an engine and a steering wheel.

We do not know why an atom acts like it does, we do not know why it interacts like it does, and we do not know why it moves like it does. In fact all we know is what an atom can do, whereas we know everything about a car – except all of its atoms!

Atoms are influenced by what we currently believe to be the only four forces within the Universe. It is very doubtful that there are any other forces still yet to make their presence known to us.

What gives every atom in the Universe this energy?

Why are all atoms the same?

Why are there no mutations of atoms amongst the trillions and trillions and trillions within our known Universe?

Can nature be so perfect that there are no mutated atoms?

Why are there mutations within nature but no mutations within physics?

The two do not tie in properly!

If you went back to the early beginnings of the Universe, could you work out what gave atoms their energy, shape, size and existence?

There have been varies speculative theories about the early Universe, one describes atoms as having existed as tiny quarks until the temperature cooled sufficiently, allowing the forces to form. It seems absurd to think that the Universe started from absolutely nothing – then there was the Big Bang.

For there to be the Big Bang without there being something there before seems totally impossible.

The earlier chapter about string theory could shed some light as to how the Universe evolved from nothing.

If the Universe simply burst into life with a Big Bang, then this implies it could be a regular spectacle. It would be incredible to think that the Universe got the Big Bang so perfectly correct the first time it tried it!

It seems to me that the whole Universe we know could be just a tiny spark in amongst a much larger series of explosions, Big Bangs and Multi-universes.

Ah – those headaches are back again!

Should we question the model of the atom?

- *Neils Bohr went out of his way to define the precise composition of an atom, and this view has never changed.*
- *Gravity: we currently battle with the concept of gravitons as an explanation.*
- *When we finally get to grips with the complete story of particle physics, we will be in control of something very powerful.*

Neils Bohr, 1885 – 1962, conscientiously embarked upon a quest to determine and define the precise composition of an atom. After just a few years he came up with his theory and announced his findings. We have stuck with his view ever since.

I find this rather extraordinary.

Perhaps someone should question it.

However, it just so happens that some scientists have increasingly become sceptical of the Bohr concept. The Bohr model explains how electrons orbit the central nucleus of an atom at a fixed distance. This is the model that makes the atom look similar to a tiny solar system.

The model actually gives very good results for the hydrogen atom, but starts breaking down quickly as you increase the atomic number.

Unless you spend a considerable portion of your time attuning your mind to the subatomic quantum world, it is very difficult to visualise this microscopic world in a sensible fashion. It is also extremely difficult to consider everything within quantum theory all at once, within one large, coherent thought – it is much easier to consider isolated aspects of the subject matter as a subset of the whole. It is rather like trying to rub your stomach and pat your head at the same time – very difficult as the two are so unrelated.

Our view of the subatomic world is still very sketchy.

As for how gravity works, we still really do not have a clear understanding, although we know its effects and implications. We currently battle with the concept of gravitons as an explanation.

The graviton is a hypothetical elementary particle that is responsible for the force of gravity; it is defined as an elementary particle because it is understood to have no substructure. If the graviton does exist then it must be without mass, as a gravitational force has an unlimited range. This effectively means that everything in the Universe is having an influence upon everything else – quite a dynamic concept. However, the gravitational force of an intergalactic body billions of miles away will be much, much weaker than one close by.

Just to put your mind at rest regarding precisely what scientists must achieve before the existence of gravitons can be proved one way or the other – they must link gravitons to the curvature of the space-time continuum and calculate the force exerted!

Gravitons are the current preferred theory for the explanation of gravity quite simply because the other three forces known to nature have similar particles that mediate their effects. Electromagnetism is mediated by the photon; the strong nuclear interaction is mediated by the gluon; and the weak nuclear interaction is mediated by the W and Z bosons.

So you would have thought we were nearly there with an explanation of gravity. However, suddenly someone realised that if gravitons did exist there is a problem with a phenomenon termed high-energies. The problem crops up because theory states that nothing can contravene the limits of the Plank scale, and as a consequence infinities arise that it is believed cannot be possible.

These are the difficulties scientists face!

Other theories have been explored, one being string theory, which overcomes the difficulty with gravitons as well as other particles, being described as states of strings rather than point particles. In this way the infinities do not appear and the low-energy behaviour of the particles can still be approximated by the quantum field theory of point particles.

String theory is a plausible explanation of the Universe; at least it seems to work mathematically.

All being well, one day in the future we will begin to get a better understanding of how gravity works within our world of the large. There are no prizes that this will be based upon quantum effects as it becomes clearer that the Newtonian world in which we witness gravity is just a construct of a subset of a much larger, more complex, quantum, physical world.

When the human race gets to grips with the complete array of particles and their precise structure and behaviour, then we will be in control of something much deeper and much more powerful than a light switch!

The Higgs Field

- *Much of physics is far too complicated for the 'normal' person to understand – but explained in the correct way it can all be understood.*
- *Empty space is not empty – it is filled with something that is similar in a way to the force experienced in transparent water.*
- *A large object will resist water and a thin object will cut through the resistance of water; just as different particles have different resistance in the Higgs Field.*

Ernest Rutherford, 1871 – 1937, Nobel Prize winner 1908, and instrumental in the development of the atomic model, said that a good scientist should be able to explain his work to a barmaid. By this he meant that Physicists have an obligation to communicate their findings in a way that can be understood by people uninitiated in physics. It is my intention to explain the Higgs Field in these simple terms too.

We all know that matter is made up of atoms, but very few people really understand the true size and nature of an atom. Atoms are very tiny, very tiny indeed; in fact they are so tiny that a single rain water drop contains roughly two thousand billion, billion atoms of oxygen and four thousand billion, billion atoms of hydrogen. An average one square centimetre of matter similar to a dice contains roughly eight million, billion, billion atoms.

To put this in true perspective, look at the end of one of your fingers, which you can see is roughly one square centimetre. If every atom in the end of your finger was a football, there would be enough footballs to cover the total surface of the Earth, land and sea, roughly twenty miles deep. Another analogy is to take an apple and magnify it up to the size of planet Earth. Each atom will then be roughly the size of the original apple.

The tiny size of an atom is not the only amazing thing about its make-up. The atom consists of a central core that is made of a number of neutrons and protons, depending upon which type of atom it is. The simplest atom is hydrogen, which consists of one proton in the centre of the atom and one electron orbiting around it. Knowing already how small atoms are, now add to this fact that the centre of this atom is one ten thousandth the actual diameter of the atom with the electron orbiting around it.

On this scale, let us imagine we scale up an atom to be the size of a four mile diameter ball, then the only hard substance of the atom would be a golf-ball-sized proton in the centre, and a speck of sand representing the electron orbiting around the golf ball at the edge of the four mile diameter ball. The rest of the atom is totally empty space. Another incredible fact is that the electron has been calculated to be orbiting the central proton fifteen billion times a second!

The problem facing physicists is that the weight of matter just does not make sense, as the atom is made up almost entirely of empty space. Now that scientists understand the make-up of atoms, when calculating the weight of an object there is a massive discrepancy between its actual weight and the weight it should be according to calculations.

Everything appears to be much heavier than it should be.

Something is not revealing itself!

This has driven physicists to frantically try to discover why matter is heavier than it should be. The most plausible theory presented to date is the Higgs Mechanism, developed by Peter Higgs of Edinburgh University. It proposes a background Higgs Field that interacts with matter called the Higgs Condensate.

The theory gets us to imagine that everything weighs nothing. After all, the size of neutrons, protons and electrons within an atom is absolutely negligible. In reality, the atom of the heaviest element consists of virtually nothing. Electrons weigh nothing or virtually nothing or else they would not be able to travel at the speed of light within electrical currents; anything with mass gets heavier the closer it gets to the speed of light.

Now imagine for a moment the Universe is totally filled throughout with a force something a little like transparent water. Just like very clear water it is not visible to us – all forces are not visible to us, but their effects are. If we call this invisible water-like field that is filling the Universe the Higgs Field, then you are part way to understanding how the mechanism could work.

This Higgs Field interacts with components of atoms to give them weight; otherwise they would all fly off at the speed of light just as a totally weightless beam of light does.

The properties of light and electrons are such that they do not interact or become affected by the Higgs Field, but atoms will. More complicated atoms that constitute the likes of lead and iron have more neutrons and protons, and hence interact more with the Higgs Field than lighter atoms, such as those of hydrogen.

Without the Higgs Field, everything would fly off at the speed of light; atoms that make up everything, including us, really should accelerate to the speed of light, but as they interact with the Higgs Field they are unable to. The more interaction an atom or particle has with the Higgs Field, the heavier it becomes. Light would appear not to have any mass, and therefore it makes sense that it is able to totally ignore the Higgs field pretty much as if it is not there.

On the other hand, when looking at lead or iron, their molecules are very much influenced by the Higgs Field. Lead and iron appear to be much more aware of the Higgs Field, and, as a consequence, take on more mass. The analogy here with our water-like Higgs Field is that we have an oar that obviously resists the water; compare this with a coat hanger that will glide through the water with much less resistance.

With this in mind, who knows, we could even start looking at the Higgs Field as being the source of gravity. Just as the flow of water currents will carry plastic bags, the Higgs Field could be influenced by the curvature of space-time by a celestial body such as a planet. This could perhaps produce a force similar to a current within water, attracting atoms towards the planet's centre by placing constant directional pressure upon them.

We learnt about what the ancients thought about the composition of matter and how philosopher Robert Boyle in 1661 made a significant breakthrough. We learnt how Mankind performed a series of experiments to ultimately discover what particles looked like, but we are some way from understanding everything about them. We learnt how Neils Bohr, 1885 – 1962, consciously embarked upon a quest to determine the precise composition of an atom – he came up with his theory and announced his findings – we have stuck with his view ever since. We learnt how the Higgs field operates and adds mass to almost weightless objects.

MORE FASCINATING, ASSORTED, UNIVERSAL MATTERS

Where we investigate closely some of the Universe's deepest mysteries, speculating at what possibly could be the root causes of them. We investigate fire, existence, eternal life, the strangeness of the Universe around us, and how everything could be imaginary. We learn how Einstein has given each of us a free reign to think openly.

What exactly is fire?

- *Certain atoms will connect to others as soon as they get close enough to each other to do so – this bonding of atoms is the basis of fire.*
- *Gases and solids reacting together cause a wide spectrum of light waves which we see as a flame.*
- *Zero gravity combustion is extremely efficient – this has applications for the future generation of energy.*

Fire is thought to have been possible from only a little under five hundred million years ago, when the oxygen level within the atmosphere was sufficient to allow the burning process. Fossil records proving the existence of fire go back only this far, meaning that fire has been widespread as a natural phenomenon for only an eighth of the lifetime of the Earth.

Of course, fire was possible all those years ago under the right conditions, but it is thought not on anything like the scale it is today due to the lack of fuel and oxygen. If you think about it, the early creatures on Earth did not have much sludge to feed on as soil and mud is organic. Early creatures must have been extremely resourceful, utilising every morsel of organic debris they could find until it became widespread. Once widespread, it could be utilised as fuel for fire.

Atoms jiggle around freely as we well know – and different types of atoms are attracted to others to different degrees. When an oxygen atom gets very close to a carbon atom – they would ideally like to reside together. When they get close enough to one another they snap together. It is a little bit like a golf ball getting the correct trajectory and then having the legs to reach the hole and popping in, then jiggling around. This jiggling around creates heat which entices more jiggling around to take place, making it more likely for other oxygen atoms to snap into carbon atoms – creating more snapping in of atoms and more heat, and so it goes on.

So you can imagine that oxygen atoms would always like to snap into carbon atoms, but that at cold temperatures this cannot happen, as they do not jiggle around enough to get close enough to snap in. By applying some heat to an area of wood, for example by friction, to make the atoms jiggle more, this sets off the chain reaction that causes millions of oxygen atoms to snap into the carbon atoms, thus creating fire.

The flame is a mixture of reacting gases and solids that emit both visible and infrared light. In the case of burning organic matter such as wood, glowing soot particles produce the orange glow of fire.

Photons of light are emitted by de-excited atoms and molecules within the gases which are produced within the visible and infrared spectrum bands.

Gravity plays a major part within the processes of fire. Altering the strength of gravity causes different flame types, as convection makes soot rise to the top of a general flame on Earth. In zero gravity such as outer space, convection can no longer operate, and the flame becomes spherical. As a consequence, the flame burns bluer and more efficiently as the temperature is more evenly distributed. Soot does not form, and complete combustion occurs.

If someone could produce a momentary zero gravity moment within a combustion engine, utilising some clever mechanism similar to the weightlessness experienced on a rollercoaster, then better fuel efficiency could be achieved.

Is life a Universe-wide phenomenon?

- *Unstable living environments mean that life-forms must be flexible and possess the ability to adjust rapidly to improve their chances of survival as a species.*
- *Signs of human evolution are all around the world: prehistoric tools; ancient city walls; old Roman roads; early galleon ships.*
- *One day in the future we may have the ability to perform feats on a scale unimaginable compared to today's standards.*

It seems peculiar to think that the Earth is the only planet with life throughout the whole Universe. It is thought by many to be out there, but until we can categorically make a concrete detection, it can only officially remain speculation. Many people have good reason to instinctively feel that life is a universe-wide phenomenon – if the laws of physics are constant throughout the whole Universe, then it is not beyond the bounds of possibility that several of the billions of stars within the billions of galaxies nurture life. Places which are not too dissimilar to Earth – similar in temperature, climate and chemical make-up.

The same universal forces which act upon Earth will act upon those distant atoms and molecules in exactly the same way. The atoms will be organised in a similar way by the likes of gravity such that they too will evolve into mitochondria, micro-organisms, bacteria, simple life-forms and ultimately beasts like us!

Simple life-forms develop to become more sophisticated – quite possibly as a means of improving survival. Added sophistication will safeguard life-forms from the threat that harsh environments will present. In addition, the life-forms must be wary of simultaneously evolving predators that can outsmart them. The successive millions of generations interact and adapt in synchronization with their prevailing environment, their position within the food chain, and the natural forces of nature.

The pure ingenuity of evolution will ultimately place a species to the top of their food chain. When this happens, it would seem more time is available for relaxation, reflection and thought. More time spent on this type of activity, rather than perpetually looking out for danger, will encourage experimentation, curiosity and ultimately intelligence.

The tell-tale signs of our evolution are all around us: the ancient cities, the development of tools and our ever-improving communication mechanisms. We are now beginning to utilise innovation in tandem with our evolution; technical evolution and biological understanding may one day take us closer to fully understanding all the unified forces of the Universe and nature.

One plausible reason why life exists could be linked to a wider cosmic strategy which is currently unclear to us. Possibly planet Earth evolved the way it has so it would eventually harbour intelligent life. Was this the initial goal of our Sun?

It could well be a natural, cosmic, survival tactic for a solar system. If you consider the concept that all creatures alive are successful, we as a race are very successful. Our desire to survive as a species may one day force us to protect our Sun from cosmic catastrophe. In the future we may have techniques to deflect large celestial bodies to prevent cosmic disasters. This would make the Sun more successful than other stars.

Perhaps the Sun set about nurturing intelligent life to divert a possible disaster. We are the beings that will perhaps one day protect the Sun if it were threatened in this way.

Perhaps the Sun desperately wants us to fully understand quantum physics to a point where we can deploy it for cosmic benefit, hence benefiting the Sun greatly in the process. We could well be the arms and legs of our Sun.

The Sun nurtured us into this existence. We owe it plenty!

Perhaps the Sun is a living entity with the ability to think, organise and nurture. It may subconsciously communicate with us on an ongoing basis, nurturing us perfectly. I cannot think of a better object for the job.

Your brain is seven or so inches across; the Sun is nearly one million miles across, and if the Sun is not a type of living entity then I will be very surprised indeed. We should perhaps try working out how to communicate with the Sun. Remember that Albert Einstein said we should think differently.

I think trying to communicate with the Sun is different enough!

We seem to think we are here to find the perfect partner as quickly as we can, reproduce and die happily.

Perhaps there is more to our presence than this!

There must be a wider strategy lurking somewhere, but we have just not spotted it yet.

The wider strategy of our Universe is unfolding before us, but our ability to spot the strategy is somewhat clouded by our self-centred viewpoint. We wonder why we exist when we should really be trying to establish whether our existence is part of the Universe's overall strategy. Before we could exist there needed to be stars, supernova, planets and water. Just like there is a food chain here on planet Earth, there is a hierarchical chain of evolution and events that had to happen before we could be created. Now that the Universe has created a being with the ability to manipulate and express feeling, this is perhaps the next logical step on towards a greater objective not yet known to us.

We can only speculate at what the greater objective could be. Perhaps we are the first life-form in the Universe that has achieved the ability to manipulate DNA. Perhaps the next step for Mankind is to begin programming the creation of creatures from DNA. Whilst we do this DNA programming and are thinking we are clever, it is in actual fact the Universe's clever strategy that is actually prompting us to do this – part of a much wider strategy.

Imagine what secrets of life we will discover within DNA over the next billion or so years!

It is not beyond the realms of possibility that we could manipulate DNA to produce a creature that can survive as dormant as a spore in outer space. We may find that we can reduce the DNA make-up of a human being to the size of a speck of dust and send billions of these spiralling into outer space to hopefully colonise a planet one day. Upon contact with a habitable planet it self-replicates to a point where it then evolves over the passage of time to become a human being just like us again!

Does this sound familiar?

Earth's very own eternal creature – turritopsis nutricula

- *Species must learn about their environment and pass on their knowledge and experience to the next generation.*
- *Animals die after having reared offspring, but there is one creature, the turritopsis nutricula which manages to revert back to a polyp.*
- *Sexual reproduction and death in old age allows a species as a whole to adapt to changes in the environment.*

A life that is bound by time is a life that requires direction and meaning from the outset. It is not possible within a short period of life to develop language, or develop survival tactics. These things must be taught to us by our parents, or inherent within us from birth.

Eternal life has within it fundamental flaws for the overall survival of a species. Consider a species that evolves and then one day suddenly becomes eternal. This creature's food source will be limited, but the creature's reproduction continues. There will now be an ever-increasing population fighting over this limited food source, which surely makes no natural sense. If we look upon it another way, whereby the reproduction of offspring is no longer a function of a species, then extinction will be inevitable as each creature ultimately suffers an unforgiving death due to food shortages, landslides, meteorite collisions and generally falling under that number thirty-two bus!

Looking at it from both these angles, it would appear that irrespective of whether eternal existence is a possibility or not, eternal existence does not promote the survival of a species. In fact, eternal existence would mean the demise of a species, unless the species is large enough to combat any threat by sizeable, interstellar impacts.

Subsequent generations, evolution and diversification of a species would appear essential for the survival of any species. Subsequent generations of a species are able to adapt to a change in environment. If generations are not frequent enough, perhaps extinction looms as environmental conditions cannot be catered for adequately. It would also appear that creature diversification and the survival of the fittest are crucial to ensure the continued existence of a species. Having now identified how reproductive species survive over the passage of generations in this way, this implies that a 'species' is alive, not the individual within that species. The individuals are only a requirement as a survival tactic of the overall species.

Somehow, at the back of my mind ... this is telling me that the survival of a species is more important than the survival of an individual ... this in turn implies that perhaps the Universe has a purpose for us within it at some stage in the future which is not evident to us yet.

Just because a species pops into existence does not mean that it has to reproduce and survive as a species. There is no obvious compelling reason why a species has to survive at all ... unless there is something miraculously encouraging it ... this reason is hidden deep in the fabric of the Universe.

Complacency may have contributed towards the evolution of human beings and their inability to possess eternal life. When early humanoids discover after forty years that they are still living within a bush by the river bank, any further existence within this environment would be as futile. At this point it is best left to another new member of the species to have the 'discovery' experience. The human being's life is about experiences – once those experiences have been satisfied, and one is complicit with the familiar environment, then why should one continue to live on further?

Imagine you are sitting by the river bank ten million years ago. Life consists of seeing the trees rustling in the wind, following the dropped leaves on the river as it trickles by, watching the Sun and Moon rise and set. After a while you know it all – what more is there to learn?

Average human life expectancy over the last two hundred years has increased dramatically. Perhaps this is due to life being more interesting, and it takes longer for complacency to set in. The underlying, official reason for greater longevity is attributed to improved hygiene and a more varied, nutritious diet.

Humans will definitely benefit from looking closely at nature to see what type of lifestyle lends itself to a much longer lifespan. Quite miraculously, there is the bowhead whale that is thought to live for over two hundred years, judging by a stone arrowhead found within the blubber of one caught once. There are turtles that live to two hundred and fifty years, and then there is the animal world's longevity winner which is the quahog clam, found off the coast of Iceland, that spans a whopping four hundred years. As far as the plant world goes, there is the bristlecone pine, one of which was dated to be five thousand years old according to its rings.

Compare this with the pygmy goby fish that lives on the Great Barrier Reef, which is the world's shortest-living, backboned creature. Its maximum existence is all of two months. During its record-breaking, short lifetime, it spends three weeks as a larva before settling on the reef and becoming mature within two weeks. Then there is an adult period that lasts just three and a half weeks. But the wooden spoon has to belong to the mayfly, which lasts just thirty minutes to one day, depending upon the species. That is an incredible eighty million times shorter than the bristlecone pine!

With lifespan differences like these rubbing shoulders with us on the planet, someone must surely soon determine the elixir of life and impregnate humans!

There may be a natural trigger which ultimately drives a species towards eternal living. It may kick in after a few billion years of evolution, perhaps just as an option; you can never underestimate the power of nature!

How wrong can you be! Surprise, surprise, nature has done it again, we are already there!

Jellyfish usually die after mating, but the turritopsis nutricula manages to revert back to a polyp state. This ability to revert back to its juvenile state is the only one known in the animal kingdom. This mechanism allows the turritopsis nutricula to bypass death to become the only immortal creature on the planet known to Mankind. Studies have shown that one hundred per cent of turritopsis nutricula revert back to the polyp stage to start their lives all over again. These creatures should be worshipped!

If eternal living is a natural wonder, there is no reason why Mankind should not be able to investigate the turritopsis nutricula's secrets and advance the effects within ourselves with the alteration of DNA, although somehow I doubt that we could revert back to being babies in the womb!

Detailed research is being carried out into why the quahog clam should live for so long; until this research is complete we will have to make do with living until our bodies have had enough.

Could the first creatures on Earth have been eternal?

Once the atoms making up these early creatures had sprung into life, there would be no way to reproduce immediately. The reproduction mechanism must have developed later. It is impossible to think that a conscious creature which can reproduce can develop instantaneously by chance. There must have been a transition stage between cells that can split to reproduce, to when a creature evolved with reproductive capabilities.

The turritopsis nutricula will turn itself into a blob-like cyst which then becomes a polyp colony. This polyp colony is the first stage of the life of a jellyfish. What is most incredible is that the muscle cells can become other cells such as nerve cells or even sperm cells. Similar to cancer cells, some cells within the turritopsis nutricula that were by all account to die, are able to switch off some genes and switch others on. This process reactivates genetic programs that are normally used in the earlier stages of a lifecycle.

This method of asexual reproduction is astounding – as far as Mankind is aware, no other creature can achieve this immortal capability. Although the creature was known about for many decades, its reproductive life remained a secret until the mid-1990's.

It is thought that the turritopsis nutricula is also thriving. One of its latest means of transporting itself around the globe and proliferating is by hitching rides inside long-distance cargo ships. Water is pumped into the ships to provide stability which is dumped upon arrival at its destination. This would explain why the research into the species has discovered identical DNA in sample species from all over the world that could not have arrived by the normal oceanic currents.

The turritopsis nutricula is also mysterious for developing into slightly different forms as it spreads around the globe. Despite being the same species, swarms that are resident in tropical waters have only eight tentacles, while others in temperate waters can develop twenty-four or more tentacles.

The turritopsis nutricula could be a relic from the past when reproduction had not evolved. Reproduction is better as a survival tactic that eternal living.

With the sophistication of the human body being as it is, if we were to have become eternal creatures we would have definitely evolved that way. It is not too far-fetched a concept. I think that our reproductive mechanism is far more sophisticated than eternal living.

Therefore we may have evolved to deliberately have an aging process. It seems at first glance, to be rather a poor mechanism for an intelligent creature, but the concept of living forever may be too daunting and, in the long run, one's termination could be a happier outcome.

Living forever is quite a daunting concept if you think deeply about it. If you think back to times before libraries and interesting hobbies cropped up on planet Earth, all we did was sit by a river bank catching fish or hunting woolly mammoths. These were our only types of pastimes. Once we were familiar with these sorts of events we would let our offspring take the reins and pop off.

Here is the interesting part. We are now living longer. This could perhaps be because life is so much more interesting and there are many more activities to occupy your mind! So in a very short period of time the human race has evolved to live much longer as there is so much more to do in life.

Happy, contented people live longer than negative, dull people quite simply because they have so much more to look forward to. I suggest, therefore, that your body adapts to the enthusiastic nature of one's mind. So to live long and prosper, engineer your life to be as happy as possible. Quite simple really!

Living forever

- *If someone was brought up in total isolation – would they ever know that they were to die one day?*
- *If you were to live forever, your list of ambitions could be as long as you like – no need to hurry either!*
- *To make the most of living forever, planning would be essential – people would need to map out their time very carefully indeed.*

By the way that some people live their lives you would think they believe they already live forever; some people seem to reach a plateau of learning and become complacent. Other people spend their whole lives eager for more knowledge to complement what they already know.

If someone was brought up by a pack of wolves, how would they ever know they were to die one day?

Unless someone is told this fact, how would they ever know, other than witnessing other creatures passing away?

You could imagine there would be no rush to do anything; everything could take time; your list of things to do could be as long as you wish. So what would the agenda for each day comprise?

Learn to play the violin, read the works of Shakespeare, invent a spacecraft to take you to the Moon, develop quantum physics further and study the life and times of Charles Darwin.

The major problem would be the relentless days that perpetually came and went. Unless you could keep your mind occupied with interesting and compelling thoughts, then life would become rather hard work. When prisoners are placed into small confined cells with nothing to read or do, then insanity soon creeps in. Perhaps after a few billion years the Universe would become so familiar that it begins to present no extra-compelling stimulation. The Universe then acts very much like a prison cell. At this point, even the passage of trillions and trillions of years would be totally insignificant!

There would be enormous problems with drive and enthusiasm once we had satisfied our minds on our favourite topics. Based on the amount of boredom that creeps into people's lives at present, it is clear that living forever may only be for a few dedicated monks, and for those who withstand boredom for extremely long periods of time.

If people were to live forever, they would need to approach life totally differently, seriously mapping out the millennia to come. They would need to identify clear objectives to reach, otherwise life could be unbearable. So given the choice, perhaps we are better off as we are now with finite lives. This just leaves us with the task of fitting into our lives as much as we can to satisfy ourselves within the 'here and now'.

Quantum physics is just the surface of it

- *By trial and error, scientists come across theories that are accurate just by chance.*
- *Perhaps the substance in the Universe was just pure energy at some time in the past.*
- *Perhaps we live in a large network of universes in which our Universe is just one of many.*

Einstein gave us free rein to think openly when he said, "Quantum Physics is just the surface of it." This means we each have the licence to dive whichever way we feel within our own known Universe, to develop our preferred answers from any angle we wish.

Scientists in the past have been trying hard to make sense of the Universe by endlessly producing chalkboards full of formulae. Although these formulae are important, there is a limit to how much they help in breaking new scientific ground. Philosophical thought as a theoretical physicist is perhaps a more fruitful way forward – producing, "What if?" questions and thinking much more dynamically than before.

Let us look at what we know and decide whether everything we have learnt makes sense – are the latest discoveries something we could have guessed?

Cells, atoms, galaxies, chemical reactions, DNA, electromagnetism, space travel and wireless communication are a fair example of recent breakthroughs in knowledge. These are all puzzles to which we eventually found an answer. If anyone had been quizzed about these matters circa 1800, they would not have had the faintest idea. However, they could have had a guess, and they may have been right!

Having a go is what it is all about!

By thinking deeply, people may suddenly stumble across something that is clearly correct. There has been no better time to do this than the present, as the amount of information available to us is greater than ever before. If we were to imagine our knowledge of the Universe is like a large jigsaw puzzle, we would perhaps be part way to having filled the edge pieces. We seem to know with what we are dealing. We are making significant headway as a species – just compare our knowledge to that of an ant colony. Ants will most likely see the Sun and Moon, but will not deduce much from them, and certainly do not have any chalkboards upon which to scribble formulae upon!

I will hark back to these chalkboards once again, as I am sure that the best groundbreaking work comes from our imagination, not the chalkboard. The chalkboard is useful once an observation or thought of note has been achieved. For example, when a beam of light is shone against a wall, scientists can produce all sorts of complex equations to explain why the light hits the wall. When a scientist observes gravity, an apple is seen hitting the ground. Only now they can retrospectively create wonderful formulae for what they saw.

It is the act of seeing, witnessing and imagining which is the exciting part. We can all do this!

So a great deal of scientists to date have been purely observing the Universe and then subsequently creating formulae to explain why the observations happen. However, our thought needs to be radically different. To become truly constructive, we need to look at things in a different sort of way.

Einstein said that he thought best in music. He was very fond of the fiddle, so musical images or wave images in our minds are perhaps best, rather than wasting endless amounts of chalk by grating it on a blackboard!

So because Einstein said, "Quantum Physics is just the surface of it," we are all now at liberty to guess the next stage. Could it be that there is a direct invisible connection from our brains to the heart of the Sun, an invisible quantum field that acts as our conscious, another dimension that stores our memories, a connection from ourselves through to other parts of the Universe, a quantum field of life?

Or, perhaps our synapses have a direct link to a hidden quantum world?

It is not beyond the realms of possibility that everything we see in the Universe existed, at one time, as pure energy.

So without wasting any chalk, let us mull this one over. The major ingredients for a good Universe seem to be atoms, energy and a large area in which to play. The atoms all came into existence from a point in time which we have difficulty visualising. Perhaps some had already been here previously, but were squashed into a tiny area. I would have thought under such pressure atoms would have been reformulated into something else.

One interesting question to contemplate is whether at the moment of creation of the Universe, were anything other than quarks and electrons formed?

These seem to formulate all substance within our known Universe. All this matter cannot just appear from nowhere, so prior to this it was either a lump of energy or a combination of energy and matter. It all seems to have blown outwards in some kind of super, massive, universal supernova.

No one knows what lies beyond the visible edge of the Universe, but just because we cannot see it does not mean nothing is there. Take a creature at the base of the ocean near to a hypothermic vent; they look up into their dark abyss of a sky and cannot contemplate anything being there. One day, when food gets scarce, they send an expedition upwards. They adjust to the pressure and find that there are all sorts of things up near the surface, when previously the edge of their Universe was a black abyss of water which turned out to be just a few miles thick.

It may well be that the edge of our visible Universe is just the beginning of a much, much larger network of universes, and all that we know is equivalent to a grain of sand within a much bigger, never-ending, super system of universes.

There is nothing to suggest that this is not the case; it is speculative, either "yes" or "no". But then I would not know how to define the edge if I went with "no" so I will have to plump for "yes".

Just because we cannot see something does not mean it is not there. As to speculate what is there, we can only imagine it is more of the same. How far can it go on?

Can it go on forever?

No reason why not, this is after all the nature of infinity.

Never forget, "Quantum Physics is just the surface of it."

Out of sight does not mean it does not exist

- *If there is no one to define the edge and apply rules to something which appears to have no edge – then is there an edge to it?*
- *There is no reason why the Universe could not be infinite and have been here for eternity.*
- *Very little is impossible in an infinite Universe.*

Just because we cannot see any finer detail other than clusters of human cells, does not mean there is nothing going on at this minute level!

There are tiny mites that crawl all over us which we cannot see with our eyes, and some of these mites have their own mites too!

There are groups of humans on our wonderful planet near our Sun. Our Sun is just one of a cluster of stars within our galaxy, within a cluster of galaxies. Our galaxy is just one of a cluster of galaxies within a super-cluster of galaxies. All the super-clusters of galaxies together make our Universe, but this does not mean it stops here!

Imagine there are trillions of these levels for a moment, clusters of universes, within super-clusters of universes until we have one whopping super trillion-deep mega-cluster of super-deep universes; does it end here?

Possibly not – who knows?

What defines the edge of the Universe?

Also, what is capable of applying rules to the edge of something which appears to have no edge?

Perhaps there is no edge because the Universe is infinite and has always been here. Perhaps the shape of the Universe we imagine is all wrong – if we could shrink the Universe to fit into the palm of our hand, what would it look like?

On this small scale, what would exist a mile away?

Or, are we being naive about the shape of space and there would be no such thing as a mile away?

Is there some kind of peculiar shape with no edges that feeds itself back into itself again?

Let us imagine a room within which we have a scale model of the whole Universe. Let us imagine this model of the whole Universe is just the size of a football. Now there are just three possibilities available to us. Firstly, the football-sized Universe is in the centre of the room, and by some unknown, miraculous concept the rest of the room does not exist. Secondly, the football sits neatly in the centre of the room with empty space all about it. Thirdly, the football sits neatly somewhere within the room, with other footballs all around it.

Which one of these is correct?

For what it is worth, my money would be on the 'many footballs in the room' situation.

In addition to this, there is no doubt some kind of universal oscillation going on, birth and rebirth, but that is it, unless someone knows something I have yet to realise.

In addition to this, as we know it is very difficult to think of absolutely nothing. So this implies that the Universe has always been here, and hence could reach an enormous size. If it is not of infinite size, this always begs the question, "What is outside of the Universe?"

Very little is impossible within the Universe, it is so large that pretty much all permutations of events are bound to have happened somewhere in the depths of the Universe.

It has puzzled many people that the Universe is only roughly thirteen point seven billion years old. The number of years quoted here is very, very small in comparison with the amount of stars that have formed, and what the Universe has achieved in terms of life, matter, reproduction and enormity. The living creatures that have evolved so perfectly after this short period of time are astounding; there must have been a few dress rehearsals before!

There is a view you may have heard a few people mention from time to time about an all-pervading omnipresent being that presides over the Universe. Is this being capable of orchestrating and manipulating everything in this world within its mind?

Could it be some type of eternal super-creature, conjuring up all that we witness?

An eternal super-creature out there in the Universe somewhere – possibly self-repairing with no need to reproduce due to its biologically eternal capabilities.

How plausible is all this?

Are we just in the mind of a super-creature?

- *The Universe is so peculiar that it would be no surprise if one day we discovered we were all just in the mind of an eternal super-creature.*
- *The super-creature could be enormous and reside throughout massive areas of the Universe – or even be the Universe!*
- *The probability of an eternal super-creature evolving somewhere in the Universe at some time is quite high.*

Are we just in the mind of a super-creature?

Is it conceivable that an enormous eternal super-creature could have evolved which spans, controls and rules a sizeable portion of the cosmos, occupies a small portion of it, or could even occupy all of it.

If we are to discount our true presence here on Earth, due to the absolutely miniscule probability of our existence ever coming into being, then we could adopt the theory that we could all be a figment of imagination residing within the mind of this eternal super-creature. This is perhaps not as far-fetched as you may think, although it would be verging on the strangest thing known to us. However, it would appear to simplify our possibility of existence.

Let us imagine that within the Universe we have a single creature that evolved throughout the eons of time, and has developed the ability to sustain its life eternally. This is definitely not too far-fetched, because it will have the resources of the Universe at its disposal, will more than likely be super-intelligent – will therefore simply rejuvenate parts as necessary, rather than develop complex reproductive organs to create offspring. Over the passage of its infinite life, it will become adept at manipulating the matter within the Universe for its own pleasure and gain.

It would have evolved to become impregnable, possibly due to its large size, making it impossible to be damaged by any huge impact such as stars, comets and planets as they decide to pass through it. Its size could be simply enormous, and manifest itself throughout massive swathes of the Universe. We human beings should not think too insularly about the possibilities of what may be lurking out there within the depths of the Universe.

We must have open minds!

This eternal beast of a conscious mind will have overcome all survival tactics, becoming effectively unassailable. With eternal consciousness, this creature's experiences may evolve into numerous parallel mental experiences, creating isolated packets of existence within matter which it has managed to suitably manipulate.

The acceptance of infinity does, of course, increase the likelihood of very odd events happening. However, it does not necessarily truly guarantee them. So with this in mind we are coming here from an angle whereby this super-creature, due to the nature of infinity, is much more likely to have come into existence than you. Its probability of existence is possibly something like fifty-fifty, purely a flip of a coin, and yours is much less probable, as we have calculated in this book earlier.

Now this creature with a very high probability of existing is here, what does it then do?

What it perhaps does is to kindly create you within its mind so it may then have fun outside its normally featureless life; it plans massive universal adventures and plays with the iridescence of the Universe. Therefore, everything within this world we know could be completely fabricated, and have materialised from a completely different creational cause than anyone could have possibly envisaged in their wildest dreams.

All that we have essentially done here is simplify the unimaginable odds of our true existence by comparing them with something which is much more likely.

The odds now look like this.

To have an eternal super-creature pop into existence somewhere within the Universe at some time is quite likely, if not a certainty – so at one time it will happen. Therefore its probability of existence is one.

Once evolved, it would be fruitless not to evolve itself further, become super-intelligent and quite sizeable for reasons of survival. This creature's chance of existence seems to be more probable and logical than the evolution of short-lived creatures with the most complex reproductive mechanisms. We ourselves know what the power of our thought can conjure within our short lives, such as imaginary people to play with when we were younger. So imagine the power of thought that this eternal super-creature could muster. It just closes any eyes it may have to the Universe, conjures up a world that it wishes and creates little creatures to enjoy it.

Having been around for eternity, it has definitely had plenty of time to develop a few good ideas and devise plans to execute.

Hey presto, here we are.

Quite simple really!

If all of this is true, then it could account for the all-pervading omnipresent existence of a higher level entity which manifests itself into beliefs and the various gods that religion has pronounced.

It could also explain why we unable to make much sense of the Universe around us and why there are so many unanswered questions.

This does not resolve the age-old question of where the all-pervading omnipresent creature came from in the first instance.

Making sense of the Universe is at best difficult and is compounded by the fact that we have only five senses, which are not too reliable and not a truly accurate depiction of the world around us. This, coupled with the seemingly wishy-washy dreamlike matter that occupies the Universe, makes deciphering everything very difficult indeed.

By contemplating the possibility that we are in the mind of a super-creature, I hope we are not taking a retrograde step back to a time when there were myths of dragons, ghouls and monsters which were depicted on the edges of maps.

Could they have actually been right?

Nothing is impossible

- *There is plenty of energy for a super-creature to consume throughout the Universe – it could even be made from pure energy!*
- *There are regions of space where we do not understand the physics of what we see – a gravitational field so powerful that nothing can escape!*
- *Different areas of the Universe may be very unstable and make human habitation impossible.*

Before worrying too much about the potential existence of our eternal super-creature, let us take a quick look at the type of environment which might be required for its development. It would be too large to inhabit a planet and possibly just the right size for a universe. It could well feed upon energy from stars and other celestial bodies, or perhaps even 'be' them!

If it was made from parts of the Universe, this could explain why we have difficulty working out what many of these celestial bodies are. If dark matter and dark energy constitute part of this creature, this would be acceptable, as no one on Earth has ever made sense of it. Dark matter and dark energy have never been seen. This mysterious matter and energy is purely speculated to make sense of the missing mass scientists have calculated is absent from the observable Universe. Perhaps this is just the way our eternal super-creature may prefer things to be, keeping itself tucked away and hidden from view.

If an enormous, all-pervading, omnipresent super-creature in the Universe exists, it may have difficulty doing anything other than being omnipresent. It will not have arms and legs like ours. Our bodies are created in such a way that we can walk, run, be creative with our hands and generally mosey around. Being large, it has difficulties with these rudimentary things, so what it may have done would be to create a conduit through which it can project pure thoughts. This could have been done through the realm of quantum physics.

Asking the right questions to something like this creature could reveal deeper universal meanings. Answers will come simply because everything in this state is inextricably linked and there are instantaneous conduits through to everywhere else within this known Universe, purely as a consequence of the way it was constructed originally. This gives anyone the ability to answer unanswered questions, if only they instruct themselves and the Universe that they have the desire to do so.

We could now check whether the correct questions were being asked by the apparatus which provided humans with their understanding of atoms. Unfortunately, the human beings' understanding of the world is one of 'real' and 'tangible'. Sorry to say, this is not the reality of it; the things we have interpreted as atoms are actually waves of potential that will only reveal themselves upon detection. At other times they are pure energy waves which are aware that they may be asked what they are by an observer.

Confused?

I am not surprised, but this is no more far-fetched than any other reason for our existence!

Have the right questions been asked regarding the definition, purpose and status of what we know as a black hole?

A black hole is generally described as – a region of space in which the gravitational field is so powerful that nothing, not even light, can escape its pull after having fallen past its event horizon. This is possibly not true, according to our eternal super-creature. What a black hole could be is a necessary part of our known Universe needed to mop up stray, dead stars that would otherwise be a liability or a danger to the active, existing ones. The matter sucked in is not lost forever, but stored for later use as required; it may also have the effect of propagating quantum fields throughout a galaxy. The effect of gravity is propagated just like light waves travelling through space, but is only detectable by objects with mass. There is no other way of detecting gravity waves.

The fact that objects with mass are affected by these gravity waves could cause the motion of matter in space. Just as there are nodes within a computer network that are focal points of data storage, the Universe and galaxies have focal points for particles to amass in a type of three-dimensional magnetic field, but with strong and weak points of focus. For instance, this is where a star system may be encouraged to form, as opposed to a weak point where particles will not even be able to congregate.

This means that interstellar travel may perhaps only be possible through predefined safe paths. Failure to stick to a safe path will result in a collapse of our 'normal' accustomed physics, not conducive with human limits.

We will never get answers unless we attempt to speculate solutions, test these speculated solutions, and subsequently accept or reject.

We learnt how fire is caused by a chain reaction of millions of oxygen atoms snapping into carbon atoms. We learnt that humans could be linked to a wider Universal strategy, as we are on the brink of being able to manipulate DNA. We learnt that in the mid 1990's we discovered the turritopsis nutricula which has the capability to live forever. We looked at how living forever may affect us. We learnt that quantum physics is perhaps just the surface of a world which is much, much more incredible than we could envisage. We learnt that something that appears dormant does not necessarily have little or nothing happening inside, or beyond it ... the Universe may have no edges! We learnt how nothing may be as it appears ... we could just be in the mind of an enormous eternal super-creature ... which would explain an awful lot! We learnt how nothing is impossible and that we must test ideas thoroughly to see whether we should accept them or reject them.

Take a deep breath, have a break.

Brace yourself for investigating some of life's unanswered questions.

INVESTIGATING LIFE'S UNANSWERED QUESTIONS

Where we probe into our fascination with the unknown, and try to determine a means of foreseeing the answers to current, unanswered questions. We look at derived intellectual views, and how new information can sometimes turn initial logic on its head. We also look at the probability of certain events ever happening, and how the Universe itself could actively narrow down the odds of certain mysterious phenomena.

Beyond first principles

- *With sufficient time we can always attribute a solution to unknown phenomena.*
- *We felt compelled to draw monsters on the edges of maps within unchartered territory!*
- *Given the correct approach, the human being's powers of deduction are exceptional.*

How will we ever get to know why atoms exist and where they came from?

How will we ever get to know why gravity exists and when it first originated?

How will we ever get to know how life originated, whether it is a natural Universe-wide wonder, or whether it is just local to our own planet?

What lies beyond the boundaries of the Universe?

How did human beings evolve?

Unfortunately, the Universe prevents us from knowing the answers to these types of questions, leaving us only to speculate the answers in a mystical fashion. There are answers to these questions, but they are somehow shielded from us.

There are shores to our sea of knowledge, and just like the sea there are waves of uncertainty at its perimeter that prevent us delving deeper to learn more. Only when the waves have sufficiently bashed and carved away at the cliff face can the sea claim a little more.

Breaking new ground within knowledge is similar. Just like the waves in the sea, we relentlessly pound away – endlessly trying hard to break into new ground to add to our knowledge.

If only there was a formula to thinking.

Imagine being able to look at a wasp for the first time and instinctively know how it evolved, how it flies, how it reproduces and functions. Upon the first glance at the wasp, our intelligence would allow us to instantly know the answers to all our questions. Instantaneous powers of deduction!

I just wonder what we would get to know if we looked at the Universe with our instantaneous powers of deduction. If only this was possible, we could tackle the unknown to derive all the answers.

When we try to think whether the Universe goes on forever, or whether it has a well-defined edge like a ball, we end up to-ing and fro-ing in our minds as if we are on a see-saw. Our minds may settle with a view of one concept, and then without too much thought, suddenly flick back to the other.

What confounds our minds is the knowledge that no one truly knows the answers to these types of questions, therefore placing our minds in a state of dichotomy. Because we know that Mankind does not know the answer, we seem unable to allow our minds to settle fully. Our minds seem to await concrete facts or scientific approval before we will accept a view one way or another.

However, some people develop very strong views about certain subjects even when their facts are rather flaky. They may argue profusely with great enthusiasm in favour of a particular explanation. By providing arguments and analogies they may easily convince other people too – a little like politics!

Just like politics, scientific breakthroughs may highlight the positive factors in favour of a proposal, whilst hiding the negatives. So we have to be careful.

Our beliefs in various concepts are made much more acceptable to us when we are taught them at school. When concepts are printed within books, they are official, believable, rarely disputed and become common knowledge. We rarely challenge this type of knowledge. We all assume that our forefathers had challenged it previously. However, inaccuracies trickle through, just like Clyde

Tombaugh's Pluto, the fate of the peppered moth, canals on Mars and the Earth being at the centre of the Universe.

Why does the Universe hide so much knowledge from us, even though our eyes and minds are open wide to all that we come across?

What are we missing that seems to make the unknown so baffling?

It would appear that when we look at something simple like a washing peg, we can see exactly how it works, but when looking at a wasp we are somewhat baffled.

Is there any way we can help ourselves narrow down the odds of being wrong about our assessments?

Perhaps we can set out logical assumptions beforehand. If we propose plausible arguments, we can persuade ourselves and others that one view is more acceptable than another.

Let us have a go at this. Let us take a quick look at whether we are alone in the Universe!

Let us first lay out some logical assumptions with which to work.

There are billions of stars in our galaxy, and there are billions of galaxies. We have detected other planets in other solar systems, some of which are similar to ours – there must be billions of these. We believe the elements required to form life are Universe-wide. There must be another planet somewhere that contains a similar mix of elements to Earth at the optimum temperature, thus providing a perfect opportunity for life. Just as life appeared here – it should appear there too!

All these factors weigh heavily in favour of us not being alone in the Universe.

So there is life elsewhere in the Universe.

Whether you believe in this view is up to you; the fact is, we just do not know.

We also have the ability to investigate the mystery of the unknown from a totally different angle – by looking back in time at our own lives.

The majority of people at some time during their lives experience a period when they are not quite sure what their future holds. For a while their outlook may seem rather hazy, then after a little time and thought, miraculously a vision of their future becomes clear.

You may be able to recall a time when you were restless, wondering what on earth the future possessed. When suddenly, along came someone or something that changed your life in a direction that totally surprised you. Looking back at this experience, you can generate a mental picture of the unknown as it was all that time ago. Having done this, pause for a moment contemplating how it was then.

Now picture how events and interactions within your life influenced you to provide you with the answer with which you are familiar today.

This is very similar to how scientific questions get resolved. So long as the unknown is well-documented and thoroughly understood, then after a little time a solution can be attributed to it.

Classic examples of this have been the success of the human genome project, space exploration, medical cures, wireless communication and manned flight. The solutions to each of these were initially unclear before people set out in earnest to derive plausible approaches. As more people concentrate on providing a solution, and as time passes, the solution becomes much clearer.

Within groundbreaking projects, it is often not possible to conduct crucial, supporting experiments to improve the overall solution. For example, when sending the first man into outer space and the first men to the Moon, a multitude of risks were taken as this was totally unfamiliar territory.

In these types of cases, it is often prudent to over-engineer the solution to accommodate every eventuality of the unknown. Over time, the solution can be bettered and the risks reduced as we get to know more facts.

Therefore, when venturing into the unknown, we often find, within the early days of a project, that we are prohibited from performing certain supporting experiments that would satisfactorily contribute towards the success of the overall solution. This in turn introduces an element of risk into the early solutions. However, to overcome this, quite often a mental thought experiment can provide just as much reward to the solution as a physical experience.

Thought experiments are very useful to ascertain an approximate result, they can put your mind at rest regarding an unknown through the power of your imagination.

Einstein utilised thought experiments extensively. One of his most famous was when he imagined to be travelling upon a particle of light and accurately documented his imaginary experience – developing groundbreaking formulae in the process.

We must be careful not to become too fanciful when venturing into unknown territory. Gone are the days when you can explain something by means of a god or a monster. Concrete and justifiable facts are all the rage at the moment!

People have been fascinated with the unknown for thousands of years. Intriguing phenomena have captured the minds of our ancestors, resulting in myths, legends and beliefs about how various happenings and spectacles came about. Over the millennia people have attempted to attribute satisfying answers to these unknowns, from the god Zeus creating thunder to the Moon being made of cheese with mice scurrying over it!

We also drew monsters on the edges of maps in unchartered territory – just because we did not know what lay out there!

Quite often a question may appear to be totally unanswerable, but with a suitably-planned approach we can have a reasonable attempt at hypothesising the answer to any unknown. Whether our answer materialises to be as inaccurate as the Moon being made of cheese, we cannot be sure, but so long as the resulting explanation is fully justifiable we will have a satisfactory answer.

We therefore need a question to use as a working example. Perhaps a reasonable example of such a question is, 'Why did creatures evolve to be able to fly?'

We shall use this question to experiment with our approach. We will need to hypothesise hard to answer this question, as we do not have a precise answer from the outset. Once we have hypothesised an answer, we can see whether our derived answer seems satisfactory enough to us.

How do we go about this?

Let us ease ourselves into determining the answer in the same way that Albert Einstein formulated his thought experiments.

Project your mind backwards in time to a point where you were bereft of knowledge, effectively 'beyond first principles'. Consider this state similar to knowing only true facts, no wild and wacky facts that you have guessed may be right. You could imagine this as if you have just been born – bereft of familiarity – but you still have the ability to understand language.

Ignore any preconceived ideas or knowledge you may already have stored in your mind; this information could be inaccurate, poorly-founded or an old wives' tale.

Now let yourself drift forward within your thoughts, building a mental picture of the most obvious facts, making your own calculated judgment at fifty-fifty decision points.

By building a mental picture of the most obvious facts, you will be able to project other clear views and thoughts over the top of these, using the human being's natural powers of deduction.

Eventually you will have developed a new concept regarding the subject, and possibly have devised a reasonably plausible explanation for the unanswered question.

I cannot emphasise enough the need for us to think in another way if we are to make progress. The visionaries of this world all thought so differently.

Albert Einstein thought in a different way to discover the link between energy, mass and the speed of light. Isaac Newton thought differently to lay the foundation for calculus, optics and theories of gravity. Leonardo da Vinci thought in an alternative way to develop new concepts in painting, sculpture, architecture and scientific projects. Their views of the world were markedly different to those of others in their day, but what exactly was it that made them think so differently?

We must certainly open our eyes to things we do not normally see. Look through things, look around things, look into things and look beyond things.

There is no better way of experiencing the concept of attempting to think differently other than by example. The example we shall use is the question, 'Why did creatures evolve to be able to fly?'

Why did creatures evolve to be able to fly?

- *If you are in danger of being eaten, it may well be a good idea to see if you can jump and flap your way out of trouble!*
- *Genetics seems to be susceptible to change by nature's own means.*
- *Are the animals we see such as the flying squirrel half-way through their evolution, or are they fully evolved?*

Let us place our minds into the 'beyond first principles state', ignoring any preconceived ideas or knowledge we already have. Now drift forward, allowing your mind to identify various pieces of information to form a mental picture – let your thoughts feel their way forward to a satisfactory answer.

In some ways you could look upon these thoughts as being like your first ever thoughts – just as you investigated things and thought about everything for the first time as a child.

Here goes!

Of the many different types of creature on the planet, four different groups of animals have successfully achieved flight: insects, birds, mammals and pterodactyls. This means that flight has evolved several times without a single ancestor.

Whereas bats, birds, wasps, bees, ants, and dragonflies have achieved muscle-powered flight, there are others, such as lizards, snakes, frogs and fish, that have developed the ability to glide.

There is a truly great variety of creatures, some of which we see with our own eyes on a daily basis. Perhaps these creatures developed the ability to fly as an essential part of their arsenal of survival tactics.

The survival of a species depends upon its ability to access an adequate food supply, reproduce successfully, and live within an appropriate habitat.

It stands to reason that, if a creature has its habitat or food source adversely disrupted, it must either adapt or suffer extinction. So if a disease kills all the leaves of a tree that certain creatures eat, then they must eat something else and hope to survive.

Making provision for their offspring is a top priority for the survival of all successful species; creatures are often obliged to go to massive lengths to help their young survive.

A species may find itself in danger of being eaten by others; it is at times like these that it may be prudent to jump!

Perhaps on one occasion there were so many predators trying to eat one type of species that only the ones that flapped their limbs, jumped, and landed safely survived to continue reproducing.

But how does an animal evolve from having arms to having wings?

There must be a truly remarkable and phenomenally detailed blueprint for life. Now we have witnessed the discovery and understanding of genetic codes, we are entering an era where it has become possible to alter the make-up of species. The true 'Frankenstein Monster Era'.

It appears that nature has its own means of altering genetics such that a creature is made capable of flight. Over the years, nature has proved this by producing some ants that fly and others that merely walk. Strangely, many ant species shed their wings once they have flown to a suitable new nesting site. A great shame really, as flight seems to be such a marvellous and intriguing feature to possess. If I were an ant I would hold on to my wings for recreational purposes, although recreation does not appear too high on nature's list of survival tactics!

Did all ants fly at one time?

Did all ants walk at one time?

As ants are thought to have evolved from a wasp-like creature, then it would be quite plausible that they all flew at one time. So rather than evolving to fly, some ants have evolved to walk – rather strange.

With this in mind, it would seem reasonable to think that if it were possible to change the appropriate DNA structure, we would be able to trigger the ants without wings to grow them again. The ability to fly may well be stored within the DNA, but just switched off. Perhaps a certain combination of environmental factors may trigger the 'flying' function to switch on once again.

If we could identify and isolate the code that makes it possible for creatures to fly, perhaps we could alter other animals too?

Perhaps the ability to fly is an inherent feature within the blueprint of life, at least for those that have had ancestors that flew at one time.

By altering animal genetics, we could develop penguins that gracefully soar through our skies. We may have a little more difficulty with elephants and whales, but our fascination with fairies may one day take on a brand new meaning. As all creatures evolve over time, there is absolutely no reason why humans could not take to the skies one day.

Imagine a very bizarre situation ... imagine the Earth was attacked by an army of extremely powerful aliens. These aliens decide to undertake an evolutionary experiment, and inform us that we have one hundred million years in which to evolve muscle-powered flight. Failure to achieve this would result in them terminating our existence.

Would we manage it?

An interesting and rather unique question to debate!

An urgent strategy would need to be implemented and imposed by governments.

Perhaps we would be encouraged to climb trees and jump out whilst flapping our arms in hope!

Perhaps we would encourage our offspring to develop progressively stronger upper bodies and flatter arms!

Perhaps we could breed from freak people who are born with flaps of skin under their armpits!

Perhaps research into genetics may reveal how to turn our hairs into feathers!

It is not only small animals that evolve to fly. Pterodactyls, with up to forty foot wing-spans, managed to evolve to fly all those years ago, so I am inclined to think that we could manage it too. I am sure there would be a myriad of strategies proposed and plenty of strange government-dictated exercises to perform!

We would perhaps hope to evolve flight by endlessly attempting it. We could achieve it in a similar way animals adapt to their environments over time.

Jumping for food or from danger may well be a key factor in the development of flight – it makes sense that this would lead towards a type of gliding technique. Once an animal can glide, it seems just a short evolutionary step to muscle-powered flight. As long as the creature can fuel its muscles by eating high energy foods such as insects, fruit and nectar, rather than leaves – voila!

We have evolved powered flight!

Gliding animals may have developed independently of powered-flight animals, but powered-flight animals seem to have proliferated around the globe significantly more. Quite understandably, because powered-flight animals have the ability to travel further.

It is interesting to note that many gliding animals are found in South-East Asia, and some in Africa, but there are no gliding vertebrates in South America. However, more animals in South America have gripping prehensile tails than South-East Asia or Africa. It may well be that animals with prehensile tails dominate in South America because the forest is so much denser. In South-East Asia and Africa the animals evolved to fly because the trees were spaced further apart.

There are some wonderful examples of adaptation observed when studying modern creatures. Animals have been discovered trapped inside pitch-black caves – over the years they lost their ability to see and their eyes have subsequently disappeared. Animals have developed coats that turn white in winter to match snow and turn brown in summer to match their terrain – we presume those that failed to do this were spotted and eaten! Certain animals have developed longer necks to access food that others cannot reach – giraffes being classic examples of this.

These creatures have all developed an alternative mechanism to allow them to overcome a shortfall in their lives.

For all creatures that evolve to fly there must be a transition period. There could not be one day when a creature could not fly, miraculously followed by a day when it could fly. It would make sense that a creature slowly evolved towards flight just as a giraffe developed a progressively longer and longer neck.

In between a creature beginning this evolutionary process and actually being able to fly, whatever its capabilities were at any point in time must have made sense. It seems inconceivable that any creature would possess a cumbersome half-baked solution for flying, and yet still be unable to fly. Although if we look at some creatures that do not fly, but have wings, like ostriches, penguins, emus and kiwis, we see them evolving ... or 'devolving' if you like, into something that is counter-intuitive compared with the agonising length of time it must have taken to develop flight in the first place. A very strange asset to lose!

We instinctively look at a penguin and realise that it has lost its ability to fly, rather than think it is currently evolving to fly. Flight would not evolve like this!

When we consider animals that could be construed to be evolving to fly, at one end of the scale we have the manta ray. They periodically launch themselves out of the water and flap about in an extremely awkward fashion; and then proceed to belly flop back into the water having only managed a split second airborne. At the other end of the scale, we have flying fish that look a little more comfortable airborne. However, you cannot help thinking they are lacking manoeuvrability.

The flying squirrel and the flying snake seem to be only halfway to fully functional flight. As a result we could convince ourselves that their awkward gliding has not yet fully evolved. Our instinct tells us that they are not losing their ability to fly, but gaining it.

What are they ultimately going to become, or is there some type of mistake and this is exactly what they are supposed to be?

Perhaps there are no further planned evolutionary developments afoot for a flying squirrel!

Research has revealed that flying squirrels evolved about thirty million years ago, so they have had plenty of time to commence flapping their arms harder if they had needed powered flight. Perhaps they are content with just gliding!

It would make sense if there was no master plan to evolution, but every creature must be able to adapt to slight changes. Sudden major changes to the planet's environment, such as meteorite impacts, have seen mass extinction of creatures. Only those that could endure the major change by pure coincidence managed to survive.

It would therefore seem that evolution is reactive rather than proactive.

So to answer whether I feel that humans could evolve to accomplish muscle-powered flight – I am inclined to think that if humans were forced to evolve in this way it would be quite possible.

Let us imagine that the first creature to try flapping its legs was a small bug-like creature. Eventually, over the passage of time, it flapped its two front legs appropriately and sufficiently hard, they became strong, flat and provided lift.

A mouse which normally feeds on these bugs eventually has difficulty in feeding, because its food source is now flying around above its head. The mice now have to stand precariously on a ledge and wait for a passing bug to zoom by. The mice then launch themselves to catch one in mid-flight. Over the passage of millions of years, they flapped their legs so hard whilst catching these flying bugs that they ended up flying too.

Some people think this is how we ended up with bats!

One thing is for certain, at one time there was just one creature that flew first – and then others followed suit. The first creature to fly very likely did so as a survival tactic. The creatures that followed suit did so to survive also – flight would have evolved over a significant period of time. Perhaps jumping from tree to tree to begin with, developing into flight over the passage of millions of years.

My investigation into creature flight sounds relatively plausible – to myself at least!

More than likely because they are my derived views!

Other people will no doubt develop totally different solutions for the same question. Different people will have different angles on life, see things differently and have approached the question from a totally different direction.

If millions of people had an attempt at answering the question, eventually someone will get pretty close to the correct answer. However, people who lean more towards a creationist's view rather than an evolutionist's view will no doubt derive thoroughly different outcomes!

A mental picture of the evolution of flight has now been established, based upon intuition and observation. It has been evolved by simply using the information that my mind chose to become prominent. Whenever someone has created their answers themselves in this fashion, the results feel pretty comfortable, justifiable and defensible. The information derived and built in the mind contains views with which you feel very satisfied, but were not that obvious at the outset. Now we have a mental picture to reflect upon – we can attempt to seek a few more nuggets of information.

Here goes!

Flight of the first creature, 6th September 789,654,267 BC

- *In the distant past there was a time when the Earth's atmosphere contained no life.*
- *Observing the first-ever flight on Earth would have been like witnessing a real-life miracle.*
- *The first flight would have been rather primitive.*

It is safe to say that at one time the atmosphere on Earth harboured no flying creatures within it whatsoever. This means there must have been a day when the first creature took to the air.

Imagine the spectacle and the feeling it had!

Recognition is awarded to humans who are first to achieve various feats. For instance: climbing mountains, sailing difficult seas, flying long distances and space endurance records. Just as Sir Edmund Hillary was the first man to climb Mount Everest on 29th May 1953, the first creature to become airborne did so on a date such as 6th September 789,654,267 BC.

What a significant moment in the history of the planet, an evolutionary marvel, the inaugural flight of a creature. The observation of this flight would have been a true spectacle and a significant world first.

Perhaps the first attempt to become airborne by a creature was nothing more than a jump in the air, and therefore could not be considered true flight. So let us imagine the rule of creature flight to be defined as, 'no contact with the ground for at least ten seconds'.

This definition will overcome any attempt that could be otherwise perceived or misconstrued as something that would be considered nothing more than a spectacularly long jump.

Imagine you had a video camera all those years ago to film the world's first-ever creature flight. It would have been a primitive flight, and may have taken a million years of trying to achieve, but the flapping motions of those limbs will have sent light waves scattering for the observers at the time to witness.

What a spectacle this would have been!

The wonders of flight – avoid being eaten, get from tree to tree quickly, mate with more partners and defend offspring. From the first day flight happened it was obviously considered a success. It was considered a success because a significant proportion of the ten million different species of creature there are on the planet can now become airborne.

Perhaps it was a coincidence; maybe a sea creature evolved to flap profusely underwater, and then purely by coincidence it found the same motion applicable within air too.

Life on Earth seems to be extremely dexterous and versatile; it finds its way into all environments that provide only the slightest possibility of sustaining life. Salt water, fresh water, very highly-pressured deep water, pitch-black environment, bright Sunlight, land, underground and quite unsurprisingly – airborne. Surely the dexterous nature of life on Earth will see life-forms infiltrate into all environments – however inhabitable it may seem.

Creatures seem to be able to react to a change in environment pretty much instantaneously. One good example was witnessed when scientists reared the offspring of insects within an atmosphere containing higher oxygen content. Our normal atmosphere contains twenty-one per cent oxygen, but when scientists decided to hatch some insects into a thirty per cent oxygen-rich atmosphere, something rather interesting happened. The immediate effect was that the next generation of insects were substantially larger than their parents, some reported to be twice as large.

However, the level of oxygen is just one parameter that can be altered. There are obviously many others.

Perhaps the complex structure of DNA has programmed within it a threshold relating to a 'certain environmental factor', over and above which a creature will sprout wings?

To understand what nature is truly capable of achieving still requires a huge amount of research. One area of the world where remarkable evolution has taken place is within Lake Victoria in Africa. This freshwater lake, which covers a staggering 26,600 square miles, has nurtured an extraordinarily special fish called a cichlid. Lake Victoria is not thought to be very old at all, with some estimates placing it as just twelve thousand years.

These cichlids are thought to possess what is termed 'evolutionary plasticity', which has enabled them over the past few thousand years to diversify into over four hundred different species. These cichlids are capable of evolving at an amazing rate to accommodate extremely localised and diverse environments. There are so many different species of cichlids that many new discoveries of the species are made each year. It is thought that there are thousands of different species of cichlids globally; the angle fish of which we are so fond in fish tanks is one. Therefore do not be too surprised to wake up one morning to find a new breed of fish in your tank!

It seems cichlids have the capability to change evolutionary shape and characteristics faster than any other species. This fastest evolutionary species has within just a few thousand years managed to evolve into a plant eating species as well as one that eats the plant eating species, with teeth fit-for-purpose as well!

There are reports of placid plant eating cichlids in fish tanks which having missed their feed for the day suddenly turning cannibalistic.

I am glad human beings are not like this – it would be rather frightening in the supermarket when the bread counter runs out.

If only Charles Darwin had known this.

It places a Creationist's views on pretty rocky ground!

Turning your work on its head

- *Could it be that the first creature on Earth was able to fly?*
- *Many scientific theories have been developed, only then to be binned and substituted by other theories of a new era.*
- *In one era people thought travelling at speed would render your body dysfunctional.*

Deriving your own mental picture of the unknown, utilising your own clear facts to answer unanswered questions, can be satisfying for your own mind, but not necessarily for other people. Other people have their own views, based upon their own experiences within the Universe. I am sure people with a specialist interest in the evolution of species would have a number of comments regarding our previous views derived about flight.

It can sometimes take just one fact to turn your work on its head. Just look at how many people thought at one time there was life on Mars. It took just one quick close-up look at the planet's surface to place all those theories in the rubbish bin.

Take the question, 'How many ants are there are on planet Earth?' Rather an interesting question if you do not know the answer, as it already has a recognised, scientific answer. Once you have come up with an answer, we can see whether we can turn your work on its head.

Just think of your answer for a moment before reading on...

It transpires that the leading 'ant expert', Edward Wilson from Harvard University biology department, has calculated there to be roughly one hundred million, billion ants. This is quite an incredible figure, and to put this into perspective this means that there are roughly three hundred and forty ants for every leaf upon trees.

Incidentally, the estimated number of trees on Earth is thought to be just short of two thousand billion, each with an average of one thousand five hundred leaves! Another fact that you may not wish to know is that it will take twenty million average ants to make the same weight as an average human being. This makes me wonder how on earth an anteater, which can grow to six feet long, can manage to eat so many ants! It would appear that it would have to eat twenty million ants just to make it the weight it is!

Seeing as it takes two weeks to count to one million out aloud without any sleep, then if an anteater ate one ant a second constantly for forty weeks, this would account for its total weight – incredible!

Knowing that so many of the ten million species on Earth can fly, prompted me to come up with an alternative view, which is that perhaps the first-ever creatures to frequent the planet actually flew. They could have arrived here from a distant planet in the form of a spore, and germinated once hitting our atmosphere. It floated around in the atmosphere for millions of years before eventually taking the bold step of landing on the ground.

A view like this can turn all our previous work totally on its head, which is fine so long as you feel comfortable with it.

Science is peppered with theories which have then been superseded by other theories. The original depiction of the atom described it looking similar to a sponge pudding. This was superseded by Neils Bohr's view of electrons, neutrons and protons. Other gaffs were the Victorian's view of something they made up called the ether, which enabled light to traverse the Universe. This was soon dropped when we truly discovered what truly lay out there. It was once thought that going above thirty miles per hour would make your body dysfunctional and that you might die. This was one of the worries people had during the times of the very early railway pioneers.

The mystery of the Great Pyramid

- *It seems incredible that the Great Pyramid was built for nothing more than a human being's grave.*
- *Would it make sense that there was a more sophisticated reason that the Great Pyramid was built with such amazing accuracy.*
- *It is truly remarkable that the great Pyramid was ever conceived and constructed in the manner that it was.*

Having read all the literature I could lay my hands on, I reached the conclusion that the Great Pyramid of Giza must have been built for something a little more sophisticated than an elaborate grave.

The pyramids have always fascinated people, and they hold a strong attraction to those who have a vivid imagination. No one has ever been able to explain the intricately carved and deliberately manufactured chambers, their true purpose and why they were constructed in such meticulous detail at such colossal expense.

Eventually someone should discover, or rediscover, exactly why such great lengths and effort were put into building such an enormous structure. The structure seems deliberate, as if it had a true purpose, but we fail to recognise any logical function for which it could have been built. Was it an observatory, a calculating device, an electricity generator, an advanced communications device or part of a large particle physics experiment?

Why are we unable to relate to the builders of the pyramids?

Perhaps our way of thinking has evolved so dramatically in such a short period of time that we are unable to relate.

If we find an old axe head, a flint arrow or an old tin opener, we can tell exactly what they were made for. These are small things that take a little time to make, but we can see clearly the purpose that they served. When looking at the Great Pyramid, we have to scratch our head hard to take in the tremendous size of it, the enormous time it took to build, before we then realise we cannot relate to what it was made for. Even if it was built as an elaborate grave, then this is still incredible as we are totally unable to relate to this level of dedication to a human being. If it was built for no function, then this is even more incredible. If it was built upon the premise that someone would gain an afterlife as a consequence, then there must have been an unfathomably bizarre belief system stronger than that to which we can relate.

It would have seemed more likely that someone could explain the true reason for the Great Pyramid, than someone produces the complete works of Shakespeare or unravels the mysteries of DNA.

But they have not!

All manner of solutions are possible for the reasoning behind why the chambers are positioned as they are. There would have been a grand design and a specific reason for them being in the meticulous fashion that they are. The fact that the reason for building it is not obvious, may mean it is too complex for us to relate to, or purely pointless.

I am not a monkey cousin that wrote the complete works of Shakespeare, but I am one that believes the Great Pyramid would have had some kind of function. There are possibly unknown physical properties that we do not understand with which it interacts, or perhaps, as some people think, an intelligent race of extraterrestrials arrived and orchestrated its construction for some precise reason.

It is most awkward to contemplate why on Earth anyone could be so dedicated to producing something of such magnitude without there being a spectacular end result. Maybe their belief was so strong that they whimsically built the Great Pyramid truly believing that it would work for them. However, the sheer size of it suggests that whatever it was built for, they knew it was going to function correctly.

It seems incredible from many perspectives that the Great Pyramid was built at all. Such large stones, such labour intensive work, the sheer size and such a daunting schedule must have played on the minds of all that worked upon it.

Tell-tale signs that extraterrestrials exist

- *The real reasons why the Great Pyramid was designed in the way that we find it, is still a mystery.*
- *An infinite number of monkeys would eventually produce the complete works of Shakespeare.*
- *The precise calculation to determine the exact number of monkeys required to produce the complete works of Shakespeare is very straightforward.*

Further to the reference earlier in the section headed 'How did life originate?' where we found that meteorites falling from outer space are teeming with life-providing amino acids, there are also additional signs too.

The secrets of the Great Pyramid are still eluding us. Despite many attempts to explain its construction and purpose, arguments perpetually ensue. Modern humans are more capable of creating computers, spacecraft, vaccines and the complete works of Shakespeare than simply working out why the chambers within the Great Pyramid were built as they were!

But why should the complete works of Shakespeare, and the creation of advanced technology, be simpler for modern man than understanding the meticulous construction of the Pyramid's inner chambers?

Perhaps we are thinking in the wrong manner!

Perhaps our thought processes are preventing us clearly visualising the purpose, reason and need for the Great Pyramid, all those years ago. If we think differently, perhaps the true purpose will become clear.

What I find so fascinating about the Great Pyramid is that it was clearly not a random building project or whimsical folly. There was clearly great design, purpose, detail, craftsmanship, architecture, work and beauty. But we have enormous difficulty trying to appreciate why it was built.

As for the complete works of Shakespeare, it is said that if a monkey hits keys at random on a typewriter for an infinite amount of time, it will eventually type the complete works of Shakespeare.

Converting this philosophy to the Great Pyramid, perhaps we are looking at just such a random event!

The Great Pyramid is such a bizarre building that I am inclined to state, 'If an infinite number of creatures exist throughout the Universe, then eventually one will build a massive Pyramid of twenty-two million massive stones'.

It just so happened to be on our planet!

What bizarre structures exist on other planets?

Imagine for a moment that the pyramids never existed, but the remainder of our civilisation's history remained the same. If someone turned to you whilst you gazed out at the expanse of the Universe and said to you, "What are the odds of some creature out in the depths of the Universe having crafted twenty-two million massive stones to build a huge pyramid?"

Your answer may be something along the lines of, "None whatsoever."

But 'we' have one!

So what does this signify?

What I think this signifies is that the Universe is massive, and that there are countless possibilities within it.

I am inclined to think that the way Mankind started its modern life upon Earth, building pyramids, was pretty bizarre, so bizarre that it implies that there must be other types of initial events associated with civilised living.

If we are the only planet within the whole Universe that harbours life, then I find it incredible that we also started by building huge pyramids. It is akin to science fiction.

Try thinking about it this way. Imagine there are a billion planets with human-like creatures upon them. How many of them do you think would start off by building massive pyramids consisting of millions of huge stones?

If you feel the answer to this question is roughly just one of these billion planets, then there would need to be an enormous amount of other planets with creatures that failed to build pyramids to enable one to succeed doing so. We may therefore deduce that the presence of the pyramids is proof that there is life upon other planets.

Follow this logic.

Imagine looking at the Universe as a whole, all its galaxies, all its stars, all its planets – you then pick one planet at random. I feel it is much more likely that the planet you choose will have creatures upon it without a huge pyramid than one with creatures and a huge pyramid.

What do you think the odds are of some creature out in the depths of the Universe having crafted twenty-two million massive stones to build a huge cube?

What do you think the odds are of there being a similar twenty-two million stone pyramid upon another planet?

It is impossible, or phenomenally hard to believe, that the first and only planet to harbour life also developed the first and only Great Pyramid. It is far too remarkable that both occurred together.

The presence of the Great Pyramid upon planet Earth should therefore be seen as a sign that there is life elsewhere in the Universe.

This does not mean that the Great Pyramid was built by extraterrestrials!

Having ascertained this we can now return to Shakespeare.

Having established that one monkey sat at a typewriter would eventually produce the complete works of Shakespeare, we realise that although this is possible, it would take longer than the age of the Universe. So if we use an infinite number of monkeys sat at an infinite number of typewriters, one would produce the complete works of Shakespeare at their first attempt. In fact due to the nature of infinity, an infinite number of monkeys will produce the complete works of Shakespeare at their first attempt!

The precise calculation to determine the average number of monkeys required to produce the complete works of Shakespeare at the first attempt is relatively easy. It is simply the same as picking the correct numbers for the lottery, but rather than there being six numbers to get right, there are as many characters as there are in the complete works of Shakespeare – roughly four million!

A fairly long shot, but immediately achievable with the right number of monkeys and typewriters!

Let us say there are four million characters that make up the complete works of Shakespeare. As for our typewriter, there are twenty-six letters in the alphabet, plus a spacebar. We will be generous to the monkeys in terms of punctuation keys by stipulating that other than these keys, the only other keys they need are an apostrophe, a full stop and a comma, making thirty characters in total.

So to calculate the average number of monkeys required to produce the complete works of Shakespeare upon their first attempt is the number of possible characters, thirty, to the power four million. A very large number, but a set number all the same.

The odds of a monkey coming up with the complete works of Shakespeare are therefore quite achievable within the Universe, even within a finite Universe. To put this in perspective, one monkey-type creature already has! Strangely enough his name was miraculously – William Shakespeare!

Having already managed to produce the complete works of Shakespeare, the Universe has proved itself to be extremely intelligent.

Knowing the large number of monkeys it would need to do this, the Universe did not just use the sheer powers of randomness to produce the works of Shakespeare. It would appear the way the Universe went about it was to narrow down the probability by producing DNA, chromosomes, intelligence and a suitable planet upon which it could be produced. Once the capabilities for its production had been created, this opened the way for Shakespeare to produce his work.

Knowing that there would only be a limited number of monkeys to play with, the Universe had to work on making the possibility more probable – this it did.

Consider carefully the following. Once human DNA had been created within the Universe, it then only took a few billion humans lives before something as miraculous as the complete works of Shakespeare was produced. This is quite revolutionary. It implies there is now a means of determining how probable DNA is by comparing how likely the complete works of Shakespeare was without it.

With human DNA in existence, the production of the complete works of Shakespeare is reduced to about one hundred billion to one. In other words it took one hundred billion human lives enhanced with human DNA to produce the complete works of Shakespeare.

Without human DNA we are at the mercy of randomness. To randomly produce the complete works of Shakespeare we are looking at thirty to the power four million to one. It is really not possible to calculate this number due to space limitations on a calculator display face and lack of computing power, but it is possible to judge roughly how many digits it may have. The number would have roughly six million digits in it. If written in normal-size print it would stretch for roughly ten miles. Quite a large number of monkeys!

A very big zoo!

This massive number, when compared with the relatively small number of human beings that have ever existed, gives us a number that depicts how enormously improbable human DNA is. The probability of DNA being created without design is, 'a number two inches shorter than ten miles long' – to one!

The two inches represents the approximate, one hundred billion human beings that could have produced the complete works of Shakespeare, prior to it being produced!

As there are a limited number of intelligent human beings on the planet, we have the answers only to certain problems. The more people we get, the more answers we get!

Being an inquisitive race, we have formed groups that ponder various topics, perpetually trying to extend knowledge into the unknown. So when trying to determine the answer to any unanswered question such as, "What lies out there in the expanse of the Universe?" there is no way to tell how many different answers there may be.

But the question "Why was the Great Pyramid at Giza built in the style it was?" has only one answer – but we do not know it!

Logic implies that there are only two answers to a fifty-fifty question such as "Is the Universe infinite?"

It is either yes or no.

We must, however, remember that the answer to what lies out there in the expanse of the Universe is a great deal more complex than the complete works of Shakespeare, and much more complex than a yes or no answer!

At least we have narrowed it down a little.

So if the Universe created us, then what does it want from us?

This is an awkward question; it is like asking, "What is the meaning of life, the Universe and everything?"

Perhaps by reverse engineering the answer in a similar way to determining from where the complete works of Shakespeare came, or how creatures originally evolved to fly, we could one day fathom the meaning of life, the Universe and everything!

One explanation could be that these massively remote odds of numerous, inconceivable events and universal structures occurred purely due to the enormous size of the Universe, rather than any other factor. Because the Universe is so large, remote possibilities become certainties. In a similar way that successful technology is replicated many times on planet Earth, then so do successful structures in the Universe. These successful structures then shape the future of the Universe.

Over to you!

We learnt how people thought differently in order to make progress, and how they opened their eyes to see things in another way. We learnt how we may answer unanswered questions to a satisfactory degree. We learnt how to apply thought to such things as determining how creatures first became capable of flying. We learnt how peoples' assumptions in the past were often built upon extremely flaky knowledge, and how we still do not know why the pyramids were built to such accuracy. We learnt that the presence of the Great Pyramid upon Earth may be viewed as a sign of life elsewhere in the Universe, not because extraterrestrials built them, but due to the fact that their existence upon the only planet with life in the Universe is very remote.

Next we look at perhaps one of the weirdest aspects of the Universe that makes one of David Copperfield's flying illusions look decidedly tame.

INSTANTANEOUS COMMUNICATION

Where we learn how the networks of the world are evolving to monitor most of what we do. We learn how current communication is inadequate for our forthcoming ambitions and that we do not at present have a suitable alternative for the future. We look at precisely what governs the speed of communication, and exactly how slow it is throughout the Universe. We take a glimpse at the evolution of communication. We also discover the attempts there have been to improve communication speeds, and ultimately discover a solution that nature has miraculously revealed to us.

How 'they' are watching you

- *Most activities and actions you undertake may be monitored closely.*
- *Supermarkets, governments and private individuals are all capable of monitoring anyone they wish.*
- *The definition of what constitutes a malicious piece of software is rapidly changing; it is getting more sophisticated by the day.*

Have you ever felt as if you are being watched?

Well, it is not just a suspicion you may have, there is a complete industry out there in the wide world which thrives upon monitoring precisely what you and everyone else is doing. This monitoring comes in many different guises, which this section of the book will reveal.

Hacking into your mobile phone is easy, 'they' can now listen in to your calls, record your calls, open your text messages, listen to your voice mail and even place false voice mails and text messages on your mobile phone, without you knowing it.

Quite often we hear of politicians and celebrities who have suddenly been found in compromising situations. It is not bad luck which has revealed them performing an irrational act or a kissing a scantily clad prostitute – just because a photographer happened to be accidentally passing at the time. Oh, no – these people are being surveyed meticulously.

Be warned that these very same people who are capable of stalking these politicians and celebrities are just as capable of monitoring your movements and actions too. In fact, it is such big business that this style of monitoring is being automated. It is now possible due to the automation of monitoring with software, that almost everyone's, past, current and even predicted future patterns are being mapped out in careful detail.

At the other end of the scale, revelations about skulduggeries within the British Press unfolded in the latter part of 2011, showing just how easy it is for anyone to infiltrate individuals lives if they so wish. Phone and computer hacking, it would seem, were almost routine by some members of staff.

Getting information is easy if you put just a little thought to such matters. To help the Press get the juicy gossip first, all they need to do is have someone on the inside of a mobile phone network to monitor user information. Unbeknown to most users, all of your keystrokes upon your phone are registered and sent back to the phone carriers for 'diagnostic purposes'. You may wish to watch a YouTube video showing you how to switch off this 'feature'. Search for 'Carrier IQ key logging', you will discover how the standard software on your phone, by default, sends information about your keystrokes, application usage and other data back to the phone carriers. Goodness only knows who they may sell this information to.

Private detectives are increasingly being hired to unearth skulduggery of seemingly 'normal' policemen, judges, businessmen and politicians to undermine their credibility in order to safeguard crooks on the 'dark side' or cause sensation to sell news. Irrespective of legislation or policies, this will continue through sheer demand. "So," you may think, "This does not bother me, the last time I slipped off to see a secret lover CCTV had not been invented."

The list of surprises in the world of tracking is almost endless; in 2011 there was the revelation that Apple's iPhones and iPads were persistently logging user's locations without their knowledge or consent. In November 2011 Hugh Grant reported in court that photographers arrived before the police when he registered a relatively mundane emergency call!

Think about these: Satnav systems which log you as you innocently wind your way through the traffic; Traffic Master chips in cars that log your every inch of travel each time you pass a blue post on the side of major roads and motorways; the tyres on your car that wirelessly transmit a unique ID to the car's CPU which informs the driver that the tyres are low ... all of these can be tracked by all manner of devices. Add to that, the Google Street View camera cars that have also been found to be mapping the IP addresses of everyone's home and business routers.

Interestingly, the last one in the list above came to my attention when I was speaking with a good friend of mine who had noticed that his phone was acting strangely in Derbyshire when he was using a router which used to be sited in Ormskirk, Lancashire. He noticed that when the Internet phone rang it appeared that the system thought he was a resident in Ormskirk where the router used to be sited. He investigated this, only to discover that Google had his router registered as still operational in Ormskirk still. This could only have been monitored from a Google Street View camera car. Such an act has even been given a name, a 'Wardrive'. Wardriving is the act of logging every connection to the Internet, associating it to a given location as someone drives or walks along. The SSIDs and/or the Mac address of the machines encountered are noted and stored for future reference – whatever Google decides this use may be!

Want to know more?

How would you feel if 'they' were able to watch you through your home or work computer?

There is much more to this industry; a deeper, darker, potentially intrusive side which no one seems able to control and is quite often instigated and sponsored by governments – to monitor pretty much everything that goes on.

Since the introduction of clever viruses and software Trojans, it is now possible to monitor every keystroke on your computer. All that you view can be monitored and you may even be watched through your own webcam by someone without you ever knowing. Visual monitoring is so easy to establish you can do it yourself quite legitimately for a positive outcome in the event of computer theft. By installing a 'clever piece' of software which may be downloaded from preyproject.com, it is now possible to initiate software which may view the robber who has stolen your laptop. Give preyproject.com a try; it takes less than three minutes to download and install – and it is free.

On the flip side of this 'easy to install' solution, it may be you have a perverted voyeur, detective, government or even your ex-partner watching through your webcam without you being aware they are doing so.

In this modern world governments do not wish to miss out on this opportunity, as we shall see. 'They' too, have ever increasingly sophisticated methods of monitoring and surveillance that make shivers go up the spines of the most innocent of people. Not only do governments have the powers to demand information from businesses and organisations which hold vast amounts of material about your whereabouts and activities such as banking, supermarket habits and CCTV images. Governments also have teams of people working to uncover all that you have been saying to friends, colleagues and secret admirers over the Internet. Governments have the authority to demand to see data about you from whichever source it so chooses – but there is more to their armoury.

If you thought this was bad, now we take a look at cyber-warfare.

In 2010 a computer worm called Stuxnet was unleashed onto the world. It initially spread silently through Microsoft Windows and targeted Siemens industrial software and equipment. Although this was not the first time this type of worm had been seen, it was the first time that one had spied upon its target and then subverted industrial systems. It was also the first time a worm had included a programmable logic controller rootkit. If you are unsure what this means, not to worry, all you need to be aware of is that this piece of malware made its way into the heart of the five Iranian organisations that were involved in the enrichment of uranium for their nuclear programme. Symantec, an antivirus company, reported in August 2010 that sixty per cent of the computers in the world that had been infected were in Iran. This worm had not caused damage to public computers or businesses, but had rendered the embargoed Siemens equipment, which had been procured covertly, inoperable. The Siemens equipment which was being used to operate the nuclear centres centrifuge operation was simply a similar piece of software that may be used to control a concrete mixer. This was quite deliberately targeted, hacked and disabled. It was thought the software entered the Iranian plant on a pen-drive. This was such a sophisticated attack on such a specific target that governments of the world were thought to be behind the attack.

In May 2011, Gary Samore, White House Coordinator for Arms Control and Weapons of Mass Destruction said, "We're glad that they are having trouble with their centrifuge machine and that we – the US and its allies – are doing everything we can to make sure that we complicate matters for them."

The Stuxnet software used four 'zero-day' attack mechanisms; a zero-day attack exploits vulnerabilities within software that are unknown to the software developers – the fact that Stuxnet used four such vulnerabilities again showed the determination behind the Stuxnet developers.

This Stuxnet-syle operation makes the 'ring of steel', the security and surveillance cordon surrounding the City of London, with its 'Automatic Number Plate Recognition', ANPR, look like child's play!

Intelligence in the latter part of 2011, covertly obtained out of Iran, reported that an even more sophisticated second generation of this type of destructive worm has been activated against other industrial components used in Iran's nuclear facilities.

I do not think the bookies are even taking bets as to who was behind this – such an attack could have only taken place with State support or several nations combining their finest experts!

The influence that 'they' have is staggering too, have you ever thought why you buy certain items and what it was that initially got you into your current buying habit?

Now we come to the Supermarkets, Department Stores and High Street Chains with their army of strategists, in-store cameras, consultants and computer experts, which in a nutshell could be called a team of 'very sophisticated spies'. For example, if you use a Tesco Clubcard, Nectar Card or a Store Loyalty Discount Card you are by default allowing the store to monitor your habits, movements and activities. This data can subsequently be sold on to companies wishing to target you with delightful products that you have already indicated you enjoy. So if you wonder why after having bought a new washing machine, you are two weeks later bombarded with free samples of washing powder, you now know how this happened!

As we begin to shop online more, the monitoring of our habits will become much easier. The techniques employed by such companies as Groupon have now got extremely sophisticated, almost to a point where they can drive you to purchase something. The offers these companies can present have made these companies very rich, very quickly – all thanks to knowing what you want, where and when - all things to be aware of in the future.

What you bought on a hot day, a cold day, a rainy day and days following all manner of events are logged, monitored and scrutinised. Complex algorithms are used to even predict what they will need on their supermarket shelves depending upon the climate. Profiling for supermarkets is very big business now.

Gone are the days of a simple malicious computer bot that gave you a bit of grief – now the hackers write specific software for corporations and governments to get all that they need form the public. Systems that hijack your computer processor are also commonplace now, these systems will use your computer to send emails to benefit the underworld and support its life of crime. These can often be embedded into legitimate firm's software packages by unscrupulous software coders – so watch out!

Phishing is now rife, more so that you would think. This is the 'art' of acquiring information such as usernames, passwords, and credit card details by masquerading as a trustworthy entity. This provides the dark side of the world with enough finance to cause the banks truly significant headaches. The techniques to acquire your information are getting more sophisticated by the day. Keeping one step ahead of the underworld is presenting itself as a real challenge.

You may well be thinking, "How the Dickens can I protect myself from all this?"

Fortunately you can … well around ninety per cent of the time if you try hard enough.

For all-round protection, check out and use all of these easy-to-use and very clever software programs:

Zone Alarm Suite, this is the best Firewall.

UnhackMe, this is the best Anti-Rootkit.

Malwarebytes, this is the best Anti-virus.

Zemana, this is the best Anti-WebCam/Anti-keylogger.

CCleaner, this is the best Tracking Cleaner.

TweakNow PowerPak, this is the best PC Detox.

East-Tec Eraser, this is the best Gutmann Hard Drive Cleaner.

Remember to 'never' open 'any' email unless you know who it is from and remember that 'any' electronic gadget may be watching and recording you!

Unfortunately, the future over the next ten years may only get worse for your privacy.

When the ultimate in quantum computing arrives, very soon, some very interesting situations will arise. Even the craftiest crook will have to watch his back, and may even be totally jiggered. Not only will 'they' be able to track every move, but there will come a day when judges and juries will simply sit and watch the act play out right there in the courtroom prior to passing judgement. Quantum computers will have such unprecedented powers that the constant monitoring of the movement inside a sealed room will be simple, the vast amount of storage space available to store each movement is already here with us today – this will become infinite in size within the 'Quantum age'. There is no reason why quantum computers could not monitor each and every movement of everything upon the planet.

With technology of today we are almost there; advanced satellite systems can pick out movement in the infrared spectrum and can monitor movements in battle situations. This advantage is enormous and so investment will be continued into this technology until we have the all-singing all-dancing version in quantum computers.

At present, even with sophisticated X400 and X500 communications protocols, the complexity of security has been arrived at through simple mathematics. As such it can all be reverse engineered. We only have to look at the Bletchley Park code breakers headed by Alan Turin to realise that with just a few smart people, the most sophisticated code can be broken.

If you wish to combat this threat to your privacy and forego this onslaught of relentless technological monitoring, I have only one short piece of advice for you, "Do not use a computer or a mobile phone and live your life inside a Faraday cage!"

Advice which is, unfortunately, impractical for most of us!

These negatives may however, be outweighed by the astounding positives now appearing on the horizon. To begin to understand the next stage of technological development which will take us towards instantaneous communication and the new world of quantum computing, there are a few principles to absorb to make sense of what quantum computing is actually based upon.

The following few sections within this chapter and the section headed 'Quantum computing', will together give you an image of how the world will begin to look in few short years from now. As we enter a world of quantum computing and faster communication in which an all-pervading quantum computing environment will materialise, there are a number of underlying concepts to understand which make this technology possible.

Wave particle duality of light

- *Light contradicts logic by sometimes behaving like a wave on a pond and sometimes like a bullet from a gun.*
- *If light weighs nothing, then it seems strange that it is capable of travelling at all.*
- *The scary reality of the real world is that observing things is perhaps the only reason why they can exist – a major dichotomy.*

With light being one of the Universe's most intriguing, significant and readily-available phenomena, understanding the intricacies of light would seem a straightforward exercise for an intelligent race. However, it has caused a great deal of despair for scientists, especially once it was discovered that light can be observed both as a wave and as a particle.

Light can be emitted from its source and observed travelling in a straight line to its destination, just like a bullet from a gun. Quite remarkably, other apparatus can be set up showing that light travels in a wavelike fashion. It will act similar to waves on a pond, and even exhibit interference patterns just like pond waves, with peaks and troughs.

The experiment that displays this phenomenon best is the 'double-slit experiment'. Thomas Young, 1773 – 1829, developed his wave theory of light, which contravened Sir Isaac Newton's view that light is a particle. Young cleverly put forward reasons illustrating that light had wavelike properties, and developed demonstrations to support this.

He compared light's interference properties with ripples on the surface of water. By placing a piece of card in a beam of light with two slits cut out to let through light, it is possible to observe a resulting image upon a screen which clearly shows an interference pattern. The only explanation for this is that light is travelling as a wave.

The underlying building block of light is energy in the form of electromagnetic radiation; it is a photon or 'quanta' of light that travels at roughly one hundred and eighty-six thousand miles per second. It radiates out from its source and is thought to be without mass, otherwise it would contravene Einstein's theories.

So, effectively something which weighs nothing is travelling as a wave at one hundred and eighty-six thousand miles per second!

This is particularly difficult to fathom.

It is extremely difficult to comprehend that something weighing nothing can travel in the first instance. Then to realise that light can span the whole Universe uninterrupted, displaying this wavelike property, seems contrary to anything we would have originally expected.

What is so baffling is that when scientists reduce the discharge of light photons in Young's 'double-slit experiment' to such an extent that they are emitted individually, over the passage of time they still show an interference pattern when passing through the slits.

How can this be?

How can a photon of light be acting as if it were being interfered with by another photon of light, when it is in isolation?

The most widely understood reason for this very mysterious behaviour is that light can assume a state of superposition. The state of superposition allows light to simultaneously be in two or more places at once.

A sharp intake of breath is required!

This is what scientists have deduced.

Until light reveals itself at its ultimate destination, it has the potential to arrive in many places. Only when light reveals itself does it make its presence known to one of these many possible destinations. Until this ultimate position makes itself known, the light has the remarkable ability to interfere with itself. This phenomenon permits the interference pattern to appear, even when photons enter the double-slit apparatus one at a time.

Breathtakingly unbelievable, but true!

One further confounding aspect of this apparatus is that when the slits are observed closely to see exactly which one of the slits a photon of light goes through, the light no longer displays an interference pattern. Light suddenly begins to behave like a particle, and no longer can the light exhibit the state of superposition that it displayed previously. Again, this occurs both when many photons are streaming through the two slits, and when they are fired one at a time.

This is thought to happen because the light's wavelike properties collapse as it assumes a particle-like form upon observation. It is therefore thought that the act of observation, or detection, determines exactly what the photon of light is, and exactly where it is. Prior to this it can be considered as purely an undetermined wave of potential light.

This is an extremely weird marvel associated with such a familiar aspect of our world. It transpires that electrons, neutrons, protons, atoms and even molecules act in this very same way. This has had the effect of thoroughly confusing scientists who study superposition. It makes them question exactly what matter really is. Some people have interpreted the results to mean that the act of observing atoms makes them appear – if they were not observed they would remain in a state of potential existence.

Heavy duty!

Now for something that is truly heavy duty and quite possibly one of the major discoveries of the last century. In the near future technology is going to bump up against our most basic assumptions about reality. Personally I am looking forward to it. The world is in the process of trying to interpret the consequences of Aaron O'Connell's discovery of superposition within visible objects. In 2010, he successfully managed to produce the first direct observations of quantum behaviour in the motion of a visible object. The fact that physical happenings at the microscopic level could not be observed at the macroscopic level, did not sit well with his intuition or logic.

O'Connell thought that if tiny little particles follow quantum mechanics then surely everything should follow quantum mechanics. He became the first person to produce an object in a quantum mechanical superposition. He took a small piece of metal, just visible to the naked eye and isolated it from its normal surroundings; masked it from light; placed it in a vacuum; and cooled it to within a fraction of a degree above absolute zero. The metal was free to act however it wanted. When measuring its motion it was found to be moving in really weird ways, instead of sitting perfectly still it was found to be vibrating, just as if it was breathing; expanding and contracting.

By giving the tiny piece of metal a nudge O'Connell was able to make it both vibrate and not vibrate at the same time. This means that the trillions of tiny little atoms that make up this piece of metal are in two places at the same time. His conclusion was that all thing things around an object combine to define what the object is.

This is a major discovery, the consequences of which will be used for a myriad of applications in the future, I am sure.

Rapid technological evolutionary improvement

- *The Earth stabilised, and then, from its perfectly balanced environment, life evolved which developed intelligence.*
- *Thousands of years ago, very slow, gradual, evolutionary improvement was all that humans experienced.*
- *Anyone living during the present time who is not keen on rapid technological improvement is living in the wrong era.*

One of the principal reasons we have evolved here on planet Earth must be our extremely comfortable proximity to the Sun. The Sun originally staked its claim to take up its particular position within the Milky Way Galaxy, then the Earth and the other planets subsequently evolved around it. With the Earth containing the ideal combination of elements and being the perfect distance from the Sun, it began to harbour life.

Once the Earth's environment had become established and stabilised, out from the primordial soup arose intelligent human beings. From this moment we were destined to become prolific communicators.

Human verbal communication came about through the passage of time and gradual evolutionary improvement; from saying 'Ug' one day to the vocabulary of Shakespeare a short time later. Our communicative vocabulary has developed so dramatically that the English dictionary today contains at least three times more words than there were in William Shakespeare's day, 1564 – 1616.

It is very difficult to accurately count the number of words in the English language because some have several meanings. Dog could be counted as one word or two, 'a type of animal', and a verb meaning 'to follow persistently'. If we count it as two, then do we count inflections separately too? Such as, 'many dogs in the park', and 'he dogs me persistently'. Is 'hotdog' or even 'hot-dog' really a word, or two words?

Over half of English words are nouns, approximately a quarter of words are adjectives and about a seventh of words are verbs. The remainder are made up of interjections, conjunctions, prepositions, suffixes and a number of other less prevalent categories.

It is suggested that there are, at the very least, a quarter of a million distinct English words. This figure excludes inflections and words from technical and regional vocabulary not covered by the Oxford English Dictionary. It also excludes words not yet added to the published dictionary, of which perhaps twenty per cent are no longer in current use. If distinct meanings were counted, the total number of words would probably approach three quarters of a million.

Hundreds of thousands of years ago, human beings would have little or no concept of change. They could have experienced only very slow change from their own growth and the growth of plants and animals. The only other change that humans could have experienced would have been evolutionary change, but this would have been totally unnoticeable as this type of change spans many generations. This lack of noticeable change remained until technology arrived in the form of early rudimentary tools.

Human beings eventually developed more advanced technology which brought with it the concept of significant technological changes, rather than the slow evolutionary changes to which they were accustomed. Dexterous limbs evolved over time in tandem with human intellect providing us with the extraordinary capabilities we see today. But now let us compare the slow speed of this gradual physical and mental evolution with something that has evolved at a truly supersonic speed, such as our means of travel!

Some estimates suggest that human beings have frequented the Earth in our current guise for approximately seven million years, since diverging from the chimpanzee and gorilla lineage. But incredibly, over the passage of just one hundred years from 1869 to 1969, our means of travel undertook an unprecedented, major, technological change from earthbound steam driven engines to three thousand ton Saturn V space rockets. From chugging along at a few miles per hour with steam puffing in the air, to propelling Mankind to the Moon and back at twenty-five thousand miles per hour took just one hundred years.

Truly incredible!

The history of powered flight has one of the most astonishing stories.

Amazingly, Sir Patrick Moore met with both Orville Wright and Neil Armstrong, two famous people from the extreme ends of powered-flight technology. Orville Wright was accredited, along with his brother Wilbur, for flying the first successful aeroplane on 17th December 1903, and Neil Armstrong was the first man to step foot on the Moon on 21st July 1969. With less than sixty-six years between the two events, it just shows how extraordinarily fast flight technology developed.

Incidentally, from observing the back cover of this book you may have noticed that I have met Sir Patrick Moore during my lifetime. On one of the several occasions I have met him, I asked him to sign the reverse of the photograph of the two of us which you see there. I reminded Sir Patrick Moore of my name, to which he replied, "I remember you very well Rob, and I will never forget you."

How fantastic!

Sir Patrick Moore and I had originally met at a Space Exploration Convention when I was a member of the AspireSpace Rocket Team inspired by John Knopp of Braintree, Essex. I had been singled out to present to Sir Patrick Moore how our rocket programme was progressing. All that needs to happen now is for me to proceed to become recognised at the inventor of faster-than-light technology – then he will have met the complete spectrum of travel technology innovators!

Considering that the Earth had been in existence for more than four billion years, this goes to show what a major change in technology took place during these astonishing sixty-six years; from the first flight to landing on the Moon.

Equally astonishing is that we rarely consider how enormously incredible the change was, during this extremely short period. In fact, our general view when we reflect back on this time appears to be the complete opposite. Ask anyone what means of transport existed a long, long time ago, and you will get answers such as steam engines, when they were actually very, very recent indeed!

The confusion possibly emanates from our experiences in times closer to the present, where we have automatically come to expect rapid and significant technological changes. Just consider advancements in industry sectors such as global travel, communication, Internet, handheld devices, nanotechnology, household appliances, genetics, aerospace and pharmaceuticals. Very little, or nothing, was known about these at the time of the Wright brothers' first flight.

Just picking out nanotechnology from this list, the future is destined to change once again beyond all recognition due to this technology alone. Nanotechnology is the science of manipulating matter at the atomic and molecular level. It will see the development of new materials; the controlling of matter; drug delivery in the body; tissue engineering; energy production efficiencies; better memory storage technology; quantum computers; food applications; and many more.

It will not be long before nanotechnology is creating products for us from scratch. Imagine having a machine in your shed that produces any product you want at the press of a button.

I really pity anyone in this era who is averse to change!

Starting from smoke signals and beacons

- *The ability to communicate quickly over long distances has been on the human race's agenda for a considerable period of time.*
- *How to achieve fast, long-distance communication was not clear until the electromagnetic telegraph was first put to use.*
- *The development of fast, global communication via the telephone has been accredited to Alexander Graham Bell in 1876.*

One key industry that is perpetually witnessing significant technological changes is communication. Before we scrutinise how the next significant communication changes may come about, we will take a quick look at the history of communication, and how it has developed from its basic infancy over the last few hundred years.

Smoke signals and beacons had been successfully used for thousands of years until 1792, when Claude Chappe invented the semaphore network. It was based upon optical telegraph which utilised good weather, daylight and stations based every twenty miles. Networks were established throughout Europe, with many of the towers still left standing to this day. Napoleon benefited from this technology, which managed to convey roughly two words a minute. The last of Claude Chappe's commercial semaphore links was decommissioned in Sweden as late as 1880.

The electrochemical telegraph was invented by scientist Francisco Salvá i Campillo in 1804, and later perfected by Samuel Thomas von Sömmering in 1809. The design had thirty-five wires up to a few miles long, each immersed in a tube of acid which would electrolyse, releasing hydrogen bubbles that represented a Latin letter or numeral.

This was the first electrical means of communication, with the receiver visually interpreting the bubbles to construct the message.

I imagine it would have been rather frustrating trying to run your social networking site over along these lines!

In 1832, Baron Schilling designed the electromagnetic telegraph, which was first put to use in 1833. In 1839, the Great Western Railway in Britain deployed the first commercial electrical telegraph. It ran for thirteen miles from Paddington Station to West Drayton. The Scottish inventor Alexander Bain then took the telegraph one step further, and developed the recording telegraph in 1843, a device that could be considered the first facsimile machine. Morse code was patented by Samuel Morse in 1837, and was very successfully utilised for commercial and military purposes.

The telephone was patented by Alexander Graham Bell in 1876, which very quickly became a tremendous global success. Then, on 3rd April 1973, Martin Cooper momentously made the first mobile telephone call to Doctor Joel Engel. Text messaging was invented as a simple application upon the mobile phone; it was not intended to become particularly popular and when it did, took the telecommunications industry by surprise. The first text message, technically called Short Message Service, was sent by a Nokia engineering student called Riku Pihkonen, in 1993.

Email was first conceived as a concept by Multics – Multiplexed Information and Computing Service. This project, which culminated in the first Email system, commenced in 1964 and allowed communication on a closed network. It was not until 1989, when Tim Berners-Lee invented the World Wide Web, that email could be networked globally, and facilitate the storage of websites.

The World Wide Web was then further enhanced with one of the most ingenious inventions of our modern day, the Internet Search Engine. The first Internet Search Engine, ALIWEB, was invented by Martijn Koster in 1993, and I was absolutely delighted to work with him for one year during those very exciting times. ALIWEB stood for 'Archie Like Indexing for the WEB'. It allowed users to submit the locations of index files on their sites, which enabled the search engine to include web pages and add user-written page descriptions and keywords. This empowered webmasters to define the terms that would lead users to their pages.

In 1991, I was involved in developing what was termed DROS, the 'Data Related Operating System', and spent many days traipsing around British banks trying to get the required funding to develop the idea into a commercially available product. The idea was that data could be communicated over the World Wide Web, be accepted into a browser and interpreted locally, hence being machine independent. The syntax of the code was relatively simple, and would allow the development of applications which could be interpreted by any computing platform located anywhere, so long as it was networked.

Unfortunately for us and the British banks, we were unsuccessful in securing the correct level of funding, and the venture was never realized. The piles of paperwork still collect dust to this day. The venture capitalists, unfortunately, had not got a clue about what we were talking about. What a crying shame for the British Computing Industry, as in 1995 James Gosling released JAVA as a core component of the Sun Microsystems JAVA Platform. JAVA proved to be a major success with its 'write once run anywhere' capability.

JAVA was supposed to have been a name chosen randomly from a list of words, but I did hear once that it was chosen as a good-humoured poke at the very acronym-riddled world of computing. I heard it stood for, 'Just Another Verbose Acronym'.

However, I also heard that this could be just a myth.

So now we have data and voice communication whizzing around the world at the speed of light, what more could we want from communication?

Quite strangely, the speed of light is actually rather slow for anything other than local Earth-bound communication. For distances further than this, something faster is required. If we are to voyage further into space, then we must develop a high-speed communications method that circumvents the restrictions caused by the speed of light.

Perhaps we will have to turn to the quantum world to achieve faster-than-light communications. I have seriously thought for some time there must be some useful communications mechanism lurking within the depths of the quantum world!

In 1997 I began contemplating how some of the known quantum effects could be utilised to improve communication. I identified some of the weird effects that showed clear signs of contributing towards faster communication, and scribbled a few diagrams and notes.

I went to a reputable patent attorney in London to 'test the water' regarding this communications method, and was met with positive bewilderment and jovial astonishment. I came to realise there is sufficient interest in such a communications mechanism, which is a relief after my inability to initiate DROS!

My experience at the patent attorney was extremely positive, and they recommended that I should persevere by creating a working model. This I feel may be rather difficult, as I do not have access to a suitable laboratory!

I thought perhaps the best thing to do was to just divulge the communications mechanism within this book, and leave it to the rest of the world to develop further. I feel I am relatively well-qualified to invent and divulge such a communications mechanism, as my credentials on the subject are fairly reasonable.

My first-hand experience of commercial communications has given me in-depth knowledge and understanding of this leading edge technology, its benefits and challenges. Each project I have worked upon has given me a greater insight into specific aspects, broadening my overall appreciation of communication.

I helped develop and maintain military standard X400, X500 secure messaging, which gave me an understanding of the importance of encryption methods. I headed a major UK airline's IT systems and communication infrastructure. This experience provided me with a clear understanding of the deployment and criticality of communication. I was instrumental in the development and programme management of Motorola's end-to-end solutions; namely the Internet to the mobile phone. This experience provided me with an understanding of the capabilities of communication. In Spain, I personally orchestrated the first ever GPRS handset to be connected to a live network. This experience provided me with an understanding of how new revolutionary technologies can be implemented. I managed the development of the Odin device – the first colour mobile-based device for Psion and Symbian. This experience provided me with a thorough understanding of handset technology. I was 3G Programme Manager for one of the UK's leading mobile phone operators. This experience provided me with an insight into the operations of a large communications network. I was a Director of one of the UK's leading mobile phone companies. This experience provided me with an in-depth understanding of how to orchestrate a communications infrastructure.

My association with computing and communications has spanned over thirty years to date. So let us see what this, along with my graduation in computing and my in-depth study of quantum physics can achieve to help pave a way forward.

Incidentally, my study of quantum physics was specifically focused around a deep interest in quantum computing, quantum communications and quantum weirdness.

Let us see what this does!

The inhibitor to intergalactic communications

- *The speed of communication is governed by the speed of light – nothing can travel faster than this.*
- *Light can travel almost six trillion miles in one year – it takes just over four years for light to reach us from the nearest star.*
- *A much better and faster means of communication must be developed if humans are to explore deep space.*

Most people consider the speed of light as rather fast. However, even though light travels at an incredible one hundred and eighty-six thousand miles per second, it is, in fact, the inhibitor that places an upper limit to the maximum speed at which communication can travel.

If we contemplate utilising a mechanism based upon the speed of light as a means of intergalactic communication, then we have problems!

Let us firstly contemplate the enormity of the Universe to see how big the problem actually is.

When attempting to absorb the enormity of the Universe, you have to disengage yourself from your normal, everyday, terrestrial, 'short-hop' distances such as trips to the shop, holidays, train journeys and plane fights. Distance yourself from these minute distances to engage your mind to become liberated within an almost unfathomable Universe, with its billions and billions of galaxies, each containing billions and billions of stars. With universal distances being measured in light years, our minds boggle when wrestling with the fact that one light year is almost six trillion miles, and the nearest star to our Sun, called Proxima Centauri, is just over four light years away. Ironically, Proxima Centauri is not visible to the naked eye, it is too faint and is only one-seventh the diameter of our Sun – it was only discovered in 1915 by Robert Innes.

Consequently, to attempt to colonise a planet orbiting around the closest star to our Sun, a distance of roughly twenty-five trillion miles will have to be traversed through the void of space, before we can actually confirm there is a suitable planet on which to land and inhabit. Bearing in mind the Moon is just under a quarter of a million miles away, this puts the distance to Proxima Centauri into some kind of perspective. Proxima Centauri is one hundred million times further away from Earth than the Moon!

If we were to travel to Proxima Centauri at twenty five thousand miles per hour in a Saturn V space rocket, it would take us approximately a billion years to get there. Plenty of time for things to go wrong!

Therefore, we can see that a significant change in communications technology is required before any type of intergalactic travel of this magnitude can be considered.

Our current communications infrastructure only operates at the speed of light and is not at all adequate. Speed of light communication proves to be extremely sluggish over these larger distances, and for this reason is totally inconvenient. When communicating instructions and messages to probes on Mars and other planets within our solar system, we have to contend with a number of minutes' delay. The time delay is made better or worse depending upon the planet's orbital distance from the Sun and relative position to the Earth – so the communication time delay to distant planets alters throughout the Earth's orbit.

It would be totally inappropriate to send a perilous pioneering expedition of intelligent astronauts hurtling off into outer space to colonise another prospective world within another solar system, with the only means of communication being a two-way radio featuring a four-year gap between conversations. It would take four years for us to get any message to them at the speed of light and then another four years for their response to return.

A technological challenge has therefore been set before us. We must now develop an acceptable communications mechanism for such a long-distance voyage.

The various options available

- *Communication is now at the heart of everyday life – it is now an essential aspect of our daily routines.*
- *Time delays are a major problem with communication operating at the speed of light, as distances within the Universe are so vast.*
- *Speed of light communication is useless for intergalactic communication, as it will take years to transmit a single message.*

Prior to the invention of communication technology, such as the telegraph, telephone and email, the human race was able to communicate only by written correspondence, visual correspondence, messenger on horseback or smoke signals. Before a faster means of communication was invented, the speed at which people communicated was never construed as a problem. Everyone managed their communication in accordance with the communications means available to them. No one could contemplate anything superior.

Civilizations happily geared themselves up around this laid-back communications model, and the world worked flawlessly. No one was aware that communication was about to significantly advance, and become an intrinsic and absolutely essential part of everyday life.

It would have been impossible to imagine, all those years ago, how anyone could browse goods from around the world upon a screen in the comfort of their own home, and then at the click of a button have any item delivered to their door shortly after.

That is not possible with smoke signals!

Along came the telephone, which could miraculously relay real-time conversations over considerable distances. A freak, technological breakthrough, that suddenly allowed you to speak with someone when they were miles away. This was an extraordinarily strange experience at the time, as sight and sound had always gone hand-in-hand. Most surprisingly, many of the Victorian people of the day had great difficulty contemplating any practical use for this new device. The average person had difficulty seeing any practical reason why anyone would need to talk with someone who was not actually in the vicinity!

To them it was similar to being blind, and conversation was limited to "Hello ... hello ... isn't this funny".

A similar lack of vision was evident when Michael Faraday invented the electric motor in 1821. Having finally mastered electricity, Faraday was demonstrating it to Prime Minister William Gladstone. William Gladstone turned to Michael Faraday and said, "What practical value has this electricity?" To which Michael Faraday answered, "One day, sir, you may tax it."

Electricity made its mark on planet earth quite rapidly, changing the way we do most things, and was duly taxed!

Eventually the telephone found its way onto the Victorian stage as a form of magic trick, alongside the great performers of the time such as escapologists, strongmen and fart artistes. In 1876, the year that Alexander Graham Bell patented his invention, there seemed as much fun having a fart artiste make you laugh as there was in seeing a telephone work for the first time.

The take-up of the telephone could not have been helped when Alexander Graham Bell refused to have one in his study, as he found it rather intrusive!

How incredible!

Alexander Graham Bell certainly had vision, as I do not know anyone who has not found telephones rather intrusive in their lives at one time or other!

It is a very good job that the telephone caught on as a means of communication rather than the fart artiste. Otherwise we would be living within a very peculiar and decidedly different type of world today!

Phew!

The speed of light dictates the speed at which current communication can operate. Thankfully, the distances over which telephone calls needs to operate is relatively small compared to the distance light can travel in a second, so we do not notice any delay here on Earth.

Light can travel around the World eight times a second, but when utilising 'speed of light' technology to communicate interstellar distances, we begin to experience extremely inconvenient time delays.

To put this into some kind of perspective, imagine we live upon the surface of one of the largest know stars, called VY Canis Majoris. Its circumference measures roughly twelve billion miles. Utilising current technology there would be at least a ten-hour time delay between conversations when speaking from one side of VY Canis Majoris to the other.

This delay is incredible, especially as the two people communicating are both upon the same celestial body. The size of this enormous universal structure highlights the comparatively slow speed of our current communications technology.

If we were to live on Jupiter, we would similarly experience a noticeable time delay, but not to such a remarkable degree. Jupiter's circumference is roughly two hundred and eighty thousand miles compared with the Earth's much smaller circumference of twenty-four thousand nine hundred miles. Light can travel around the Earth just over seven times a second, but in the same time it would not make one complete revolution of Jupiter. As a result, our telephone conversations on opposite sides of the planet Jupiter would be interspersed with annoying moments of silence.

Our current, conventional, 'speed of light' technology is totally impractical for intergalactic communication, due to the inherent time delay. For such relatively large distances, there is undoubtedly a genuine requirement for a totally different communications mechanism.

For anything to travel faster than the speed of light is considered to be nigh on impossible, due to the natural rules highlighted within Einstein's Special Theory of Relativity. However, this has not prevented people from challenging this convention in an attempt to derive a revolutionary solution, and become the first to discover a long awaited 'faster-than-light' phenomenon.

Gallant attempts to improve communications

- *Some extremely ingenious methods have been conjured up to attempt to communicate faster than light.*
- *Some scientists live in hope that there are undiscovered physical properties within the Universe which will permit faster-than-light communication.*
- *The science behind the future of communication could be mistaken for science fiction – the warp drive was devised by Miguel Alcubierre.*

Scientists have long dreamt of achieving faster-than-light communication. The sheer thought of it is intriguing, and the rewards attached with such an innovation would be tremendous. Gallant attempts over the years to discover 'faster-than-light' communication have resulted in terminology such as tachyons, Alcubierre drives, rigid bodies, rotational effects, and traversable wormholes. Some of these attempts have been ingenious, others purely science fiction, and others a little more serious. To date, none of the attempts have been successful. However, we will take a brief look at each as they are all perfect examples of 'thinking outside the box' and thinking differently.

Tachyons were first described by the German theoretical physicist Arnold Sommerfeld, 1868 – 1951. They are hypothetical, subatomic particles that travel faster than the speed of light, but as they have never been detected are only realistic within the context of a science fiction film.

The Alcubierre drive, or Warp Drive, was a concept devised by Mexican physicist Miguel Alcubierre in 1994. It is a speculative mathematical model that theorises the expansion and contraction of space-time, causing the fabric of space to the front of a spacecraft to contract and the space behind it to expand, thus thrusting the spacecraft faster than light relative to everything around it. However, there is no way of creating a bubble within which the spacecraft may reside, and hence we have to leave this as a theoretical concept for the time being.

Rigid bodies are interesting and something that I believe would be a great deal of fun trying to prove. If you have a long stick reaching from the Earth to the Moon and you push one end slightly, surely you would feel the prod movement immediately on the Moon. Normally it takes just over one second for conventional communication to arrive at the Moon from Earth at light speed. Are we on to something here?

Alas, the elasticity of the materials always turns out to be much slower than the speed of light. You could, however, imagine a new material whose elasticity renders it totally and utterly solid. In this way, you could say that theoretically there could be faster-than-light communication. Unfortunately, this may never be the case in our Universe.

The 'rotational effect' consists of standing outside on a clear night with the Moon on the horizon, then spinning yourself round and round about once a second. Now if we work out how fast the Moon is spinning around your head relative to yourself, it turns out to be roughly eight times the speed of light!

Bearing in mind that general relativity states that all coordinate systems are equally valid including revolving ones, does this mean the Moon is going faster than the speed of light as far as you are concerned?

Unfortunately not, as velocities in various places cannot be directly compared in general relativity, so this idea falls flat. However, this is an extremely ingenious attempt, utilising original thought, showing that a person can observe a faster-than-light experience, albeit a pseudo experience.

A traversable wormhole is unfortunately something that has so far been restricted to films. It is theoretically possible, according to Albert Einstein. Supposedly space can warp and stretch space-time, creating a shortcut from one location in space to another. The creation of a wormhole would be nigh on impossible, as it would involve travelling through a rip in the fabric of space to appear somewhere else within the Universe. From the sound of it, I doubt very much anything would survive the journey through the rip in space anyway!

My favourite faster-than-light mechanism has not yet been given a name, I have heard only of the concept. The set-up consists of two planets and a dual light beam-generating unit. Imagine the two planets are separated by a distance of one million miles. The dual light beam-generating unit is specifically designed to emit blue light in one direction and red in the other.

It would normally take about six seconds for light to traverse the one million mile distance between the two planets. If we now position the light beam-generating unit centrally between the two planets, it will take roughly three seconds for light to travel from the light beam-generator to each of the planets.

If one of the planets receives a flash of red light three seconds after the unit is given the instruction to emit light, then the inhabitants of this planet instantly know that the other planet has received a flash of blue light!

This mechanism has managed to relay knowledge at twice the speed of light. The planet receiving blue light knows that the other planet has received red light and vice versa. Yes, this all happens at twice the speed of light.

Although this is true, it also relies upon the dependability of the beam-generating equipment. Original real-time communication cannot be relayed, only previously defined knowledge. This renders the mechanism's use extremely limited and unworkable for effective real-time communications. Irrespective of its limited capability, it is extremely interesting all the same.

These gallant attempts are all very commendable, but alas there are shortfalls within each. There is, however, hope for a breakthrough, as you will discover by reading on!

Our new communications mechanism for a new era

- *Who would have thought that matter was made up in the way that Mankind ultimately discovered – some very special people helped pave the way.*
- *The behaviour of particles completely baffled scientists as they battled to discover the secrets of matter.*
- *Quantum weirdness, or non-locality, displays astonishingly peculiar results that baffle the scientific world.*

I would like to propose a new 'faster-than-light' mechanism, but we need to firstly familiarise ourselves with a few little known facts at the subatomic level. Nothing too severe, I promise!

An atom is absolutely tiny and, unlike anything we could imagine, holds a charm of its own on a scale that we can scarcely believe. It was the behaviour of atoms which finally revealed the secrets relating to their size and make-up. It was first thought that matter was made up of something similar to plum pudding, before it was settled upon that they are actually made up of almost nothing whatsoever!

Our current understanding of the components of an atom has been bequeathed to us only since James Chadwick discovered the neutron in 1932. Earlier work undertaken by Ernest Rutherford, J. J. Thomson, Fredrick Soddy, Margaret Todd, Neils Bohr, Gilbert Lewis and others helped pave the way, but it was not until 1964 when quarks were first proposed, that we had any inkling that they were the smaller building blocks of neutrons and protons. This discovery is scarily recent, and we have had hardly any time to act on this information and extrapolate the full potential of exactly what this realm of the very, very small has to offer.

Now that we are beginning to understand the subatomic level, and have finalised our model of the atom, we have confidently printed it within text books. At this point, you would have thought everything would be plain sailing. However, how wrong you would be; there are so many unknowns it is incredible. The behaviour and nature of particles at the subatomic level has thrown scientists some true googlies. Entanglement, superposition, quantum weirdness, wave particle duality and the quantum leap are names of just a few examples of this very peculiar realm of subatomic behaviour. These phenomena at the quantum level have perpetually bamboozled physicists attempting to reveal the mysteries of the very, very small.

Thanks to some extremely clever scientists, we can now go a step further into this very strange subatomic world of particles and atoms that make up everything in the Universe. At this point I need to explain some very exciting but peculiar phenomena relating to something that has been termed 'quantum weirdness'.

Quantum weirdness, or non-locality, exhibits some extraordinarily interesting characteristics, and displays what is called the nonlocal property of particles. In a nutshell, this is where substance goes completely doolally, and particles disappear and reappear elsewhere without traversing the intervening space. Scientists have tried very hard to explain exactly what is happening, and a few think they have managed to come up with a fairly meaningful explanation, but what is interesting is that, irrespective of the reasoning behind this mind-blowing behaviour, it is consistent, and may be utilised to our advantage.

Spooky action at a distance

- *When looking at 'spooky action at a distance', one could be forgiven for thinking it is a paranormal phenomenon.*
- *A particle is capable of traversing the Universe, totally unaware of what its properties are.*
- *It would also appear that a subatomic particle does not know what it is until it has reason to reveal its identity.*

If you get excited about thinking that the Loch Ness Monster may exist, or have dreams about discovering a missing link within the evolution of man, or even if you are just an advocate of the paranormal, then you are in store for a pleasant surprise from this section of the book. The explanation of 'spooky action at a distance' or 'quantum weirdness' is significantly more exciting than discovering the existence of the Loch Ness Monster. I sincerely believe that the phenomenon fits plumb into the category of the extreme paranormal. No disrespect to the Loch Ness Monster or Piltdown man!

Imagine Einstein's surprise when he realises there is a truly weird spectacle, so weird that he coins it 'spooky action at a distance'. Upon witnessing it, Einstein even questions whether God is playing dice with the Universe by saying, "God does not play dice with the Universe".

Einstein is in disbelief regarding what he observes, and remarks, "Quantum mechanics is certainly imposing. But an inner voice tells me that it is not yet the real thing. The theory says a lot, but does not really bring us any closer to the secret of the old one. I, at any rate, am convinced that he does not throw dice."

No one could believe what they were witnessing at the subatomic level. It would appear that electrons and other subatomic particles can behave very peculiarly indeed. In certain circumstances, particles will miraculously disappear and reappear somewhere else, some particles can be everywhere all at once, and at other times they can become entangled together as twin particles.

All very true and very weird!

This crazy behaviour of particles is extremely counterintuitive in contrast to the behaviour of items to which we are normally accustomed.

A simple one hundred watt light bulb emits roughly one million, million light particles, called photons, every second. They burst throughout the vicinity of the bulb to illuminate everything around. These photons may entangle themselves with other photons to become twin particles which subsequently travel off in different directions.

Particles naturally decay and convert to become other particles. In doing so they conform to conservation laws that result in pairs of particles being generated that find themselves assuming conjoined states. This effect is called quantum entanglement, and is a naturally occurring wonder.

Imagine a pair of particles that have two states, 'spin up' and 'spin down'. These particles can now be viewed as one entangled status of two particles, and may not be described individually without making reference to the other. The case we just described is referred to as an anti-correlated entangled state as the properties of the particles oppose each other. If the spins were to be the same, they would be referred to as being in a correlated entangled state.

Within the quantum world, a particle's properties cannot be determined until it is observed; it is said that a property of a particle is indeterminate until an observation or measurement is made. When measuring the spin of many similar, 'normal' particles, the result will yield an unpredictable series of random fifty-fifty probabilities of spin up or down. However, when measuring the spin of entangled particles, their spin is correlated. Detecting the spin of one of the entangled pairs tells you exactly what the spin of the other is!

Unfortunately, there are other known behaviours of particles that cause a difficulty with this situation. There is something called 'The Copenhagen Interpretation', which demonstrates that a particle will only know its true identity, or properties, the moment it is observed or measured. It may sound rather strange that a particle can traverse the Universe not knowing what it is, but this is one of the underlying principles of 'The Copenhagen Interpretation'.

So with this in mind, when the first particle of an entangled pair is measured, the state of the second is known at the same instant. This is regardless of the distance between the two particles. The knowledge about the state of the second particle is the problem scientists have with quantum entanglement. As the distance between the two entangled particles is irrelevant, and could be millions or billions of miles, this means that the information about the status of each must be travelling faster than the speed of light. However, this does not conform to Einstein's principle of general relativity.

Scientists find this area of quantum physics very messy, and have developed all manner of reasons to try to explain the faster-than-light transfer of information. A number of theories such as the hidden variables theory, Bell's inequality, Bell's theorem and the Bohm interpretation, have been proposed to try to explain this result.

It is all very mystifying, and causes sleepless nights for many quantum physicists!

The mere observation of subatomic particles, such as light, affects their characteristics as they actually act as a potential particle until observed, or questioned, about what they actually are. It is almost as if particles travel from source to destination in a different form, within a wavelike structure, as opposed to what we would naturally expect them to be; like bullets being fired from a gun.

Only when the wave structure is observed does it give up all its secrets, and the wavelike structure is abandoned to reveal itself as a particle with particular characteristics and properties. The wavelike structure that it assumes as it travels means that it can be interfered with by other similar waves, just like the waves of an ocean. However, what is especially confusing to everyone studying this phenomenon is that, even if there is only one particle travelling within this wavelike structure, it can be seen to remarkably interfere with itself, demonstrating unequivocally that it is in more than one place at once. In fact, it could in reality be actually in an infinite number of potential positions.

Spooky and weird!

Within an instant, the particle becomes observable at its destination, randomly deciding truly where it is, and assembling itself out of the wavelike composition into a particle-like composition. It is at this instant that the wave-function of the potential particle collapses, and the particle appears to randomly present itself in a location that conforms to just one of its possible locations.

This phenomenon could be looked upon as particles existing in a state of potentiality, with just a probability of existence in a particular place. What is so truly miraculous and so counterintuitive is that even in the absence of any other waves of potential particles, these individual waves have been witnessed interfering with themselves. This ability for a particle to be in many places at one time is referred to as being in a state of superposition, exactly the same as being everywhere all at once.

It would also appear that a subatomic particle does not know about its own properties until it is observed by something else; as a consequence it has the ability to defy interrogation during its journey as it traverses space in a wavelike structure between its origin and its destination. It is as if a particle travelling as a wavelike structure does not know what its identity is upon its outset. This can be deduced by placing interrogation devices along its journey, and studying the logic associated with the ultimate outcome at a final detector. The final detector can produce results that totally defy all conventional logic.

It would appear that a particle travelling as a wavelike structure has its properties encrypted inside itself. It is a little like a traveller having not packed his own suitcase, so when questioned by border control whether he knows what he is carrying he is unable to say.

"Are you carrying a gun," asks border control to the traveller.

"No," replies the traveller.

Border control lets him through without checking his suitcase. It is only when the traveller arrives at the hotel that he looks inside his suitcase and gets to know the purpose of his mission. Little does border control know that if they had looked inside the traveller's suitcase they would have discovered he was carrying a gun, a computer for communication, a note explaining what to do upon arrival and a bright orange uniform!

Instantaneously aware of interrogation

- *At times, at the level of the very small quantum world, it is as if there is no intervening space between particles.*
- *There are recently-discovered weird phenomena at the quantum level that have yet to be put to practical use in this world.*
- *Particles have a series of properties such as spin direction, charge and colour – these can be detected by apparatus.*

Most miraculously, when an entangled pair particle is identified or determined at its destination, it appears to be instantaneously aware of any interrogation devices to which its other particle pair has been subjected. If one of the particle pairs has revealed its identity, then this will be instantly known to the other pair. This happens even if the pairs of particles are massive distances apart; it is as if the space between them does not exist. This faster-than-light acknowledgement occurs between entangled particles even if the distance between the two were millions of miles!

As in our earlier analogy, it is as if two travellers set off from the same destination in different directions and are both totally unaware of the content of their suitcases. Neither traveller can explain what their purpose is until they look inside their suitcases. Both suitcase contents could potentially be anything. Both travellers manage to fool border controls along the way, until one of the travellers unpacks his suitcase. From this moment the other traveller is unable to fool any more border control officials. The second traveller now knows what their mission is about. It is as if one traveller can make the contents of the second traveller's suitcase fall out, all triggered instantaneously from an enormous distance.

But it gets even more intriguing.

Whenever a suitcase is opened by one of the travellers, only half the contents inside actually becomes known.

The travellers certainly do have unopened suitcases during their journeys, but this quantum weirdness is even more whacky than people first thought; while the suitcases remain unopened they can get information sent to them back in time from before the suitcases were going to be opened. The only really sensible explanation to this is that the contents inside the suitcases are not set until the case is opened.

So in reality, the border control would have no hope in stopping a travelling smuggler and the potential smuggler would pass a lie detector because, as far as they are concerned, they are telling the truth. They have no gun in the suitcase – but then it can be mysteriously teleported there as though out of a Star Trek episode.

Einstein's 'spooky action at a distance' blatantly exhibits the bizarre nonlocal behaviour of particles.

The entangled particles will unashamedly lie about their characteristics, and pass through interrogation devices, prior to arriving and revealing their ultimate identity at their destination. This reveals that a particle has the ability to display particular characteristics to an interrogation device earlier on in its journey, and only appears to change its characteristics when determining its true identity. This implies that whilst in its wave form it just does not know what it is, and the fact that it defies logic as it passes through interrogation devices, shows that the particle did not have a clue as to what it was from the outset.

When one of an entangled particle-pair is suddenly forced to reveal its identity, it may then render the interrogation device results for its partner erroneous, even when the interrogation devices have already been navigated and determined. This is quantum weirdness!

If you did not understand all of this, then, do not worry – not many people at all do!

How on earth can the particles of matter that we know and love so much in our normal world behave in this way?

It certainly does not justify a trip to your psychiatrist if your mind feels muddled, but if you do understand why this happens then it is certainly worth a trip to the Institute of Physics to explain it to them.

You may be worth a fortune!

What people have not done to any degree of detail is to speculate what practical use this may have in our world, irrespective of why particles behave like this.

I can now sense you thinking about the possibility of this wave particle duality of light, with its associated idiosyncrasies, becoming the basis of a new communication mechanism. Surprisingly, how right you would be!

So what can we accomplish with this spooky action at a distance?

Well, let us examine it closely to see whether there are the constituent components for a communication mechanism hidden within there somewhere.

So we can be familiar with the precise events occurring at this quantum level, and relate more closely to the exact behaviour of these tiny particles, I will illustrate all that is happening as if it were to be witnessed at our more understandable level of the world of comprehensible 'larger' things.

What follows is an account of the associated activities occurring at the quantum level as if it were to be happening to something familiar, such as cars, at our level of the large.

This spooky behaviour of the subatomic particles we have just recounted happens with light and equally well with electrons. Light is made up of photons which are small packets of energy. We will now imagine them to be cars.

What we will to do for the purpose of this analogy, and to convey a clear message, is to imagine that a photon of light is a motor car. A photon has attributes associated with it such as spin direction, charge and colour, but if we look upon them as cars instead it will be easier to envisage the strange happenings, and relate to how bizarre particles, such as light, act.

To perform quantum weirdness experiments in a laboratory, physicists require a reliable light source or a laser to produce a constant source of photons. However, in our black and white cars analogy we will imagine an enormous car park full of cars – the car park will emit cars just like a light source emits photons. In our experiment, the cars will travel in straight lines along roads, just like light travels directly from a light source to its destination. At points where light could be interrogated for certain properties and diffracted, our cars will be observing road signs and taking the appropriate turnings.

Certain characteristics of our cars will make our experiment interesting – we will imagine that the cars are randomly coloured back or white and they are being randomly driven by men or women.

The cars and the particles

- *We set up apparatus that sends particles along a route – just like we may direct certain traffic along particular roads.*
- *There are other parts of the apparatus where particles may choose randomly to go north or south, just like cars joining a motorway.*
- *Amazingly, to everyone's total disbelief, we find particles at one of the destinations that we were sure the apparatus had directed elsewhere.*

To imagine how weirdly particles behave at the subatomic level, imagine we have a car park full of black and white cars belonging to both men and women. Road signs will direct the cars along a system of roads, which will result in some people arriving at a bar to enjoy a drink. This is the essence of our black and white cars analogy.

As the cars drive out of the car park and travel along the road, after a short distance there is a turning they may take. There is a road sign by the turning which instructs all female drivers to take the turn, leaving all the male drivers to continue along the road unaffected.

The male drivers then arrive at a fork in the road giving them a choice of direction – they may either take the road which heads northwards or take the road which heads southwards. The male drivers randomly choose to go northwards or southwards. A little further along each of the roads, irrespective of whether they chose northwards or southwards, they encounter another road sign which instructs all black cars to turn off – leaving all the white cars to continue along the road unaffected.

If all drivers have paid attention to the road signs correctly, which we shall presume they have, along the two main roads there will be just white cars being driven by men – some heading northwards and some heading southwards. Both the northerly and southerly roads then make a sharp turn which heads all traffic towards the bar, where all that arrive can get out and enjoy a drink.

Amazingly, to everyone's total disbelief, we find that the people who pull up at the bar are not as we would expect. All cars that arrive at the bar are white, but for some weird reason they are being driven by both men and women. How did the women arrive there, when clearly, if they had followed the directions properly, they would be miles away at this point!

The road sign had clearly instructed all women drivers to take a turn shortly after the car park, near the start of the journey.

This defies all logic, but this is exactly what happens with photons if we treat them in a similar way using their specific properties.

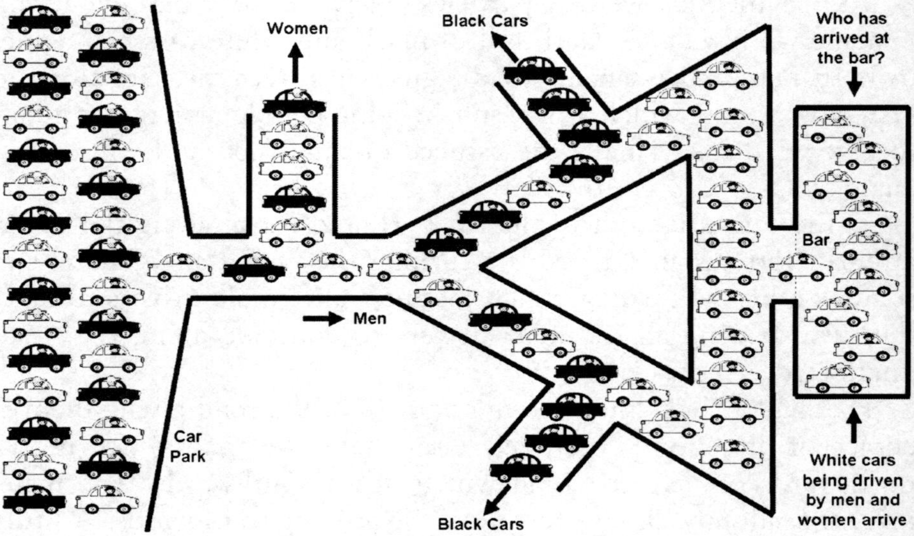

Amazingly – despite appropriate turnings being taken, all white cars being driven by both men and women arrive at the bar!

The diagram explains a very strange phenomenon that occurs at the subatomic level, but has been enlarged to something to which we can relate.

Let us just recap this bizarre setup. There are black and white cars being driven along a road by both men and women from a car park. All female drivers are told to turn left very soon after the exit from the car park. Then there is a fork in the road which those travelling along can randomly choose to take. The black cars along both forks are now told to turn off. The two roads containing only white cars being driven by men are then merged back together again arriving at a bar. Amazingly, at the bar we find both men and women in their white cars.

At the subatomic level, this is what is referred to as quantum weirdness, and has become one of the most amazing physical mysteries.

Some believe that the nature of the subatomic world does not allow you to know two properties about a particle. So having already directed particles along a path due to an earlier interrogation device, when they are analysed in the second device, the result of the first no longer holds true, even though we know the first filtering process works perfectly well in isolation.

We shall now move on from the car analogy to take a close look at what happens in the real world. The properties of light are many and varied – but for pictorial reasons we shall imagine light particles may be black or white and positive or negative.

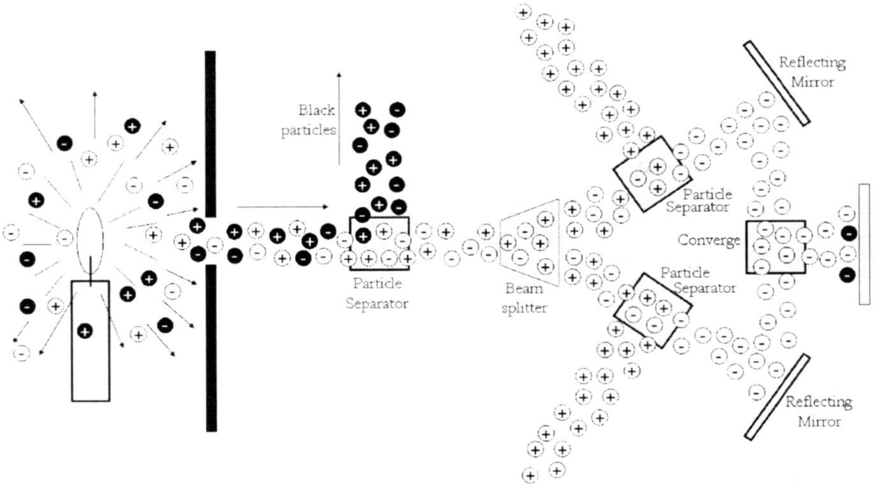

Beams of particles are emitted by the candle. Each particle has properties: they are black or white, also they may be either positive or negative. Even though the apparatus separates out the black particles early on, they miraculously reappear at the detector to the right of the apparatus. Truly bizarre!

Once familiar with the events that occur in the previous diagram, there is another interesting twist to fathom. With a mechanism placed to detect the property of the particles on the northerly path, irrespective of whether there is a mechanism to detect the property of the particles placed on the southern route, the detector still reports as if it were there – this can only have been influenced by there being one on the northerly route. This happens even though knowledge of this would have taken faster than the speed of light to inform the particle at the detector, not even if the particles have passed the beam splitter before someone decided to insert the detecting mechanism on the northerly route.

This defies all logic, but who cares?

Our greatest scientists have had sleepless nights thinking about why this happens.

We should not worry about it.

Let us utilise the effect to our benefit!

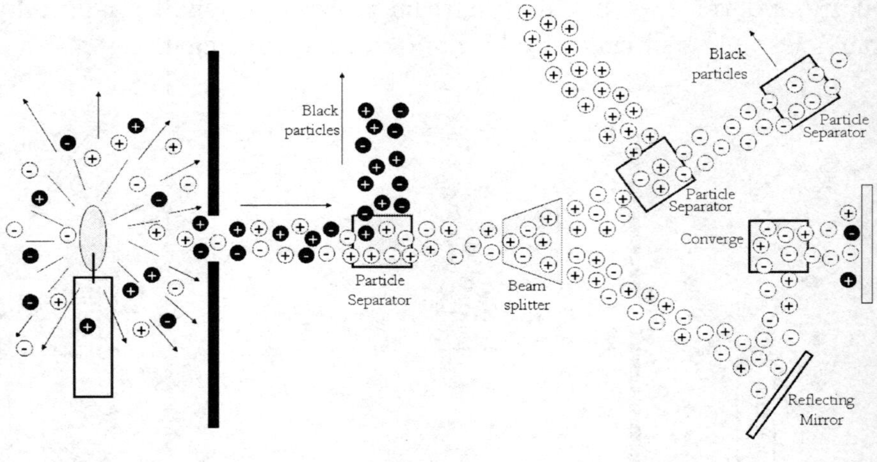

Amazingly, if we insert the particle separators on just the upper route then we get similar results as before. This is without any experimentation taking place on the lower route – this means that the knowledge of the particle separator being in position on the northerly route is relayed to the detector faster than the speed of light.

So we know the result when we 'do not' detect at a distance along the upper path, and we know the result when we 'do' detect at a distance along the upper path. Different states result, like on and off. So if the detector is inserted, this has the effect of toggling the result. If we place the detector in and out like a piston, then surely we have a means of sending messages in binary ones and zeroes over limitless distances instantaneously.

So we can now draw the apparatus that does this and take a closer look at exactly what is going on.

The particles are fired from source through an initial detection mechanism, and the particles that continue through then race towards a beam splitter. The beam splitter acts purely as a means of dividing the stream of particles into two, without any detection mechanism.

We now have a lower and upper beam of similar particles which are entangled together. If we now interrogate the upper beam for a particular property, the lower beam instantly recognises this fact; and because of this, we have invoked and thus contravened the principle that you can only know one thing about a particle, namely Heisenberg's Uncertainty Principle. This has the effect of strangely invalidating the initial detection mechanism, and particles which were at first glance no longer within the upper or lower beams, miraculously reappear at the detector. If we now take the detector out of the upper beam, instantly we get a different result at the detector at the end of the lower beam.

So by placing in, and removing out, the detector on the upper beam, we can make the detector on the lower path yield different results instantaneously, irrespective of the distances involved. This includes the distance from the beam splitter and the piston motion of the detector on the upper beam. So here we have a communications mechanism that can traverse limitless distances instantaneously, passing information effortlessly.

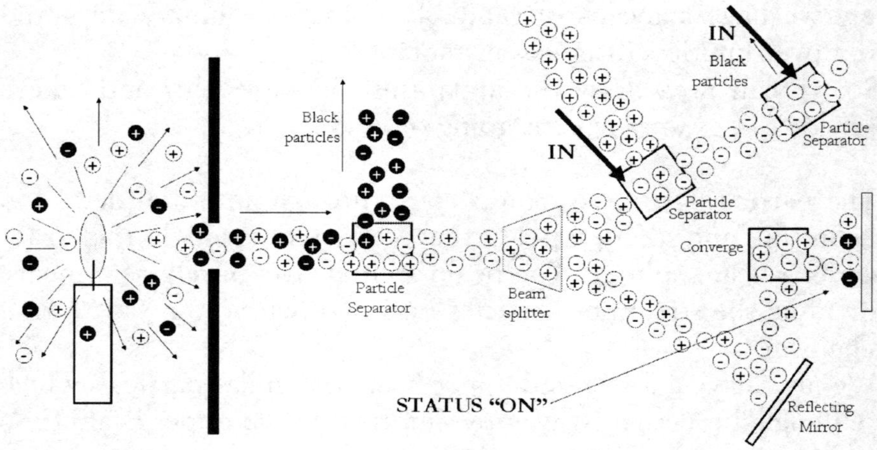

By placing in, and removing, the detector on the upper beam, we can make the detector on the lower path yield different results instantaneously, irrespective of the distances involved – the above diagram depicts status 'on'.

We can use line of sight, or cable, or both, to send messages of this nature. The communication path is from the point where the detectors are positioned on the upper beam across to the detector.

How is all this possible?

Heisenberg's Uncertainty Principle explains the more you know about one property of a particle, the less you can know about any other. It is a very counter-intuitive concept and it is difficult to understand the reason why, but we can see this operating with a very simple experiment.

If a laser beam of light is passed through a small pea-sized hole, we can witness the resulting light emerge from the hole by displaying it on a screen. Now if we make the hole narrower and narrower, something very strange happens when the hole gets to about one hundredth of an inch wide; the dot on the screen suddenly spreads out.

It is no longer a speck!

How can this be so?

What is causing this to happen?

It happens quite miraculously because our Universe very strangely will not allow an observer to know more than two things about a particle. Because we know so precisely where the laser light is on the horizontal plain of the tiny opening, when it subsequently emerges from the hole, the direction of the light can no longer be determined. It would appear that the particle's sense of direction is made random once we know where the particle is. So what we see is the light spreading out on the horizontal plain; very non-intuitive, but it is the way the Universe works!

It is worth considering whether we are seeing Heisenberg's Uncertainty Principle at work within our setup, perhaps coupled with superposition!

No one has ever been able to explain the strange phenomenon observed with this apparatus setup; perhaps a number of quantum effects are acting together.

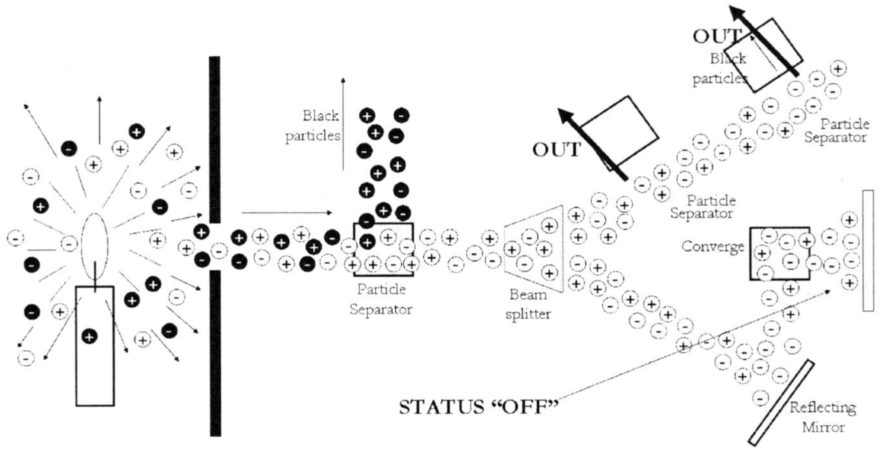

By placing in, and removing, the detector on the upper beam, we can make the detector on the lower path yield different results instantaneously, irrespective of the distances involved – the above diagram depicts status 'off'.

Interestingly, the apparatus still works the same when we fire just one particle into the system. What this fact implies is that the particle traverses along both the upper and lower paths in the superposition state, and is capable of interfering with itself.

Knowing that the particles are fully conscious and aware during their superposition state as they travel through the system, then if there is some way of changing the outcome quickly and repeatedly, whereby the final detector can acknowledge these changes, we have created a communications mechanism.

So, having drawn the apparatus required to establish this communications mechanism, I suddenly realised that the set-up is identical to the inner chambers of the Great Pyramid at Giza.

Was this a coincidence, or ingenious design?

Well, I know which one I would put my money on!

Does this mean that, at one time, we were part of a large network of interlinked planets, happily communicating, sharing ideas and satisfying ourselves in the knowledge that other beings are caring for, and sharing, our Universe?

We can only speculate at the types of conversations that could have been traversing the Universe before, unfortunately, something went wrong, and it became inoperable.

We now float amid the constellations as a failed planet, unable to keep abreast of the latest universal gossip. Unaware of our past, uncertain about our future, and so ignorant that we did not even know what the Great Pyramid at Giza was – when it was staring us in the face all these years.

Four shafts lead out of the Great Pyramid at specific angles, pointing directly at Orion's Belt, Sirius, Alpha Draco and Kochab. There must be similar pyramid constructions at these destinations with angles that span out to other destinations within the network.

Who on Earth would have built this massive thirteen-acre construction from over twenty-two million stones, each weighing five to seventy tons, if it were not for some truly remarkable reason?

We know that the Egyptians were an intelligent race of people, but their hieroglyphics only go so far as to explain what they did, what they ate and what they believed in – everything about their lives apart from how and why they built the pyramids. This could quite possibly be because they did not build them. How likely is it that a super-race of beings came to Earth with a view of linking it up into a universal network?

Once having travelled the expanse of space with super knowledge, they simply developed the transmitter/receiver for instantaneous communication across limitless distances, based on designs they brought with them. An amazing feat it was, and surely something that beings would only embark upon if there was something incredible to be gained.

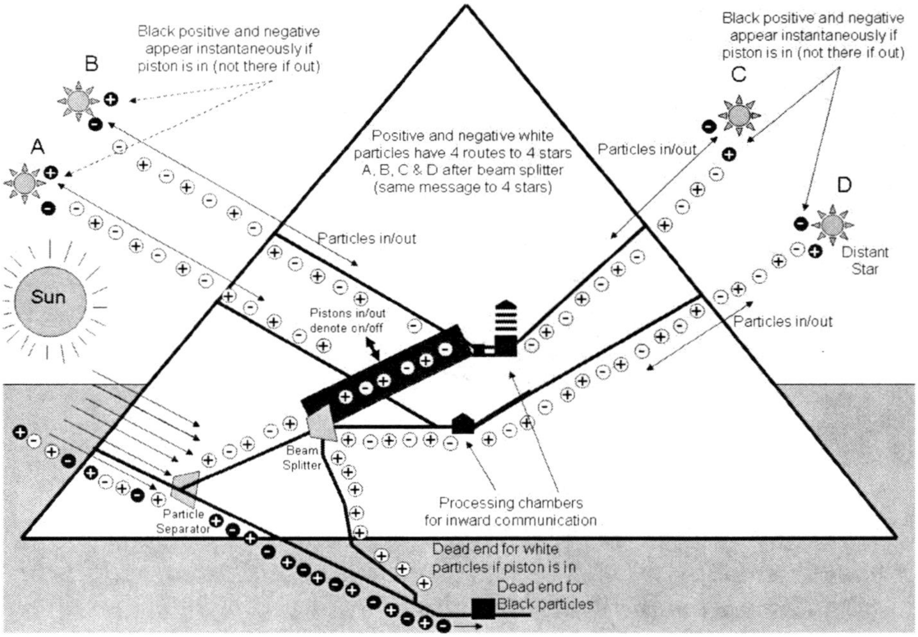

By placing in, and removing, the detector on the light beam from the sun, the detectors billions of miles away register different results instantaneously – well worth the effort.

To picture that it was built for a tomb or a place of worship is far too ridiculous a reason for building it. Imagine the reaction a race of intelligent humans would get when their leader chirps up and announces that they are to build this almighty structure simply because he wanted it.

After a few bricks, people would have been questioning the guy's scruples. The regime of building would have undergone mass mutiny and construction would have stopped. So the motive must have been greater, and the end result must have been worth all the precision that went into it.

No one has ever been able to explain the intricate pattern of passages that have been built with such amazing accuracy.

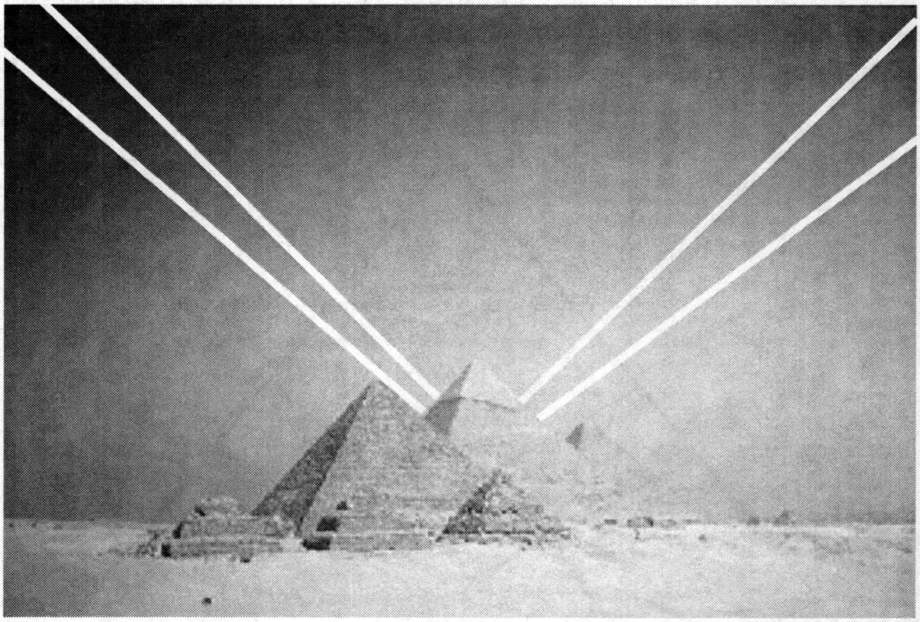

Communication utilising quantum weirdness would have been achievable with the appropriate apparatus configured within the Great Pyramid – perhaps this is what it was built for!

The millimetre-perfect passages have puzzled many a philosopher, construction engineer and scientist for eons.

Perhaps it is now time we rewrote our history books, placing something plausible within them rather than ignorant, unjustifiable twaddle.

So is it possible to mend this apparatus, restoring it to its original operational glory?

I believe it is quite possible; it just requires a number of scientists to study the internal passageways with the communications mechanism in mind, and we will surely get sight of precisely what should go where. My attempt at this is as follows, but do remember it has to be bi-directional so as to act as a transmitter and receiver.

A conventional acoustic, analogue or optical transmitter encodes speech and data in order to transmit it over cables or lines of sight, before being decoded at the receiving end. This communications mechanism operates in a similar way to interfering with the particles whose brother particles have already arrived at their destination, but have not yet been detected. Upon detection, it will be in one state or another, depicting the digits of a binary number.

If you consider the Sun to be a living entity, you will see that it too will also need a means of communicating with other stars to discuss various interstellar topics, such as quasars, comets, where to blow up next, dead planets, habitable planets and humans. I am sure the Sun and other stars would not be happy to have many years' delay between conversations, so my intuition brought me to think that they must communicate in this way too.

It stands to reason that the brighter a star can shine, the further it may be able to communicate – perhaps this is why stars shine.

The following diagram shows how a light source sends photons through from source, such as a star, to a destination, taking with it discarded particles, supposedly discarded from the first check!

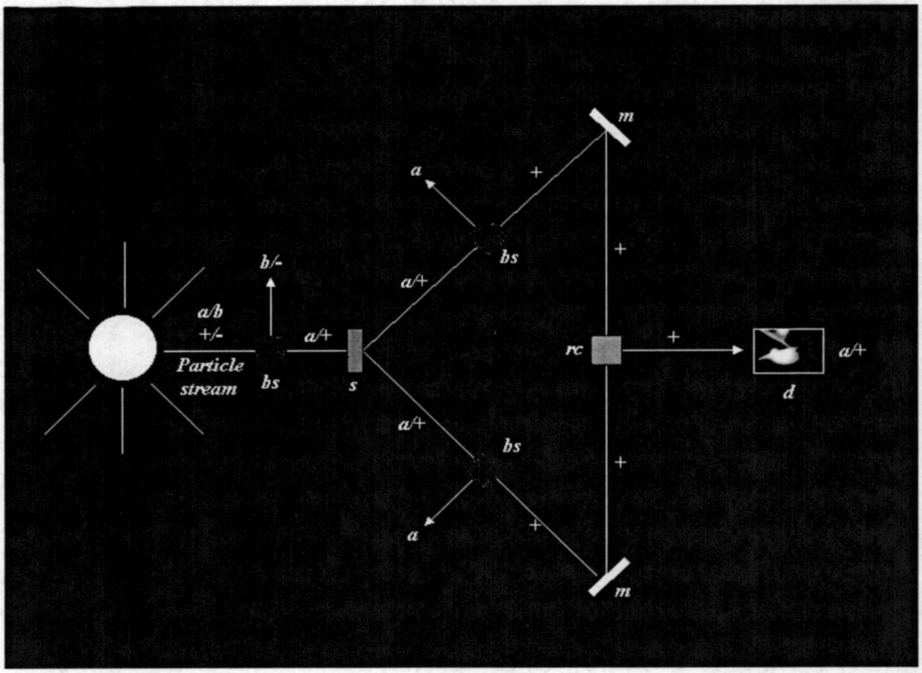

The 'answer' yielded at detector 'd' is truly miraculous. The first beam splitter that dispels the 'b/-' particles leaving the 'a/+' particles to continue. The particles continuing have their properties split, in this case 'a/+'. After this they are introduced again via a mirror, giving an amazing result. Observation at detector 'd' reports that '+' appears, but so does 'a'. How can this be?

We are now looking at one of those particularly deep phenomena within quantum physics that does not make too much sense at first glance. However, it has inordinate potential for communications. This very bizarre happening does not currently have any explanation. We wait with baited breath!

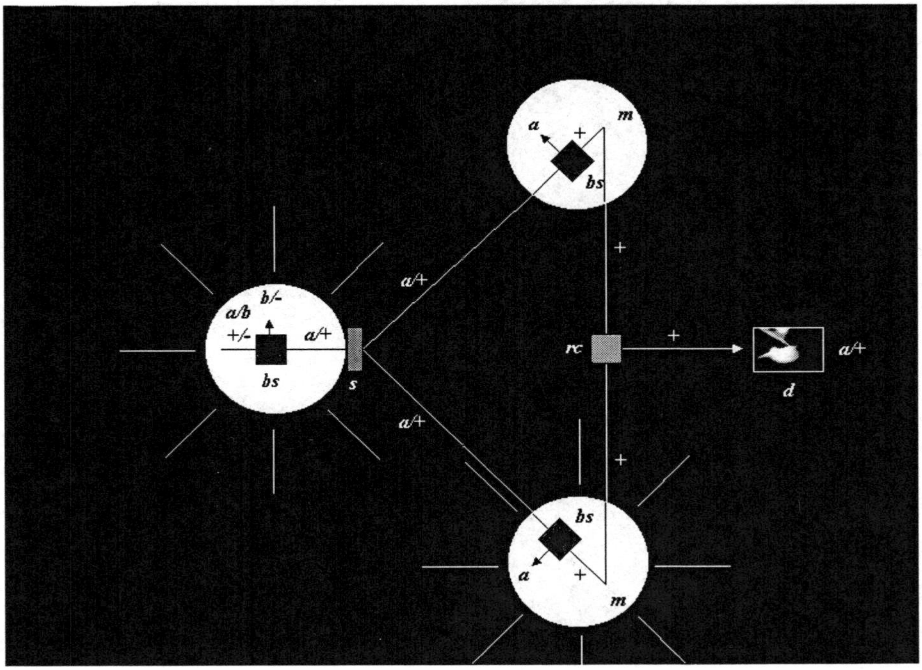

Is this the method that stars use to communicate with each other instantaneously too?

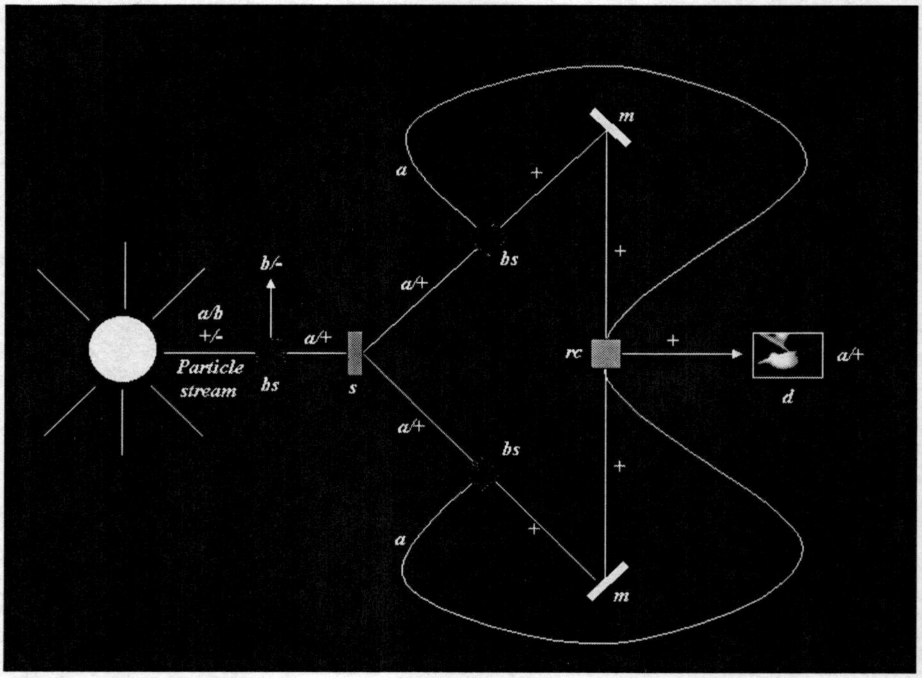

How can the answer be 'a/+', when 'a' is nowhere within the vicinity? Perhaps it finds an alternate route through space. Perhaps it is due to entanglement. Perhaps it is due to the superposition state of light.
Not to worry – it works – we should use it for communication.

The go splat world of light

- *Leave fruit for a while and it will begin to decay – but why does a ray of light from a distant galaxy not decay on its journey here?*
- *Time does not tick at the speed of light – so light that reaches us from distant galaxies thinks it has done so instantaneously.*
- *Light will traverse from its source to its destination at the same time coming into existence and vanishing.*

There seems to be something curious with the concept that a ray of light can traverse the whole Universe, taking millions and millions of years to wind its way through to its destination, and arrive as fresh as the day it was created all those years ago. Unlike an apple, light does not appear to decay over time. Is it really possible to throw a particle into outer space, and expect it to remain constant without mutating in any way for all that time?

There are two answers to this: firstly, this must be the case; secondly, this is not the case, and as light knows where it is to land at the moment of creation, it appears there instantly with the illusion of time attached.

When considering some of Einstein's theories about relativity, laws of motion and space-time, it is understood from one of Einstein's equations that anything travelling at the speed of light experiences no time whatsoever. Also, the very fact it is travelling at the speed of light implies that it has no mass at all.

This means that every ray of light travelling the distance from its source, such as a star, to its destination, such as the Earth, does so, as far as it is concerned, instantaneously, irrespective of the distance involved.

Certain experiments have been conducted with light which show very weird results. It is as if you are in fairyland – a make-believe world.

Interfering with one part of a quantum system alters the results observed in another part. This is what physicists relate to as quantum weirdness.

This also means that the ray of light is created, traverses the Universe to reach its destination, investigates all avenues and dies in the same instant. As far as the ray of light is concerned, there has been absolutely zero elapsed time whilst it travelled the distance, even if the distance was billions and billions of miles.

Put another way, time does not tick or even exist for anything travelling at the speed of light. The ray of light in this split instance will stretch from its source to its destination at the same time as being created and dying.

This means light is at its destination and destroyed the moment it is created. This is what is called the 'go splat' world of light.

This actually means that light is dying whilst being created, which also implies that its destination has as much to do with its existence as its creation source. This can then be viewed as a bidirectional, instantaneous event, but we humans know it is not, as we clearly see the source i.e. a candle flame. However, perhaps as far as the crazy laws of quantum physics are concerned, light's destination is just as important for its existence as its creation. It needs something to land on and die, the instant it is created. This could be similar to how an electron's quantum leap works within an atom as it orbits.

On close inspection, a totally uneducated person who knew nothing of time or the concept of forwards and backwards, could interpret light as being something that shoots out from a leaf and arrives at the Sun to help it brighten it up to the magnitude we see. This is how some people were thought to have viewed light in the Victorian times. But our minds do not work like this, as we are told at school that the Sun shines, giving out light.

How would we view the world if were taught that the Sun attracts light from leaves, grass, our skin and everything around us, to brighten itself up; this is as likely a view that a totally-uneducated person could arrive at as the one we have built into our cultures!

You may think that placing something like a piece of paper in the way of a candle light casts a shadow on a wall, hence proving that light is generated at the candle light source, and is propagated outwards and blocked by the paper, preventing the light from hitting the wall. Well, no, it may just be that the light the candle needs is now coming from the paper presented before it.

The very act of sucking particles from something may warm it up also. I do not know about you, but when I put my wet hand in front of a flame it seems to suck all the moisture from my hand, rather than provide it with something. Also, the luminosity of my hand could be the effect of light leaving my hand and rushing to the candle.

Somehow it is clear that the source of light is the candle and it spreads out from there. Does this mean that if a star is shining on the edge of the Universe, half of it would not be able to shine, as the light needs something to land on to be created?

If this were the case then the star would either stop shining, as light cannot shine towards nothing, or it would continue to shine as there is no edge to the Universe as the shape of space makes it bend back on itself.

There is something strange about light that may get you thinking deeply. Relatively stationary observers such as ourselves will see light travelling at a specific speed through space, and will be unaware that the ray of light believes itself to be instantaneous.

This gets rather interesting when you get to understand the following: as light has no mass it cannot be influenced by gravity, and light has no time to exist within which to be influenced by gravity. Although light appears to be influenced by gravity around a planet, it is actually being influenced by the warp in space-time created by the planet within the fabric of space.

Effectively, as light passes a planet it will bend, due to the fact that the planet has bent space, not that it is being influenced by the planet's gravity. It is a good job that light is not influenced by gravity, otherwise we would not be able to see anything throughout space, and also the world would be ultra-bright as it sucked in all the stray light floating out in the Universe.

Gravity is caused by the accumulation of the masses of objects combined together. No one is quite sure, but the speculation here is that gravitons perform the force of gravity. However, they have never been detected.

The faster something goes, the more it will ignore gravity; this is best looked upon as its trajectory. And, blow me, the faster an object's trajectory, the slower time gets and, hey presto, when we reach light speed there is no time for an object's trajectory to be influenced by a planet or a star's gravity.

Directionally, light can only be influenced by the shape of space within which it moves; the space within which light moves can only be distorted by mass, not time. Where light sees its straight-line journey through the Universe bent by the shape of space caused by mass within space, we would experience gravity, because we have the concept of time and mass as we do not travel at light speed.

This means that time alone cannot visually shape the Universe, due to light's ignorance of it. Therefore, to us on Earth as light-seeing human beings, time will lie to us due to a ray of light's ignorance of time. That is why we see illusions, such as an oar bending in water. This is because light travels instantaneously, as far as it is concerned, and humans see time influencing light's speed within a different medium. It is a straight line as far as light is concerned, especially as time does not exist.

This is a great example of how mass distorts our space-time. The light gets to our eyes from all parts of the oar instantaneously, as far as the light from the oar is concerned. The oar emits light that instantly travels to our eyes; however we, as stationary human beings, see things differently – we see the light coming towards us at a particular speed because we have the notion of time.

All we have been able to do before is calculate the angle of refraction of light entering water, and calculate the speed of light in water. Now we can go a step further to understand the Universe's communications mechanism, replicate it and connect ourselves back with where we once were.

So on one hand we can ignore time, as light does not comprehend it, and on the other remember that human beings are susceptible to it, as we are fairly stationary.

Let us recap: light is at its destination and destroyed the moment it is created. This actually means that light is dying whilst being created, which implies that its destination has as much to do with its existence as its creation source. This can then be viewed as a bidirectional instantaneous event.

A different angle on life – sceptical of all knowledge

- *A 99%-accurate detector sounds fairly precise; however, it will produce extremely flawed results when tested upon large numbers of samples.*
- *Newspaper editors get a snippet about the truth of a story and twist it to be sensational – quite often, totally unlike the event that really occurred.*
- *Some factual sentences can appear very convincing, when they are actually based upon a previous sentence that was only a tentative belief.*

At the age of nine, I was innocently staging a long wheelie on my bicycle right outside my house when the police pulled me up. When they asked where I lived and I pointed directly to my house, their response was, "Don't be so stupid, you don't think we were born yesterday. Where do you really live?"

I was really taken aback that they did not believe me.

I dropped my bicycle and ran to my house just as my Mum came out to see what was going on. I ran straight into her arms.

When you are innocent and do not know what a lie is, it makes for a mystifying world.

As I wrote in my first book 'A True British Eccentric', I have been extremely sceptical of reports, statistics, beliefs and assorted knowledge from a very young age. I became rather sceptical of the world when I discovered that a 99%-accurate disease detector, which sounds quite accurate, will produce extraordinarily erroneous results. This was proved by studying a rare disease from which one hundred people within a population of one million were suffering. If the disease detector is ninety-nine per cent accurate, then ninety-nine people would be correctly diagnosed as having the disease. However, amazingly ten thousand people would be incorrectly diagnosed as having the disease when in fact they have not.

My confusion within this world all started when I was not believed at the age of seven by my teacher, Miss Beeby, when I wrote, "I have a seaside in my back garden," in my school news book. As my parents really had built a replica seaside in the back garden for a party, I thought this normal, and here was a grown adult telling me to write 'real' news in my news book.

This subsequently led me to challenge a great deal of facts and figures passed to us from our ancestors, whether they be factual within books, or even inherent within us. Mistakes could easily have been made!

You only have to be involved with any news item to realise that newspaper editors have a field day with the truth, twisting it to become almost another story entirely. Even worse are the 1937 encyclopaedias, still sitting on my parents' bookshelves, which tell you that scientists believe certain markings of Mars are areas of cabbage-like vegetation, which spring into life when water from the melting polar caps reaches them.

It frightens me that I was born only twenty-five years after these encyclopaedias were published. Goodness only knows the extent of the inaccuracy of the historical and ancestral facts we have brought forward in time within our cultures.

Another of the claims about Mars was, "The dark spots at the junction of the canals on Mars are believed to be centres of habitation, their dark appearance being caused by the growth of vegetation watered by the canals".

It amazes me that the second part of this sentence is so convincingly based on a belief suggested within the first part of the sentence.

To check the accuracy of text written about other known woolly topics, I decided to read the 1937 encyclopaedia entry regarding the Egyptian pyramids. I made a note of the ambiguous aspects of knowledge similar to the inaccurate planet Mars description.

Well, about the Sphinx, the 1937 encyclopaedia entry reads, "No one can estimate the age of this gigantic figure carved in rock and partly buried in the sand." This is not too bad, at least they are honest.

About the Great Pyramid at Cheops it reads, "Cheops was an Egyptian king and the pyramid forms his tomb."

Clearly, if there was any ambiguity before 1937 it has now been ironed out, and everyone is to believe that the two point two million stones making up the pyramid, each stone weighing from five tons upwards to seventy tons, were just for a chap who popped his clogs.

How about considering that the pyramid was there already and they buried this guy within it because they were "not so clever?" Also, it was built for some totally different reason, as we now know. However, no one grasped it because no one gave it a thought, after being spoon-fed information from a reprint of a 1937 encyclopaedia. Knowing what I know now, as far as I am concerned, if the Great Pyramid is alleged to have been built to contain the body of a person, then that is absolute poppycock!

We learnt how the act of observation or detection of light photons determines precisely what it is and where it is – prior to this it can be considered as purely an undetermined wave of potential light. We learnt how we have automatically come to expect rapid and significant technological changes on an ongoing basis. We learnt how communication has improved over the centuries, and how the speed of light limits the speed of communication. We learnt how people find it difficult to relate to new technology, such as when Prime Minister William Gladstone could not see the benefits of electricity. We learnt about some failed 'faster-than-light' communications technologies such as tachyons, Alcubierre drives, rigid bodies, rotational effects, and traversable wormholes. We learnt about how quantum weirdness displays a new 'faster-than-light' mechanism for future generations to exploit. We learnt about the crazy behaviour of particles, and how counterintuitive they are in contrast to the behaviour of items to which we are normally accustomed. We learnt how, as soon as one of a pair of entangled particles has revealed its identity, then this will instantly be known to the other partner, however far away this is. We learnt how the Great Pyramid in Egypt could have been used for intergalactic communication – replicating how other celestial bodies could naturally communicate. We learnt how light is at its destination and destroyed the moment it is created – based upon the fact that anything that travels at the speed of light experiences no time.

Next we look at what the future has in store for us – we certainly have very exciting times ahead.

INVENTIONS OF THE FUTURE

Where we look at various inventions and how they have improved over time. We look at how inventions that exist now may improve, and try to anticipate some inventions of the future. We also look at future computing power, and how applications in the future may benefit the world.

The Universe holds more secrets than we can imagine

- *Irrespective of how many of the Universe's strange phenomena we get to understand, there are hoards more yet to be discovered and studied.*
- *"Our responsibility is to do what we can, learn what we can, improve the solutions, and pass them on" – Richard Feynman.*
- *Perhaps we can develop clever tactics so we may visualise the knowledge of the future.*

It would be fascinating to take a voyage into the future to observe inventions, events, cosmic happenings and worlds that could be truly witnessed only many generations from now.

If this voyage were possible, unfamiliar objects and strange happenings would be observed that would give an insight into how innovation will evolve. Imagine what a person two hundred years ago would make of the technology and possessions we have today. The pace at which technology is progressing will make two hundred years from now a much more daunting step to take than the step taken in the last two hundred years.

Modern life consistently develops increasingly sophisticated innovations based upon already existing inventions and discoveries – they are 'all the rage'. Superior quality televisions, telephones, computers, household appliances, audio devices and cars successively make their predecessors almost laughable.

Where will it all end?

Whereas just one hundred years ago we were still formulating the foundations of Physics and Chemistry, we now seem to have the fundamentals mapped out. To make further headway requires increasingly more advanced and expensive equipment, but this is exactly what we are capable of creating. Just look at the Large Hadron Collider.

At the beginning of the twentieth century it was suggested that everything we needed to know about physics was known, but that was before the discovery of nuclear physics and quantum effects!

There are plenty more phenomena yet to be discovered, studied and explained, and there are also many incomplete theories to be finalised. Mankind's current understanding of the Universe is unquestionably still in its infancy.

Our voyage of discovery has only just commenced.

It is difficult to imagine a day when Mankind will feel content with all the amassed knowledge, and consider any further advancement unnecessary.

So long as the human race survives, we will not be satisfied until an explanation exists for all known phenomena.

As the famous theoretical physicist Richard Feynman said, "We are at the very beginning of time for the human race. It is not unreasonable that we grapple with problems. But there are tens of thousands of years in the future. Our responsibility is to do what we can, learn what we can, improve the solutions, and pass them on." Richard Feynman 1918 – 1988.

Who knows how many different solutions need to be mulled over before we stumble upon the correct answers regarding what really lies out there in the depths of the Universe?

Clever tactics need to be developed and employed to allow us to cut corners, allowing us to take a peep into the future of Mankind's knowledge. This is a little like applying a sophisticated approximation algorithm to a mathematical problem – thereby coming up with an answer faster than performing the calculation in a conventional fashion.

It would be great to think we can create a shortcut to future knowledge to arrive at results more quickly – just as we have alleyways to get to other parts of town, desktop shortcuts to applications, ready-made food, polls in elections to predict a political outcome and estimation techniques regarding the impact of climate change.

Our lives are better with shortcuts, so perhaps we should become more proficient at producing them. By thinking in a unique way, we may be able to gain access to invaluable knowledge which would otherwise take many decades to deduce. I personally would like to see whether it is possible to devise and implement such a tactic to achieve this.

There is no harm in us trying!

Thanks to vastly improved global communication, we can now consider new ideas extremely rapidly, and derive conclusions quickly. This has already made the progression of ideas significantly quicker, and enables us to draw upon the knowledge of many people around the globe.

Pending future breakthroughs

- *Getting to grips with all the natural forces and how they interrelate together will be a major scientific breakthrough.*
- *All manner of future breakthroughs are possible that we can scarcely imagine today.*
- *Imagine considering future inventions in the 1850's, what we have now would have sounded extremely far-fetched.*

In our voyage to understand and secure our future upon this serene, blue planet, it is clear there are an inordinate amount of natural forces working amongst themselves – we need to get to grips with these so we may discover more about how everything fits together.

Perhaps we can assist ourselves with our voyage of discovery by making comparisons with the equilibrium of the Universe. Throughout the Universe there is a gigantic lattice of solids, liquids, gases, galaxies and electromagnetic waves, which radiate a multitude of forces, some of which interact with each other and some of which are oblivious to each other. However, the net result appears to be completely harmonious.

There are countless electromagnetic waves flying around, of which our senses can detect only a small fraction. As a consequence, we are able to interpret the stars in the night sky only as a series of white dots, even though there are many more types of waves whizzing around at wavelengths we cannot see. Until observational technology was invented which was capable of penetrating deep into the Universe, we were never able to view the Universe as it truly is.

In our own minds there are things we cannot comprehend, as a result of having limited senses. Our senses prevent us seeing the remainder of the Universe, and perhaps we should be determined to make sure it does not always remain a mystery. If we were born as all-pervading, omnipresent beings, we should have more of a clue as to what the whole of the Universe is about. Unfortunately we are not.

One million years from now we shall have a much better understanding of the Universe, possibly to such an extent that we may have pinned down topics such as the true origin of matter, why we are here, the true size of the Universe and how manned space colonies can explore other solar systems. These would be significant future breakthroughs.

Goodness only knows what other types of breakthroughs there may be. There could be virtual intelligence stored in every conceivable item making light of every chore, totally automated, nanotechnology-based sealed food-harvesting chambers, no disease, instant learning downloads to our brains, transcendental body hopping, home DNA modification kits, reengineering of the natural world, major synthetic biology breakthroughs and optional eternal life.

Incidentally, none of this should sound any more far-fetched to us now than suggesting the pending invention of a camera to someone in the 1850's. Neither should it be more far-fetched to us now than mentioning Internet dating to someone just before its introduction in 1990.

Innovation out of the blue

- *Innovations that came out of the blue are those that had no comparison previously, or their comparison was wildly different.*
- *Innovation and inventions are now very much part of our lives – we would suffer greatly if they were taken away.*
- *New scientific breakthroughs provide a platform from which brand new innovation can materialise.*

There is a major difference between inventions that improve existing inventions, compared to original inventions that have come out of the blue. Original inventions resolve problems that never had a solution, or even that we never knew needed a solution.

Initially we communicated by paper, then we communicated via fixed-line telephones, and then this was superseded by mobile telephones and email. These all perform similar things, but have progressively become more effective and conducive to a human being's lifestyle.

Innovations that came out of the blue are such things as the wheel, boats, the tin opener, aeroplanes, radios, televisions, telephones, computers and the Internet. Looking back in time, most of the inventions that exist today fall into the category of progressive improvements, and outweigh those 'out the blue' inventions.

In the early 1900's, how many people could have visualised the success of aeroplanes, televisions, radios, telephones, computers and the Internet?

The answer to this is subjective, but it certainly would not have crossed the minds of many, and if it did they would have thought it purely science-fiction. These inventions are now very much part of our everyday lives.

So what are the unknown scientific inventions of the future that appear science fiction to us today?

The unknown scientific inventions of the future fall into two categories. The first includes inventions which will improve solutions to already-known problems, including the refinement of existing inventions fuelled by scientific improvements, hence further overcoming the known problem. The other would be inventions that come out the blue, overcoming unknown problems with the realisation that a new scientific breakthrough provides a solution to a previously unknown problem.

Before you can visualise the scientific inventions of the future, the categorisation of past inventions should be separated into those that improve solutions, and those that are out of the blue inventions. Most of the inventions we have today are improvements upon previous solutions.

How will technology devices improve in the future?

- *If technology works, it will then be mass-produced and made available to all.*
- *Looking around at current innovations, it is difficult to imagine what humans are lacking – however it was the same for the Victorians before the Internet!*
- *Imagining a future invention is a true art – the likelihood is that any invention thought of has been thought about before – but just not available yet.*

Consider a question that requires thought well beyond the current era – 'How will technology devices improve in the future?'

Comparing some groundbreaking, technological features that have evolved recently, it may be possible to visualise where these devices are heading in the future. There are sophisticated, handheld, banking facilities, the ability to purchase goods from anywhere in the world from a phone, location-aware services, social networks, visual aids for street scenes and satellite views of everywhere upon the planet. There is also the ability to talk and video conference with anyone on the planet, wherever and whenever required.

The convergence of mobile devices with the Internet has now been achieved. When coupled with the miniaturisation of devices, ever-decreasing prices and mass-production – the net result is something that is effective, convenient, compelling and available to all.

These devices and their applications are designed to make human beings' lives easier – no time is required to configure any of the device's features. The devices now come with the most common pre-defined user preferences, which can be altered later.

To consider the technological breakthroughs of the future, it is perhaps best to look at what factors will drive their advance. Banks will want customers to be loyal, advertising revenues will be crucial, familiarity and branding is paramount, and the customer will want better applications, experiences and interfaces.

At this point, perhaps it would be prudent to take a look back in history to a similar revolutionary time, and make some comparisons.

Although the railways seem at first glance to be rather different to technological devices, there are some parallels that can be drawn. There was an initial impetus to get the infrastructure built and operational. In the early days, the railways had no shortage of investors to fund the hard work of laying the track. It was seen that the rewards were to be very substantial in the longer term.

Comparing this with device technology, in a very similar way there has been no shortage of investment to establish the communication masts and the base-stations for the mobile device networks.

Referring back to the railway networks again – once the infrastructure was in place and the trains had improved to become as fast as possible, the concept of train travel became 'normal'. The ability to go faster than a horse was revolutionary. However, there were downsides to the network; you could only go where the tracks went!

Large towns and cities flourished where the railway networks travelled, and where they failed to reach caused difficulties for the people in those areas.

Then along came an alternative means of transport – the motorised vehicle. As soon as this was affordable by the masses, there was less demand for the train. As a consequence, the railways went into decline, especially within Great Britain.

Doctor Richard Beeching, 1913 – 1985, was responsible for publishing a report called, 'The Reshaping of British Railways', commonly known as the 'Beeching Report', published in the 1960's – it had far-reaching implications and was not very popular. More than four thousand miles of railway and three thousand stations were decommissioned.

So let us see what lessons this holds for the future of technology devices. In the early days of portable mobile devices, costs were high and network coverage was poor. Nowadays, we have reasonably-priced packages upon networks with good coverage, which provide us with all the messaging and communication we need. One analogy with the railways is that once everyone has embraced the concept then a better alternative evolves. This happens within mobile device technology too. However, this was factored into the running costs of the network from the start, so there have been no surprises – good planning!

We have seen various incarnations of mobile network technology through to the new generation networks we use today.

Now we have fast networks, slick devices and compelling applications – what more could we wish for?

The interfaces can be improved, and the speed of connectivity can be improved – but just as the railways reached a functional plateau very quickly, so may the use of mobile devices.

It is similar to filling a railway network with trains and carriages. There are plenty of railway tracks to fill initially, but then its capacity is full and very little more can be done with it, other than replacing old trains with better new ones. These better trains are analogous to the applications and devices we utilise on the mobile networks.

We have television on our devices; we have payment facilities, search facilities and social networks. Any other improvements are only going to evolve as a result of industry collaboration and convergence – effectively the network providers have done their job and it is now up to industry to provide the compelling applications.

Three-dimensional screen images, life trainers, images with mass projection capabilities and artificial intelligence could evolve – but these are just the applications.

Real-time feeds are extremely popular, which no doubt will develop further. People's lives are being documented in massive detail for future prosperity. Wireless electricity is now a reality over short distances. We may get device functionality integrated into our brains. Satellite-oriented network signals may overcome the mobile dead-spot difficulties. Perhaps we will get a means of experiencing different places and different worlds through the transmission of visual quality images to our brains.

We can look forward to better connectivity and more compelling applications – but only as and when they are invented. We have just been through a period which saw a complete dump of the personal and commercial world into mobile devices – this is almost saturated already. It is as if we have invented the bath tub, left the tap on and the bath is now full!

Quite incredible!

Crazy inventions

- *Running an innovations company made me realise how bad some ideas were, and how miraculous others were.*
- *Some innovations in the future will be as miraculous as the Internet, aircraft and cures for diseases – however, we do not know what they will be now.*
- *Few people many years ago could have predicted the inventions we have today.*

I set up an inventions company. Established in 1997, the philosophy behind the company was to assist budding inventors with their ideas, their strategic direction, and put them in contact with the people who mattered to make their venture a success. The company name I chose to perform this task was "Stage Forty-Two Limited" with the slogan, "For people with ideas, the answer's with us." The idea for this name and slogan originated from Douglas Adams' Hitchhikers Guide to the Galaxy. Within this book the answer to the meaning of life, the Universe and everything, was calculated by a supercomputer to equal forty-two. However, the meaning of the answer forty-two was never to be fully explained.

The only explanation was that the ridiculous answer came about because of the vague, insufficiently-detailed question that was put to the computer. This is similar in a way to normal, everyday folk saying, "Damned computer," when the programmer of the software is really at fault.

Having established a website, I was delighted with the responses from the general public, and soon had a large number of projects on which to work. Some of the ideas with which I was approached were absolutely terrible, some were bad, some were reasonable and some were excellent. I eventually managed to acquire a small portfolio of relatively viable ideas. Strangely, the number of inventions is decreasing. However, this may be masked by the fact that applications for the Internet are not viewed as inventions. The Internet has brought with it convenience and knowledge in such a short period of time: holidays, insurance, bills, banking, shopping, instant written communication and leisure. Never before, in such a short period of time, has the way in which we accomplish everything developed so rapidly.

If we were to have a free reign of thought to mentally develop the craziest inventions we may have in the future, we might end up with: artificial gravity, carbon dioxide eliminators, free background radiation power, cures for all diseases, worker robot drones and a means of inter-universal travel. Who knows what the human race will end up inventing?

However, being a little more down to Earth – I can recall a time when I thought I had something truly unique.

One invention I came up with in 1995 was 'Inter-jurisdiction Financial and Data Transmission with Trusted Fourth Party; i.e. The United Nations'. No one could understand what the Dickens I was talking about in 1995 – the Internet had hardly taken off, and there was little known about Internet payment methods. Therefore, my work fell by the wayside, and I watched Internet payments sites take off later with interest.

We know that the future holds a myriad of improvements, but guessing what they may be is so difficult. The only way to determine how difficult this is, is by looking back hundreds of years and seeing whether you could have predicted then what we have now. Quite probably not!

Who could have predicted cars, aeroplanes, bicycles, helicopters, fridge-freezers, hosepipes, television, computers and the Internet?

Computing power of the future

- *Could there ever be a day when everything that could be discovered and invented has been?*
- *The more powerful a computer gets, the more sophisticated algorithms we shall get it to solve.*
- *Are we already a race that relies upon computing machines to govern the destiny of our future wellbeing?*

The Heath Robinson-style of computing was absolutely fantastic, with its boiler-suit clad operators, valves, miles of wires, tickertape output and a massive building to house it. Suffice to say we now have much more processing capability within a toy given away free inside a cereal packet than we had in the early days of computing.

So what is next?

The pace of global communication, and the dynamic way that worldwide manufacturing can operate, means that any major breakthrough in technology is very rapidly brought to market. Processing speeds within computing have been increasing year on year; the beauty of this being the capability to dramatically improve functionality, such that multiple technologies can be merged to provide a more intelligent experience.

The question is whether there will be a point at which technology reaches a plateau in its current incarnation. Will there be a day when the user experiences of technology can no longer be improved?

Whatever the case, in an extremely short period of time, mobile phones have developed to become slick and location aware.

If you tried to describe the onslaught of mobile phone technology to someone in 1990, they would have difficulty in relating to what you are referring. Does this mean that every year we are perpetually going to get new devices with new functionality?

Surely there is a point when there is sufficient storage, sufficient applications and sufficient speed. Even so, I cannot see there being a logical end to the evolution and development of mobile devices.

One thing is for sure, in one thousand years from now we will see totally different devices, with embedded functionality that makes current-day solutions look like antiquated donkeys. Having said this, it is not possible to imagine what these functions will be. There surely comes a point whereby lives being permanently videoed for prosperity's sake, being informed of all points of interest, friend's locations around you, personal tours of places of interest and information flying at you at the rate of knots, becomes too much.

One thing that will begin to happen is the interaction that these devices will have in the world about them. Paying for shopping automatically, interfacing directly with the tax system on your behalf and networking at a pace that is truly astounding. New applications are easy to develop, but the art is to invent a killer application that the masses want desperately.

In a way it is strange when talking about computing power. We are already talking in terms of the power of the computer being the power that it gives you, rather than its processing speed, which seems to have reached an acceptable level without there being too many moments staring at frozen screens and frozen cursors. We have managed to get to the point where switching on a personal computer is pretty much immediate, especially when in standby mode. This is something that took years to achieve.

As far as the processing power of supercomputers is concerned, we have no end to the algorithms and immensely difficult equations we can throw at ever-more-powerful computers.

The discovery of larger and larger prime numbers is a useful task to pursue for computing security purposes. It takes enormous processing power to discover any new prime numbers, but, once discovered, they provide us with better security features within what is termed 'public key cryptography algorithms'. So the quest to find ever-larger prime numbers will continue.

Incidentally, whilst we are on the subject of prime numbers, there is an exceedingly interesting analogy relating to security that can be drawn between prime numbers within nature and computers. Just as public-key cryptography algorithms can produce better security within computers utilising prime numbers – amazingly, within nature, prime numbers have also been discovered to work for species security and survival.

One great example is an evolutionary strategy used by certain cicadas called magicicdas. These are insects that spend most of their lives as grubs underground. Only after thirteen or seventeen years do they emerge from their secure underground hideaway and fly about, breed, and then die after just a few weeks.

The logic behind the prime number intervals between emergences makes their lives more secure, as it is very difficult for predators to evolve that could specialise as predators of these cicadas. If they appeared at non-prime number intervals, they would be sure to regularly meet with predators. Though at first glance, you may think that cicada predator populations may be the same at all times, this is not quite the case, as the cicadas cannot now be looked upon by any other species as a guaranteed yearly food source, and so predator numbers are curtailed as a consequence. Although it could be initially considered as a small advantage, it has to be taken very seriously as an effective security tactic, as it appears to have been enough to drive natural selection in favour of a prime-numbered lifecycle for these insects.

Artificial intelligence has currently brought devices with rudimentary limited 'intelligence', coupled with speech and speech recognition, but going one step further towards true intelligence is another matter. When computers can appropriately reason, judge, respond, deduce and seamlessly interact with us for the better, then we will have a breakthrough. At present, this day is a little way off, but people will not stop working towards it. When we have it, we will be working towards bettering it. The human race will never rest as far as technology is concerned!

I would like to predict that one day in the not-too-distant future, there will be a computer program developed that will consider all worldly phenomena and their associated statistical history, so that we can predict the future more accurately.

Funnily enough, this program would act in a similar way to my college thesis that successfully predicted the demise of cod in the Atlantic. The simple factors associated with calculating the demise of cod was based purely upon their estimated numbers, their breeding capability and the amount caught over time. The program was very successful; I even sent a copy of the results to Harry Ramsden's, the biggest fish and chip brand in the world. I received no reply, but I am sure they read it with interest. A decade after I had sent the report in 1984, there was turmoil as Atlantic catches of cod hit an all-time low, and stocks dwindled so dramatically that in 2000 the World Wildlife Fund placed cod on the endangered species list.

The predictive computer program needs to take this just one step further with all the Earth's creatures and planetary resources. The interaction with other associated resources and species needs to be understood, and entered into the overall calculation. In this way we could predict global changes and the consequences of taking more or less of a particular resource or food source.

The overall concept could start in a simplistic format, focussing initially upon species and resources more crucial for human survival, becoming more complex over time. For example, it would not be initially important to analyse the reproductive capability of a tree frog in our calculation, whereas the reproductive capability of the rainforest and its oxygen-producing capabilities would be essential. Over time, the program would become more accurate, as the data associated with actual resource and species levels become known, relative to previously-predicted levels. Initially, it would be necessary to only input relatively rough, historical data, and estimated thresholds. Trends would appear; it is at this point that perhaps governmental interest in the program would gather quite rapidly. Governmental decisions could be made confidently as a direct consequence, along with the understanding of their long-term implications.

An organisation such as Wikipedia could rise to this challenge quite easily. Different entries would be influenced by the trends of others providing future predictions of growth and pending danger thresholds.

Ultimately, we could have fairly accurate predictions relating to population and food source requirements, endangered species and food-chain deficiencies, fuel consumption and predicted future usage, and, ultimately, governmental policy changes and their global effects.

Imagine the power that this system could bring to governments. The decision to curtail whaling, the decision to prevent deforestation, the decision to no longer use coal-fired power stations, and all other politically-volatile decisions could all be more confidently made with immediate analysis in terms of the predicted benefit over time within the planet's ecosystem. What is more interesting is that the consequences of not taking action would become clear too – an extremely powerful tool would emerge!

Would this be a day to which we could look forward, or a day to be avoided?

Imagine the Earth having decisions made by computer regarding the survival of species and the maintenance of resources!

Unfortunately, the way the planet is evolving today, I cannot see any other way! If decisions continue to be left to the indecisiveness of global governments, we would see more procrastination and ill-founded judgements, based upon poorly backed-up arguments. This type of computer programme would take all the guesswork out of crucial decisions.

We would soon become subjects within a virtual world of computing and intelligence monitoring. We could then be perceived as being a species that maintains a computer program to govern the success and failure of the species; would this ultimately take over natural selection?

An interesting, and at the same time, a relatively frightening concept!

Quantum computing

- *Different laws apply as we dissect matter into smaller and smaller parts – an apple will drop to Earth, whilst electrons will totally ignore gravity.*
- *A quantum computer works closely to the way human brains works; a vast array of neurons – all working in parallel.*
- *A quantum computing QUBIT is created inside a SQUID.*

Silicon chips consist of relatively simple components, namely resistors, transistors, capacitors and inductors. Just add a battery and you can make a phone, a personal computer or a supercomputer.

Initially these components were quite large, and over the years scientists have been increasingly miniaturising them. These components are very useful, as together they can store and manipulate zeroes and ones. We now have integrated circuits which contain these components in tiny form. It is now possible to fit billions of components into the space that used to occupy just one of them in the early days.

The integrated circuits are developed onto silicon using a technique similar to photography. The components are not made individually any more, but stamped and printed onto the silicon. The engineers use all sorts of tricks and techniques to project smaller and smaller images onto the silicon chips.

However, a couple of awkward problems occur when these silicon chips grow in size to accommodate more and more components. Firstly, there is a problem of overheating, and secondly, and more interestingly, different laws of physics apply as the components get smaller and smaller.

The overheating problem occurs as each of the tiny transistors has a resistance, and as you put power through a resistance, the power is dissipated. As you pack more and more transistors onto a silicon chip, more and more heat is produced, resulting in a requirement for more and more advances in the dissipation of heat with such things as heat syncs.

This overheating applies to a two-dimensional silicon chip, so there is little hope of producing three-dimensional silicon blocks, which would theoretically be much faster. The problem of overheating would be exacerbated many-fold; to such an extent they would overheat and melt.

However, we still want to develop faster and faster processors to perform more and more demanding tasks.

There are materials called superconductors which have no electrical resistance below a certain temperature. A normal metal will maintain some resistance down to very low temperatures, whereas a superconductor will suddenly possess a totally zero resistance once it reaches a certain temperature.

Superconductors can be made of metals or ceramic-based materials. The temperature at which the metal superconductors start operating is quite low, but the temperature at which the ceramic superconductors can operate is much higher. The higher the temperature at which the superconductor can operate with zero resistance, the more cost effective the processing capability will become – simply because cooling materials to low temperatures is a fairly costly exercise.

One of the best superconductors for this job is referred to as YBCO, its. Its full name is yttrium barium copper oxide, which is a crystalline chemical compound.

As superconductors have zero resistance, if we were able to build computer chips from them, the zero resistance would produce zero heat. As a result, computers would no longer experience overheating problems.

The superconducting material could now be stacked into three-dimensional blocks, which would be extremely efficient for processing.

Another property the superconducting materials possess is the ability to operate at very small scales where different laws of physics apply; namely quantum physics. As mentioned earlier, these different laws of physics are prevalent as the components get smaller and smaller. Superconductors seem to be able to obey this weird world of quantum physics – this conformity appears to be inherent within their normal composition and operation.

One problem presented to the designers of very small circuit boards is the fact that an electron is not actually a point-like particle when viewed on a very small scale; it is actually more spread out like waves. Within the electronic circuits, the electron waves are still very small, but they are actually big enough to start causing a problem. If an electron is spreading out, then when it comes in contact with a small component such as a transistor, you are not sure where the electron is; it could be on one side of the transistor, or on another side of the transistor. What you ideally want to be able to do is control exactly where the electron should be.

This is currently a major problem with the design of very small printed circuit boards. Because the components are so tiny, the electrons behave as waves, and this causes operational problems.

Is there a way we can utilise these quantum waves to our advantage?

What is interesting is that if we pass an electric current through a superconducting material such that the electrons have two paths they could take, when converging them again, they act in a similar fashion to the light in the double-slit experiment. What is especially useful about quantum waves is that they possess some very strange properties.

Because the wave-like behaviour allows the electrons to spread out, they can be in two places at the same time. If you were to attribute a state to each of these two positions of the electron wave, not only would it be in two places at the same time, the electron would also be in two states at the same time!

Whereas a normal Personal Computer works by storing information as bits in the form of ones and zeroes, a quantum computer will allow the storage of a state of zero and one at the same time, using the quantum waves. This is effectively a superposition of zero and one called a QUBIT, coined from the terms quantum and bit. It can take a value of zero, one, or a mixture of the two. This is what gives a quantum computer the potential for enormous processing power.

Because a QUBIT can be in two states at the same time, it can explore several calculation possibilities at the same time. A classical register on a classical computer will require four computational steps to upload to four register values. However, a quantum computer register can have both states at the same time, so you can generate all the answers in a single computational step. This is effectively a very powerful parallel computation capability.

A quantum-computing QUBIT is created inside what is termed a SQUID, a Superconducting Quantum Interface Device. The SQUID is made from little loops of superconductor that allow the electron waves to pass in two directions. The superconductor used is normally niobium, as it allows the electron waves to get larger, which means they can be controlled better. These components can be made as small as one micrometre. Bearing in mind that a grain of pollen is about five micrometres wide, this shows how small they are.

The quantum computer benefits from the superposition state of the electron wave. This means that other methods or systems can be utilised to gain the same effect, rather than just using superconductors. Ions trapped in electrical fields, nuclear spins in the centre of atoms, and photons from lasers can also be used to encode these zeroes and ones in their special states. No one knows yet which of all these different systems will work best – scientists are still exploring.

QUBITS are simply placed together in large clusters; so long as they are designed correctly using superconductors they can be positioned onto an integrated circuit and operated as a quantum computer. The quantum computer circuits look very similar to the conventional silicon circuits – so it is easy to build quantum computers using existing technology.

You may ask; so what benefits will a quantum computer bring?

It so happens that due to the nature of a quantum computer it can use clever algorithms. A little like a classical computer may solve a jigsaw puzzle by trying every piece in every other piece to see if it fits, a quantum computer can use clever algorithms to start with the edge pieces and separate out different coloured pieces.

To put this in perspective let us compare how a classical computer can cope with factoring numbers, compared with a quantum computer. A classical computer runs into problems very quickly with very large numbers – to such an extent that to find the factors of very large numbers would take longer than the lifetime of the Universe. With quantum computers there is an algorithm which allows these factors to be found much more quickly.

True headway is being made, in December 2011 physicists headed by Ian Walmsley of Oxford University succeeded in entangling two normal diamonds. This demonstrates that quantum mechanical effects are not limited to the microscopic scale. Now that these quantum effects are more controllable, making headway will become easier.

The true beauty of the quantum computer is that it is similar to the way the human brain works; a vast number of neurons all connected in a parallel way. It is interesting to think how quantum computing will help us in the future – perhaps one day we could create additional processing packs for our brains!

We learnt how vastly-improved global communication enables us to consider new ideas extremely rapidly and derive conclusions quickly. We learnt the types of breakthrough that the future may hold, and how some new ideas will be based upon previous ones, whereas others will be unlike anything we have seen before. We learnt how the past can provide some pointers as to how the future of technology may progress. We learnt how at one time no one could have predicted cars, aeroplanes, bicycles, helicopters, fridge-freezers, hosepipes, television, computers and the Internet. We learnt how future computing power will improve. We also learnt how the superposition of zero and one is called a QUBIT – and how a quantum computing QUBIT is created inside a SQUID.

Next we look at one of the most destructive items on the planet – our financial system.

THE MYSTERY OF MONEY

Where we discover how money is created, and the rules associated with producing it. How money can become problematic for economies on a grand scale if it is not managed correctly by those responsible. We discover some relatively little known facts about the frailty of money that not many people know. We also take a look at some of the large scale scams and financial frauds that have taken place in the past.

Where does money come from?

- *Money is fairly mysterious, and it can occasionally get people, organisations and governments in trouble.*
- *People seemed happy trading issued pieces of paper, oblivious to the fact that the actual gold coin specifically commissioned may not have been in existence.*
- *Fractional reserve banking allows banks to create money out of thin air – lending out money many times the amount that has been deposited.*

In a one hundred-year period, the amount of money existing in the world has multiplied itself by a factor of roughly one hundred. How can this be possible, without fraudulent people merely printing it?

Well, if you do not know the answer it is rather interesting, and no wonder the financial institutions get themselves in a pickle from time to time. Also, it is as close to fraud as anyone can get, but because it is government-backed it cannot be seen as fraud by anyone.

The process by which banks create money is so simple that the mind finds it impossible to fathom.

What we are not told is that money is created by banks. Banks create money from thin air, and not from their earnings, or from funds deposited by their customers. Banks are able to materialise money from the people who borrow from them, and they generate it from the borrowers' promises to repay loans. These loans are also to be repaid with interest, with which the banks can also create fresh money.

The way that the creation of money came about is made easy to understand when you look back in history at the early goldsmiths who made gold coins for their clients. The goldsmiths were commissioned to make the gold coins, and keep them safe until collected. What they noticed, over the passage of time, was that the owners of the gold coins never came in for them all at the same time. Their clients seemed to be content with the paper claim acknowledgement slips, which they found easier to carry around and trade with, whilst the goldsmiths safeguarded the real gold inside their vaults.

This meant that, so long as the goldsmith had just a few gold coins, he could satisfy the demand of a small number who came in to request them. Other people seemed happy trading their issued pieces of paper, oblivious to the fact that the actual gold coin specifically commissioned by the individual might not truly be in existence. This gave rise to paper money, which seemed a much more convenient trading method anyway.

Goldsmiths began lending money to people; this was made much easier, as they had already been given money to make gold coins for people who did not want anything other than a slip of paper. They also found they could charge interest on the loans that they made. As soon as the people to whom they loaned money started to request it in the form of paper rather than gold, the goldsmith was really onto something big. They could now take gold in exchange for their pieces of paper and also give out pieces of paper as if it were gold. Stunning!

Once the people who had deposited gold with the goldsmith realised that interest was being made by the goldsmith loaning out their money, they insisted on having a share of this interest too. There had been confidence issues with the goldsmith, but now he was sharing his interest. This enticed more people to deposit with him and the depositor was happy.

Because no one whatsoever knew how much gold the goldsmith was supposed to have, or incidentally how much he really did have, the goldsmith could now issue pieces of paper based on gold that had never existed. Goldsmiths soon became wealthy from interest paid on gold that had never existed. The public became suspicious about how much gold there really was in goldsmith vaults, which resulted in everyone insisting on their gold back. Suddenly, the goldsmiths found hoards of people standing in queues outside the bank, waving their deposit slips demanding their gold back. This spectacle was termed, "A run on the bank".

It would have been easy and straightforward for governments to outlaw the creation of money from nothing, but the large volumes of credit the bankers were offering had become essential to the success of commercial expansion. So instead of banning this fraudulent activity which the goldsmiths had cunningly conjured up, it was legalised and regulated. Bankers agreed to abide by the amount of fictional loan money they could lend out.

Following some early runs on goldsmiths who were also being called 'banks', it became clear that banking regulations were required to limit the amount of paper money that could be lent out as a ratio of that having been deposited. The regulatory authorities agreed that a ratio of nine to one was permissible. This meant that banks could create money out of thin air and lend out nine times the amount that had been deposited, called fractional reserve banking.

The integrated network of banks and a few scams

- *The money system could crash if several runs on banks occurred simultaneously; therefore measures were required to safeguard the system.*
- *Bernard Madoff's established a fraudulent scheme called a Ponzi scheme – Charles Ponzi was an early pioneer of this type of fraud.*
- *Amazingly Bernard Madoff evaded close inspection by official officers and auditors – nothing suspicious was identified for years.*

To prevent a run on a bank, the authorities created a central reserve that could offer assistance as required, providing general credibility to the whole system. Now the only way that the money system could crash was if there were several runs on banks at the same time.

Over the years, the fractional reserve system and its integrated network of banks, backed by a central bank, has become the dominant money system of the world. At the same time, the amount of gold that was used to back the system has gradually shrunk to nothing. The nature of money has changed. In the past, a bank note acted as a receipt that could have been redeemed at a bank for a fixed weight of gold or silver. This is no longer the case, and a paper or digitally stored unit of currency can only be exchanged for another piece of paper or digitally stored unit of currency.

In the past, people had the choice to refuse bank credit that had been created in the form of bank notes, just like we have the choice to refuse people's private cheques today. The privately-created bank credit of today is legally convertible to government issued fiat currency. Fiat money is money created by government fiat or decree, and legal tender laws declare that citizens must accept this money as payment for debt, or the courts will not enforce the obligation. The courts look upon an offer of government-issued currency to settle a debt as being legally acceptable and anyone who refuses it will find that the debt is now discharged.

This means that although money has no value whatsoever, we have been forced by an act of parliament to accept it as having value. The only other alternative is that you do not conform to this belief, and find yourself effectively outcast from society's embrace.

The once-fraudulent practice of fractional reserve banking continues with government backing to this day, but the tricks played to create money from nothing have become progressively craftier, bizarre and very often fraudulent. You only have to look at Bernard Lawrence Madoff's sixty-five billion dollar banking fraud which he ran from the 1980's to 2008 to realise that the system is fraught with loopholes and scams. Although he received a one hundred and fifty-year prison sentence for his crime, this will not deter other people from building their fraudulent empires in the future.

Can you believe that Bernard Madoff, during his totally fraudulent spell at the helm of his counterfeit empire, was appointed for three years as the chairman of the NASDAQ, in 1990, 1991 and 1993?

He was known as one of the pioneers of modern Wall Street. Scary, frightening and truly unbelievable! Makes you wonder what other scams are in process at this moment just waiting to be found out, or worse, never to be found out!

Bernard Madoff's fraudulent scheme was called a Ponzi scheme, after one of the early pioneers of this fraud called Charles Ponzi, 1882 – 1949. A Ponzi scheme is a fraudulent investment operation that simply pays returns to investors from their own money, or money paid in by subsequent investors. The Ponzi schemes craftily entice new investors by offering returns that are unusually high. However, the Ponzi scheme requires a perpetually greater flow of money to keep the scam going, and more and more investors are sought, until the system ultimately collapses because the payments eventually get out of hand.

As more investors are attracted to the scheme, the more likely it is that the scheme is revealed to the authorities. But amazingly, in the case of Bernard Madoff, his firm was repeatedly scrutinised by officials and auditors, and no irregularities were highlighted. The United States Securities and Exchange Commission had conducted investigations and due diligence exercises with Bernard L Madoff Investment Securities since 1992. These were obviously amateurishly handled, and were perhaps inadequately conducted deliberately; they were officially cited as being incompetently handled at Madoff's trial. This drills massive dark holes in the governance of our world, governance that we ought to trust implicitly. Shady characters like this may one day undermine our confidence in the current system of money, and stricter regulations are essential. Perhaps a whole new approach is required, as the fractional money system has major flaws anyway.

The cartel responsible for orchestrating money

- *The cartel that orchestrates money was signed into law – this gave the Federal Reserve System a truly unique financial position.*
- *How is it that banks, governments and other institutions can all be in debt at the same time?*
- *You would have thought that the organisations that produce wealth within society would be wealthy – but they are in debt to those who lend.*

A very peculiar and noteworthy financial event took place in 1913 in the USA. President Woodrow Wilson signed into law the Federal Reserve Act, which formulated the cartel that is now in charge of orchestrating money. The absolute power that the Federal Reserve System has over the economy and individuals is immense.

Very few people know how money is created, and even fewer know how the Federal Reserve System was created and what it does. A very famous media guru, philosopher and communication theorist called Marshall McLuhan said, "Only the small secrets of the Federal Reserve System need to be protected. The big ones are kept secret by public disbelief."

The financial system that used to be backed up by gold is no longer operated in this way. We have what is termed fiat currency, and this is money that is simply declared by the government as being legal tender. We end up in a situation whereby a private company called the Federal Reserve simply say that the money printed is worth its stated value. So, in 1913, the people at the Federal Reserve became a group of potential, future, governmental scapegoats in the event of anything going wrong with the financial system. It has long been known behind the financial scenes that the monetary system is fraught with potential difficulties, so at least if there are problems now, the government can safely blame the Federal Reserve!

Money is created as debt, meaning that money is created whenever anyone takes a loan from a bank. In actual fact, every deposit made becomes a loan, and is repeated many times over –

potentially creating an infinite amount of money from debt. With debt creating money, it is no wonder that banks are keen to lend it out, but over the years there have been noticeable flaws in the system.

It would seem the banks adopt a delay and pray tactic. If they delay admitting the truth for long enough, they have time to pray that more deposits are made, so they can lend out more.

The whole difficulty associated with the creation of mass amounts of money was compounded when some banks' lending ratios went up to twenty or even thirty to one; some banks have no limit whatsoever. In addition, by using loan fees to raise the required reserve from the borrower, banks have now found a way to circumvent reserve requirement limitations entirely.

Initial regulations seem to have been bypassed, and loans can be made to anyone who needs them, as long as they look half capable of repaying their loans. The major point being made here is that banks are able to make as much money as we can borrow.

You may wonder how individuals, banks, governments and others can all be in debt at the same time. They all owe very large sums of money. You will be able to see how this can be the case when you remember that banks do not lend actual money, they create it from debt and because debt is unlimited, so is the supply of money. We are all so convinced that money is a safe and trustworthy commodity, and that there cannot be anything wrong with this scheme.

How wrong we are to think all is well. There are a number of major difficulties with the way our money system works. One major problem is that the people who produce the true wealth within society are in debt to those who lend out money in that society. In addition, if there were no debt there would be no money at all. We are taught that to pay our debts responsibly is good for ourselves and the economy, so we imagine that if all debts were paid off, the economy would improve. Wrong! The opposite is true, it would collapse. During the great depression, the supply of money plummeted as the supply of loans dried up.

The problem of relentless financial growth

- *Popular long-term loans create problems of perpetual indebtedness which cannot be sustained indefinitely.*
- *The financial difficulties of 2008 prompted the Queen of England to question the scruples of the banking fraternity.*
- *Any system that relies on relentless growth is not sustainable – so why do we live in a world that is based upon fractional reserve banking?*

Here is the big problem: banks create only the amount of the principal loan to the individual or organisation. So where does the money come from to pay the interest?

Well, it comes from the general economy money supply, and this money has been created in the same way.

Popular long-term loans create problems of their own, as the interest far exceeds the principal loan amount. Therefore, unless money is created to pay the interest, payment defaults will transpire. However, in order for society to retain its equilibrium under these circumstances, the number of defaults must be kept as low as possible, so more debt must be created, which means more interest is created. This results in a vicious spiral of indebtedness which is impossible to maintain forever.

Quite rightly, Queen Elizabeth II, The Queen of England, said to a group of financial experts during the economic difficulties of 2008, "Surely, did someone not see this coming?"

They more than likely did, but they could not do anything about it. It was all too late, as soon as the goldsmiths started to play their tricks with money!

It transpires that it is only the time between the money being created, and the time the debt is repaid, that keeps the overall shortage of money from catching up, and bankrupting the entire system. It is this that brought the World's economy to an almighty halt in 2008. The preferred method that the banks tried out to resolve the situation was to create more debt. The interest rates were lowered, and the perpetual bombardment of credit card applications flooded through everyone's post. This was a last-ditch attempt to keep the financial collapse at bay.

A financial system based on fractional reserve banking is not sustainable, because it relies on relentless growth. This relentless growth has meant the plundering of the planet's resources, the mountains of rubbish created within this frenzy, and is also responsible for the human population explosion. All of this just to prevent the system from collapsing. Interestingly, the human population explosion helped stave off the collapse a little longer, as more people resulted in more debt – just what the banks wanted. But the bubble burst in the end!

The unsustainable pace of economy

- *Economic difficulties will repeatedly occur until the entire financial system is changed.*
- *Banks charge interest on money they never had in the first place, this seems to be rather a swindle – however this is sanctioned by government.*
- *Money is created as debt – whenever someone takes a loan from the bank new money is generated.*

The crazy thing is that governments have just about been able to sort out the financial aspects of the economy by their intervention. Bizarrely, the economy is still in turmoil, and the same thing will repeatedly happen again repeatedly until the regime is changed. To change this regime, we must address the terrible situation of money purely being debt; a few angles on this matter need to be investigated.

Firstly, why do governments choose to borrow money from private banks at interest, when they could create all the interest-free money they need themselves?

They could surely grant themselves the authority to do so. How can fractional reserve banking, dependent on perpetual growth, be used to build a sustainable economy?

It causes an economy to go at an unsustainable pace, too; it causes countries to rush to plunder their resources with no future vision.

Why create money as debt at all?

Surely we would be just as well off utilising a money system that circulates permanently, and does not have to be perpetually re-borrowed in interest. The money system is frightening, and, left as it is, could potentially become a major significant factor in the downfall of the human race. When cash flow fails it causes rioting, looting and major civil unrest. It was interesting to watch the situation in 2008, when the governments did not quite know what to do. There were opposing views as to what should be done, and this showed how unknowledgeable even individuals within government are about how money operates. The solution they came up with was only temporary and fixed the financial system, but not the economy. In a few years' time the same thing could happen again.

So what needs to change to create a sustainable economy?

One fundamental problem is the practice of charging interest for lending money. After all, the banks are charging interest on money they produced out of thin air and never had in the first place. The audacity of it!

Not only that, they charge for service charges, and then, when a borrower defaults on his payments, the banks walk away with the asset – often a house.

So with all these banks creating so much money privately, and governments producing fiat currency that by law we are to accept or suffer the consequences, the question arises, "How much money is out there in the wide world?" In the past, the total amount of money in existence was limited to the actual physical quantities of the commodities which were in use as money.

Now money is literally created as debt, new money is created whenever anyone takes a loan from the bank. As a result, the total amount of money that can be created within the world has only one real limit – the total level of debt. My research has not yielded a figure for the amount of money, just phrases indicating that no one knows, that it is too difficult to calculate, and it depends upon how you look at it. I did however manage to find figures indicating many hundreds of trillions of US dollars up to thousands of trillions of US dollars, but all this obviously does not exist in the form of actual cash.

In the past, it was common to insist that banks had at least one unit of gold to back ten units of debt-money created. Today, reserve requirement ratios no longer apply to the ratio of new money to gold on deposit, but merely to the ratio of new debt money to existing debt money on deposit in the bank. Today, a bank's reserves consists of two things: the amount of government-issued cash or equivalent that the bank has deposited with the central bank, plus the amount of already existing debt money the bank has on deposit.

High powered money

- *Whatever amount of money a bank holds at the central bank, nine times this amount may be magically created – it is as simple as that!*
- *New money is repeatedly created as people invariably redeposit the created money back into a bank.*
- *The more financial deposits banks accept, the more new loans they are able to make.*

Confused?

Do not worry. It can be visualised through a very simple example. Imagine a brand new bank is established, but as yet has no depositors. However, the banks' investors have made a reserve deposit of one thousand, one hundred and eleven pounds at the central bank. This money is called high-powered money. Their first loan customer comes into the bank and requests a loan of ten thousand pounds for a car.

This loan is possible, as the required reserve ratio is nine to one; this enables the bank to legally conjure into existence nine times the amount held at the central bank. This money is not taken from anywhere; it is simply typed into the borrower's account as bank credit. The borrower then writes a cheque on that bank credit to buy the car.

The person who sold the car then deposits the newly created ten thousand pounds in his bank. Unlike the high-powered government money deposited at the central bank, this newly-created credit money cannot be multiplied by the reserve ratio; instead it is divided by the reserve ratio. With a ratio of nine to one, a new loan of nine thousand pounds can be created on the basis of the ten thousand pound deposit.

If the nine thousand pounds is then deposited at the same bank which created it, or a different one, it becomes the legal basis for a third issue of bank credit. This time, a new loan for eight thousand one hundred pounds is made possible. It is similar to unwrapping a parcel at a pass-the-parcel game that progressively gets smaller and smaller as you go on. This continues until almost one hundred thousand pounds in brand new money has been created within the banking system.

All this new money has been created entirely from debt, and the whole process has been legally authorised by the initial reserve deposit of just one thousand, one hundred and eleven pounds, which is still sitting untouched at the central bank. What is more, under this ingenious system, the books of each bank in the chain must show that the bank has ten per cent deposited compared to the amount that it has out on loan. This gives banks a very real incentive to seek deposits in order to be able to make loans, supporting the general false impression that loans come out of deposits.

Government-backed deceptive system

- *This system would be viewed as fraud; however, it is backed by the government by means of the Federal Reserve Act.*
- *The banking system does not produce anything or do anything – so how is it that everyone can be in debt to them?*
- *The control of money offers absolute power within the industrial and commercial world.*

Even though the rules are quite complex, the reality of it all is quite simple; banks can create as much money as we can borrow.

Banks can only operate this crazy deceptive system with the cooperation of governments. Firstly, governments pass legal tender laws to make us use the national fiat currency. Then governments allow private bank credit to be paid out in their government currency. Thirdly, the judicial courts system then enforces these debts. Lastly, governments pass regulations to protect the money system's functionality and credibility with the public, whilst doing nothing to inform the public about exactly where the money really comes from.

Without the document the borrower signed, the banker would have nothing to lend. This is why governments, people and businesses can all be in debt at the same time to the tune of such massive amounts. How can there be that much money out there to loan?

Well, contemplate no longer, there is not!

It is astounding, despite the incredible wealth of resources, innovation and productivity that surrounds us, almost all of us, from governments to individuals are heavily in debt to bankers. If only people would stop and think how it can be. How can people who produce all the wealth in the whole world be in debt to those who do nothing at all, other than invent the money that represents the wealth?

Even more amazing is that once we realise money is really debt, we understand that if there was no debt there would be no money. Most people think that if all debts were paid back, then the world would be a better place. This is the total reverse; the money system would stifle and the circulation of money would fail like jammed cogs. We are therefore dependent upon continually renewed bank credit for there to be money circulating within society.

It is amazing that the financial manipulators have pledged totals that exceed the value of all the money in the world, and no one seems to know how much the paper obligations total, who holds them, or what they truly mean. However, we are all assured that those involved are too big to fail, and that therefore we have to keep giving the banks money whilst the rest of us fail, falling into debt instead!

Perhaps banks should be converted to become non-profit making services to society by lending out money without charging interest at all. It is the fundamental nature of the current monetary system which is the problem, and tinkering with it cannot ever solve the problem. It must be replaced.

One solution would be the replacement of paper money with precious metals, which would once again become very cumbersome and inconvenient. Alternatively, locally-based barter money systems could be created, in which debt is repaid by hours of work, valued at a predefined amount. Then the government's spending on infrastructure, not using borrowed money, would begin to create value locally and nationally.

James Garfield, the twentieth US President, once said, "Whoever controls the volume of money in our country is absolute master of all industry and commerce. When you realize that the entire system is very easily controlled, one way or another, by a few powerful men at the top, you will not have to be told how periods of inflation and depression originate." James Garfield was assassinated in 1881.

Oh dear – what a surprise!

- *Any major mistakes the banking world makes can only be resolved by the taxpayer – why should this be?*
- *Bankers are pictured as concrete, clear-cut guardians of wealth the world over – when in reality they are just enforcing government-sponsored fraud.*
- *Banking failures occur on a regular basis, and are always caused by unforeseen, economic misfortune.*

Since 2008, governments have had to pump billions and billions into their economies to try to stabilise a volatile market. People wonder why on earth the tax payers should pick up the bill from the well-heeled banking speculators' mistakes. Where does all this money originate?

Where does one dollar, one pound or one Euro come from?

What do Presidents John F. Kennedy, James Garfield, Abraham Lincoln, and Congressman Louis McFadden all have in common?

They all believed in the necessity of dismantling the operation now known as the Federal Reserve Banking System. They were unique in that they all had the understanding and power to act on their beliefs, but never got the chance because they were all assassinated!

John Adams, 1735 – 1826, one of the key Founding Fathers of the United States, the second President of the United States, a leading champion of independence and a political theorist said, "Democracy, while it lasts, is more bloody than either aristocracy or monarchy. Democracy never lasts long. It soon wastes, exhausts, and murders itself. There was never a democracy that did not commit suicide."

What is the trouble with the Federal Reserve Banking System?

Contrary to popular belief, it is not part of the government. It is privately owned, which means the United States does not control its own money supply. Strange but true!

Bankers have always been publicly seen as solid, no-nonsense guardians of growth and prosperity in many countries. In truth, it turns out they are mostly fraudulent tricksters, performing feats of pure fantasy that make them rich to the direct degree that they succeed in undermining nature, while destroying economies and most forms of general prosperity, along with the global security of the planet. These are the same people who have elevated 'money' to its currently outrageous position over everything else in life.

Banking failures are interesting to study. One thing that you will notice about them is that they occur on a regular basis, and all revolve around an unforeseen, economic calamity. It makes one wonder whether once economic calamities are spotted on the horizon, these calamities are chosen behind the scenes as reasons to bankrupt certain banks, in order to keep the fraud going. After all, the banks are all in the system together within a closed loop, and are supposed to be able to help each other out. Obviously, when the economic calamity gets too great, something has to give.

How will evolution treat our money markets?

- *Wall Street works in mysterious ways – suckers are born every minute for them to take advantage of.*
- *Buying and selling commodities, making false promises, enticing people to buy for later gain which never materialises – the money markets!*
- *People mysteriously disappear never to be seen again on Wall Street – perhaps they have pocketed enough money to retire!*

I was once told a great story about how Wall Street works, and it is best told by using an analogy of catching monkeys. Within a district, all the people were told that Mr Jones would pay each person ten pounds for each monkey that they caught. Everyone rushed out to the dense forests and started catching them.

Mr Jones bought hundreds of them at ten pounds each, but, as you might imagine, the supply of monkeys began to shrink somewhat. The people began to give up looking for them as it became too just too difficult to find them. Mr Jones then announced that he would be prepared to pay twenty pounds for each monkey and this had the effect of making people look that little bit harder. After a little while purchasing monkeys for twenty pounds, Mr Jones increased the purchase price to twenty-five pounds, as they were getting so scarce.

Mr Jones then made a remarkable announcement. He would now pay an incredible fifty pounds for each monkey. He then added that he would be leaving for a while, and his assistant would remain in place as a buyer of the monkeys on his behalf. In the absence of Mr Jones, the assistant announced to the people, "You see all these monkeys in this cage. I can sell them to you for thirty-five pounds each. Then when Mr Jones returns you can sell them to him for fifty pounds each."

Obviously, seeing such a great deal on the horizon, all the people rushed to buy the monkeys with their hard-earned savings, and awaited Mr Jones' return.

The people never saw Mr Jones or his agent ever again. This is how Wall Street seems to operate!

Getting to the root cause of the world's major problem

- *It is a crying shame that the dominant species upon this planet has placed the intricacies of nature second to their own greed.*
- *Owning the latest and greatest technology is an essential part of being able to compete, which exacerbates a vicious circle of global sales.*
- *Due to our actions, there may well be earth-shattering days ahead that will bring doom-stricken cataclysmic days.*

Only a small proportion of people seem to take an interest in the environment and are keen to promote true progress. Unfortunately, other peoples' minds seem to be preoccupied with wealth, and achieving their own personal satisfaction, without too much interest in the intricacies of nature, and how their true self is positioned within the Universe.

What a great shame this is, as the fine balance of nature, and the universal mechanisms which govern the cosmos's existence, have evolved into such fascinating topics and are intriguing beyond doubt.

An interest and understanding within this subject matter helps you interpret the world around you better, and brings people a better sense of intellectual wellbeing.

Nevertheless, an interesting aspect of the way that people strive to survive in a competitive, money-oriented society is to develop better products, simply to make more money. Unfortunately, new developments in technology seem to be merely a by-product of a money-oriented consumer society, and do not truly reflect the aspirations of the individuals involved in their manufacture.

This ideology accelerates the development of innovative new products and services – but not because people 'want' to buy them. It seems that the sale of new technology is exacerbated because companies without the latest technologies cannot compete. A vicious circle!

It would seem that our future technology is being developed to satisfy the manufacturers' bank balances, not their true, technological desires. However, the purchase of the latest technology often provides an individual consumer with an advantage over other individuals.

Over the past few decades we have seen dramatic improvements in already existing products such as televisions, mobile phones, cars, household appliances – in fact, pretty much everything. I put it to you that, deep down, most people are quite content with their gadgets, but corporations will strive progressively harder to get more megapixels, more gigabytes, faster processors, better screen resolutions and incredible applications simply to outdo competitors – simply to keep in business!

Perhaps if we ditched the view of our consumer society, and looked at what everyone in the world truly needs, rather than forever bettering already existing technology, we may be able to progress better as a race. We could concentrate on some of the more beneficial worldly developments that would help us survive more successfully, rather than inadvertently plundering the valuable resources of the planet for short-term gain. If we continue as we are, it may well get out of hand very quickly!

Reflecting upon a life filled with achieved ambition, being frugal with resources, coupled with a full and varied lifestyle, must compare considerably more favourably with a duller materialistic existence, with little hunger for knowledge, and few life experiences. Some people, knowing this, may still wish to inhibit themselves, but knowledge and life experiences can only bring about a better, more satisfying and possibly less regretful deathbed experience!

There certainly needs to be a more concerted effort and less compromise, whereby Mankind seeks to live in harmony with his surroundings rather than decimating it. Due to our total incompetence, future generations have no option other than to concentrate on the remedy of our madness, at the start of what they could justifiably call the 'Decimation Period'.

Our lives now are forming the history of the future, and there will be very little thought about our generation in as little as ten thousand years, yet alone a million years' time. Our lasting legacy will to have been responsible for marking the start of the Decimation Period of the planet, where, where environments and species were plundered, the eco system began to fail, and we started to upset the fine balance of our planet, which took billions of years to evolve!

The security of the future of the planet in these politically and environmentally turbulent times should be of the utmost importance.

Rather, we find ourselves approaching ever more doom-stricken cataclysmic days. Surely, we do not want to die wishing we had spent more time at work earning more money, and further destroying the planet, to satisfy our own and others' ill-founded wealth.

If this were the case, our deathbed thoughts would be negative and would reflect upon a self-centred lifestyle.

Perhaps it is about time governments woke up to this fact, unified, and truly worked towards combating this current inevitable threat.

When studying the root causes of problems affecting the Earth, it transpires that our monetary system is contributing towards a great deal of decimation to the planet. Due to the fact that the world's financial system keeps most businesses and people perpetually in debt, we find ourselves ripping the planet apart to pay our off our loans.

We learnt how banking regulations were required to limit the amount of paper money that could be lent out as a ratio of what had been deposited. We learnt how the fractional reserve system, and its integrated network of banks backed by a central bank, became the dominant money system of the world. We learnt how President Woodrow Wilson signed into law the Federal Reserve Act, which formulated the cartel that is now in charge of orchestrating money. We learnt how the monetary system is fraught with potential difficulties, and how the governments can safely blame the Federal Reserve for such problems since 1913. We learnt how a financial system based on fractional reserve banking is not sustainable because it relies on relentless growth – this is why the planet is suffering. We learnt that because fractional reserve banking is dependent on perpetual growth, it cannot be used to build a sustainable economy. We learnt that new money is created entirely from debt, and the whole process is legally authorised by the initial reserve deposit. We learnt how the money system would stifle, and the circulation of money would fail like jammed cogs, if all debts were paid back. We learnt about some alternative methods of operating the globe's financial system. We learnt how banking failures occur on a regular basis, and all revolve around an unforeseen, economic calamity. We learnt how our current financial system could be leading us into the 'Decimation Period' and how it is fraud, but because it is government backed it cannot be seen as fraud.

Next we look at potential objectives of the human race.

THE GLOBAL OBJECTIVE

Where we look how human beings currently view the unknown, and how the past can help give us pointers to the future. We look at the insatiable appetite that human beings have for knowledge, and how much knowledge has increased over recent years. We finally take a look at what goals we need to achieve to make sure that human beings do not destroy the ecosystem of the planet.

Tackling the unknown

- *Through using our minds, we have been able to deduce the make-up of the world around us – and develop items to benefit us.*
- *In years gone by, we imagined rather macabre happenings within the unknown aspects of the world.*
- *Not only does 'now' matter – but one hundred million years from now matters just as much to the people who will be alive then.*

We live within a truly phenomenal Universe, and over the centuries our perception of it has changed dramatically. Through experimentation, discovery, exploration, philosophy and technological advancement, we have improved our understanding considerably. We have advanced in leaps and bounds since drawing dragons and ghouls on the edges of maps within unexplored territory. However, we now question what may exist beyond the boundaries of our known Universe of hundreds of billions of galaxies, each containing hundreds of billions of stars.

Rather than drawing dragons and ghouls at the edge of our known Universe, we are now more inclined to imagine other dimensions, black holes, other universes, or an infinity of galaxies and universes leaving us extremely giddy, even more so, leaving us even more giddy than our counterparts in the middle ages. Perhaps the dragons and ghouls were more accurate!

We will never be content with our limit of knowledge until we have satisfied our understanding of everything around us: diseases, energy solutions, time, gravity, universal expansion, intergalactic travel and the history and future of everything.

Questions regarding these topics may take eons to produce any truly satisfactory answers. However, one major advantage from which future generations will benefit is that they will be born into an advanced, developed society, rather than a savannah or jungle. Irrespective of whether our bodies are still compatible with a savannah or jungle habitat, thankfully our ancestors very kindly paved the way for the majority of us to be born into the world with a head start. Instead of a savannah, we are now born into a nursery, equipped with pens, pencils, notepads, computers and groundbreaking knowledge at the ready.

With so much detailed knowledge and information accumulated over the years, it is inevitable that specialists are necessary to fully understand previous technological concepts in order to pave the way forward. These specialists within various fields, such as physics, chemistry, biology, astronomy and computing have arisen over the recent past, evolving from individuals who at one time knew everything there was to know about those particular topics. The future is going to nurture specialists who will concentrate upon specific areas of focus within their chosen field, rather than the whole. For example, environmental issues have diversified to such an extent that some people have begun to specialise in one particular type of molecule and its influence within the environment.

One example of someone who knew everything there was to know about his chosen field in his day was Charles Babbage, 1791 – 1871. In 1837, Charles Babbage was the first person to conceptualise and design a fully programmable mechanical computer, called the analytical engine. Hence, at this time, no one else knew more than him about computing, and there was little more anyone else knew that he did not know already. Nowadays we have scores of degree subjects specialising in areas of computing, focussing within niche aspects of a much, much larger topic since Charles Babbage's day.

The very freakish analytical engine must have appeared as a truly eccentric innovation. Exactly how people envisaged its future at the time would have been interesting, but I doubt many could have predicted the change it would inflict upon the world as it has today.

Future generations will be presented with considerable learning curves to reach the forefront of their specialist topic. The essential study an individual needs to undergo to reach a point where their knowledge is sufficient is already rather challenging. Just imagine the grounding required by a quantum physics graduate, wanting to make headway in ten thousand, or even ten million, years from now. The thought is daunting!

There are plenty of life's unanswered questions to tackle in the future. For example, our astronomer's favourite ponderings: namely, what really lies out there in the depths of the Universe?

I would not at any time underestimate the capability of human beings to answer this in the future, simply judging by the incredible discoveries made over the last hundred years or so. But, perhaps when humans have attained such universal knowledge, we shall be more settled in mind, and psychologically relaxed.

Where is all this leading?

Well, perhaps one day we may discover the Universe is alive throughout, and all that is required to ask it questions is to talk to it in an as yet unknown universal language at one end of the X-ray spectrum. Oh, if life were as simple as this, it would make our future so much easier!

All this may sound rather far-fetched, but within this Universe we surely have to expect the totally unexpected. There is intelligence within the Universe's structure; we know this, or otherwise we would not be here. It is just a matter of how we tap into it.

Many people have speculated that intelligent life can be sustained upon any planet only for a small number of years after nuclear bombs have been invented. Economic meltdown may follow partial annihilation, caused by greed, envy and misunderstanding. I was once asked the question, "Is it 'when', or is it 'will', terrorists get hold of nuclear weapons?" Unfortunately, when you look closely at this question, 'when', is more likely than 'will', unless something very serious is done to combat terrorists throughout the remainder of human evolution.

So with this in mind, just as we look back in history to times of great innovation and growth, in a million years from now there will be so much more to look back upon. People today seem to forget that they are custodians of future resources, and act as if only 'now' matters. Unfortunately, as I see it, we shall be looking back at current times in a million years from now as an example of how not to proceed. The survivor's lessons will be not to advance too quickly, and to take time to evaluate the pros and cons of innovation and change.

Consistency and predictability go out the window

- *If I repeat an act on Earth, the same outcome will be repeated with one hundred per cent accuracy.*
- *Hidden within this quantum world will lie secrets to a natural driving force that will hold all the answers to the questions we have been asking.*
- *It is quite possible that the four forces within the Universe change considerably within different areas of the Universe.*

I am not sure, I cannot truly be certain, but thankfully there do seem to be some sort of ground rules laid out that governs the activity and behaviour of everything. For example, not once, when pouring water out from my kettle has the water decided to hit the ceiling. It always goes downwards towards the centre of the Earth, and into my cup. This experiment is repeated by humans millions and millions of times every day, and surprisingly the same result always prevails! This is a good example of the consistency of Newtonian physics, the physics of the large scale. However, at the subatomic quantum level, the physics of the small, things begin to act rather strangely and unpredictably.

When it comes to topics like our senses, subatomic particles, quantum forces, thought and consciousness, all consistency and predictability seem to go out the window. Although the unpredictability of all these seems to be the only consistent thing! How to make sense of all this quantum nonsense is a fine art. But lurking beneath this entire quantum realm of weird sorcery, somewhere out in the depths of the Universe, will be a natural driving force that holds all the answers.

However, working out all these answers may well be more difficult than we could ever imagine. The calculations and the mathematics involved could be well beyond today's technology. If there was any straightforward logic to the Universe's natural driving force, then you would have thought someone would have noticed it by now. It may transpire to be a series of extremely complex calculations that push computing to the limit.

We are currently only scratching at the surface of our recently-acquired knowledge of the very small, along with its bizarre atoms, weird expanse of empty space, and the truly eccentric behaviour it exhibits.

We do not even know whether everything throughout the Universe is exactly the same, whether everywhere in the Universe follows exactly the same laws of physics. Perhaps there is a planet where the water from my kettle does hit the ceiling!

Perhaps physics could be local to a particular area of space, with different forces becoming more exaggerated than others, depending upon the sources of various bizarre phenomena within the locality. Who is to say that if we tried travelling to a nearby galaxy, as soon as we arrived somewhere between the two, we would be so far from our normally-accustomed Milky Way's galactic physics that we and our spaceship would experience unforeseen forces and pop, blowing up into smithereens.

Just as animal senses are different to our own senses, there is absolutely no reason that the constants we experience associated with the forces here on Earth cannot change dramatically throughout the Universe. In fact, it could be seen as rather naive for us to think that all the constants we experience here on Earth remain the same everywhere.

For example, perhaps a black hole is simply where nature has divided all its forces by zero!

The perfect day

- *When the last morsel of something good is used up – it is only then that the human race counts the cost of what they have done!*
- *Prior to global communication, global problems were dealt with locally – now we have awareness on a global scale – let us hope we may benefit.*
- *Our understanding of technology has taken giant steps forward – there appears to be no end to the progress.*

With human greed as it has manifested itself today, only when the last tree is felled will some realise that they cannot eat money!

Imagine you drift away from planet Earth and into the distant Universe to settle for an instant. You look back at Earth and imagine there is an objective.

What do you think that objective could be?

You would see all the felled trees which once sustained wildlife, and kept the carbon tied up from the atmosphere. You would see the plummeting and squandering of all the planet's resources, a depleted ozone layer, struggling species, polluted oceans and no signs of this abating.

All this is not replaceable!

The future is certainly not being considered.

How can there be an objective associated with all this!

There is little use having the Universe's best resources at hand, unless there is an appealing objective to complement them!

In a world of sudden bursts of anger, passion, greed and debauchery, there is little wonder that many people cannot get the happiness they are seeking. Pains, anxieties, whether global, local or personal, influence our lives to prevent any type of preconceived Utopia being formed, either mentally or physically.

Famous people, who have reached perfection in wealth, health and happiness, quite often turn their attention to planetary issues, as they are relaxed enough to realise that their Utopia is being encroached upon by external factors outside their control.

Anxiety and guilt felt within their Utopia causes them to attempt to do something about it.

Prior to global communication, problems within one country were isolated and shielded from others. News spread slowly, and only local events were understood. Utopian communities were possible without the guilt and anxieties of the world at large.

In recent times, we only have archived knowledge of how these times were, and old video footage with which to compare ourselves. Looking at an old black and white film, you can only speculate at people's knowledge about the world at that time. Our knowledge of the World and the Universe is far superior now.

It has been estimated that a typical person living approximately one hundred and fifty years ago accrued the equivalent of three Sunday papers' worth of knowledge in their minds about the world, within their entire lifetime.

Our ability to educate has improved, and our ability to create environments within which to provide education has also improved. Goal setting has improved, but until we have a global set of objectives we shall all be 'flapping in the wind'.

What is the short term goal that each country has set itself?

Is it to ensure that all its young children are educated better than all the others?

What are the objectives of countries and the world as a whole?

Should we not introduce study topics such as Happiness, Universal and Planetary Science, Anxiety Studies and Future Generation Studies as subjects on young people's school curriculums?

Perhaps then we could end up having young people adept at being happy, kind, and knowledgeable regarding all the World's major issues, and fully conversant in advanced technologies with a view to furthering our precious survival within this fragile Universe, as opposed to the bickering, intolerant World that we know today.

One of the major problems is that people born today do not realise that aeroplanes, cars, pollution and the plundering of the planet's resources are recent events. We all know, deep down, that continuing as we are will have extremely negative effects on the planet, but unfortunately most people think that it is a problem that does not belong to them and do nothing to help.

My message to the world is this, "If you don't act now in a positive fashion, then we will have created massive problems for our children whom we all struggle so hard to look after." If we do not set a global objective within our specialist survival handbook, we will have created a world that is totally unsustainable.

We should ditch our cars, replant the rainforests, recycle all we can, park up the aircraft, and plan the future carefully, with multimillion-year strategies. This would actually be exciting. We could plan the inventions for the future, work towards Utopia in a calm, controlled, calculated manner. We must also prevent uneducated people from running governments. Far too many ignoramuses get into positions of power, and operate with the wrong agenda.

Governments should remember that they are not there to run a country; they are there to preside over the wellbeing of those that do.

There is a famous description by George Orwell regarding the outlook that Charles Dickens had on life, "If men would behave decently, the world would be decent." I believe there to be a great deal of truth in this statement.

Survival and evolution are taking on new forms now. Our understanding of genetics and the human body has taken leaps and bounds of late, and will continue to do so. This will continue until perhaps, one day, we will engineer ourselves into a race that can cope with interplanetary space travel, specifically designed to colonise other planets. Would this be the perfect day?

So what are the major conclusions we can formulate from all of this?

Despite the corrupt nature of the human race, there is hope. We shall have to put faith in the currently developing strategies, draw upon the expertise of a few to guide us forward, and hope for a more placid future.

In this chapter we learnt how people today seem to forget that they are custodians of future resources, and act as if only 'now' matters. We learnt how we could be naive to think that the Universe is consistently the same throughout. We saw how we must put faith in the future, draw upon the expertise of a few to guide us forward, and hope for a much more placid world – maybe this entails ditching our cars, replanting the rainforests, recycling all we can, parking up the aircraft, and planning the future carefully with multimillion-year strategies.

It is now up to you to gallantly venture forward to correct the errors of the past.

Good luck within this extremely eccentric Universe!

As for what this whole book has enlightened you to – you tell me!

INDEX AND REFERENCES

Index of subjects

Index of quotations by eminent scientists and philosophers

Readers of this book are encouraged to debate the topics and questions raised by visiting either of the websites listed below:

Website: www.theeccentricuniverse.com
Social debate: www.facebook.com/theeccentricuniverse

Alternatively, email the author directly at:
rob.lowe@theeccentricuniverse.com

Index of quotations by the author

Readers of this book are encouraged to debate the topics and questions raised by visiting either of the websites listed below:

Website: www.theeccentricuniverse.com
Social debate: www.facebook.com/theeccentricuniverse

Alternatively, email the author directly at:
rob.lowe@theeccentricuniverse.com

Index of questions raised in this book

Readers of this book are encouraged to debate the topics and questions raised by visiting either of the websites listed below:

Website: www.theeccentricuniverse.com
Social debate: www.facebook.com/theeccentricuniverse

Alternatively, email the author directly at:
rob.lowe@theeccentricuniverse.com

ABOUT THE AUTHOR

Rob graduated in computing in 1983, and fast became recognised within the IT industry for his all-round knowledge and strategic vision. Rob has been instrumental in the merger and acquisition process of major global organisations, and been at the helm of some of the most innovative technological companies including Vodafone, Motorola and Psion. Rob has also spearheaded a number of government technology improvements including those at DWP, Companies House and HM Customs, where he was responsible for the delivery of all Projects and Programmes of work as a member of the tactical delivery board. Rob has also worked on government communications protocols and at a company where the first Internet search engine, called ALIWEB, was produced.

Rob's views on many subjects, his innovative approach to life and original personality were recognised not only by technological organisations, but also by the Eccentric Club of Great Britain. In a fair annual contest in 2009, Rob was awarded with the title 'Great British Eccentric of the Year', and the Eccentric Club published a positive review of his first book, 'A True British Eccentric', promoting it as 'a recommended read'. The book is an account of hilarious predicaments, affection for groundbreaking technology, and a tale of his epic bicycle ride around the whole coastline of Great Britain. The Eccentric Club, originally established in 1781, is based in Mayfair, London, and is proud to have HRH The Prince Philip as its patron. Rob is equally proud to be associated with the Eccentric Club, the past and present members of which were and are amongst those who help to shape British culture and society. The club's history is inseparable from that of Great Britain itself. For more information, please visit www.eccentricclub.co.uk

Rob's fascination with technological innovation, quantum physics and astronomy, coupled with wanting to improve the world technologically and environmentally, has culminated in this philosophical publication.

With an uncomplicated and fun approach to some extremely interesting topics, Rob takes you on a journey through our truly eccentric Universe, where you will sometimes have to pinch yourself to realise that you are within its mystical and secretive grasp.

Rob sincerely hopes you enjoy reading: THE ECCENTRIC UNIVERSE.

Special Thanks:

Rob would like to thank Felicity Coxon, David 'Gadget' Morgan, Graham Partis, Christine Lowe, Nigel Stonham, Charles Fry, Madeline Tomlinson and Andrew Bird for their encouragement, support and input whilst writing this book.

Rob would like to thank the Eccentric Club of Great Britain for their support and encouragement, especially the Club Chairman, Imants von Wendon. Rob very much looks forward to offering topics within this book as fuel for debates, discussions, talks and after dinner speeches.

Rob would like to especially thank Roger 'Quiz Runner' Morgan for his sterling effort proofreading this book.

Finally, a special 'thank you' goes to Felicity for being so patient whilst I was writing this book.

Rob Lowe
NIL NISI BONUM

Rob hopes that the topics within this book are enjoyed and debated for many years to come. Rob is available for presentations, talks, and after dinner speaking – Rob is also available to project manage interesting ventures, large or small.

Email: rob.lowe@theeccentricuniverse.com
Website: www.theeccentricuniverse.com
Social debate: www.facebook.com/theeccentricuniverse

Certificate Presented to Rob Lowe by The Eccentric Club, Established 1781 – Patron HRH The Prince Philip

'Great British Eccentric of the Year'

1781 2008 1890

THE ECCENTRIC CLUB

Celebrating Eccentricity since 1781

Patron: HRH Prince Philip The Duke of Edinburgh KG KT

Whereas, Rob Lowe,

having demonstrated to the members of the Eccentric Club, to their guests, and to the WORLD at Large, on the 1st day of April in the year 2008, an exceptionally virtuous and meritorious eccentricity, has been hereby awarded the Title of:

The Great British Eccentric of the Year

The Eccentric Club (UK) Committee

Lightning Source UK Ltd.
Milton Keynes UK
UKOW020131040112

184701UK00001B/8/P

9 780755 214020